VOLTERRA FUNCTIONS

VOLTERRA FUNCTIONS

ALEXANDER APELBLAT

Nova Science Publishers, Inc.
New York

For permission to use material from this book please contact us:
Telephone 631-231-7269; Fax 631-231-8175
Web Site: http://www.novapublishers.com

LIBRARY OF CONGRESS CATALOGING-IN-PUBLICATION DATA
Apelblat, Alexander.
Volterra functions / Alexander Apelblat.
p. cm.
Includes bibliographical references and index.
ISBN 978-1-60456-169-2 (hardcover : alk. paper)
1. Volterra equations. I. Title.
QA431.A664 2008
515'.45--dc22 2008045972

Published by Nova Science Publishers, Inc. ✦ New York

To the memory of Serge Colombo

CONTENTS

PREFACE

The Volterra functions appeared at the beginning of the second decade of twentieth century in the works of famous mathematicians Srinivasa Ramanujan, Jacques Touchard, Vito Volterra and Edmund Landau. Surprisingly, they considered them in a variety of fields of mathematics, namely in the theory of definite integrals, integral equations and prime numbers. The functions bear the name of the Italian mathematician Vito Volterra (1860-1940) who together with Joseph Pèrés brought them to an independent, mature status as solutions of the integral equations of convolution type with the logarithmic kernels. Between 1943-1953, the Volterra functions have aroused special interest of several French mathematicians (Pierre Humbert, Louis Poli, Maurice Parodi, Serge Colombo and Pierre Barrucand) who investigated them in the context of the Laplace transformation where the functions play an important role as direct and inverse transforms. After this period, a gradual decline was observed in the research activity associated with the Volterra functions. The last review dealing with properties of the Volterra functions *"Étude des Transcendantes Intervenant Dans la Résolution des Équations Intégrales de Volterra a Noyaux Logarithmiques"* has been published as a brochure by Serge Colombo in 1964 and it has a rather limited scope. Since then, nearly five decades had passed and it became evident that the Volterra functions are not only involved in the theory of convolution integral equations and integral transforms, but also in a large variety of problems associated with the properties of other special functions. Therefore, a new and significantly enlarged review on the subject is desirable.

The Volterra functions came to the author's attention rather early, through the involvement with operational calculus, but in different contexts and over different periods of time. At first, during the 1985-1989 period when the author found that the infinite integrals with respect to the order of the Bessel and related functions can be developed into series of the Volterra functions and some infinite integrals of gamma and polygamma functions are expressible in terms of the Volterra functions. However, the main preoccupation with these functions started after the author met for the first time in the summer of 1994 Professor Serge Colombo. This happened rather incidentally, the author intention was, being then on sabbatical leave in the Alberta University, Edmonton, Canada, to write a book (still unfinished) on the Laplace transformation. In the book, the planned emphasis was on mathematical techniques of problem solving with relations to properties of special functions rather than, as traditionally, on the theory of the Laplace transform and the solution of differential equations. In order to prepare material on "lesser known" special functions, the author wrote to the Catholic University of Louvain, Belgium, where Colombo served as professor during the 1965-1981 period, asking to obtain the brochure on the Volterra functions published there. To the author's surprise, together with the desired brochure (his last copy) the author received a letter from him, from the author,s retirement place - Antony, near Paris. He was very pleased that there was someone who remembered and still has interest in his old works and he invited the author to meet him. In such way started the author's personal and professional relations.

During the authors' first meeting in 1994 and the few that followed since, the author found Serge Colombo (already in his late eighties) a very intelligent, open-minded and warm person, in full physical and mental capacity. The authors felt very good together in spite of more than twenty five years difference in age having a common language and interests in mathematics. He continued to work, as he told the author, devoting about twenty hours per week on several subjects in mathematical physics (gas dynamics and relativistic electromagnetic theory). Some kind of a repeated ritual took place during the authors' meetings. In advance, he was making reservation for a lunch in historical restaurants, famous for the excellent French food. During these lunches the authors would discuss general plans for the next year, but this part was usually short (the authors continue to deliberate their problems later walking to a nearby café and by mail) and the main part of the time was devoted to his reminiscences and many anecdotes from the life of mathematicians that he knew personally and how he valued their works. The list of them included such names as J. Hadamar, P. Lévy, P. Humbert, B. van der Pol, L. Schwartz, F.G. Tricomi, A. Erdélyi and others from his generation. One of the questions that the author was curious to know was why suddenly at the beginning of the fifties practically all activity in the domain of operational calculus and the Volterra functions was interrupted. He attributed this to a lack of interest after the death of Pierre Humbert in November 1953, who was the leading figure in the field. However, the main reason was the overwhelming reign of the Bourbaki trend in French mathematical life that made it difficult and in some cases impossible to find a suitable position and support in academic institutions. He himself was forced to leave France, initially for Saigon, Vietnam and later to Louvain, Belgium. Actually, the 1964 brochure was reprinted because it was published at first in 1963 in Léopoldville, Congo-Zaire. The author sensed that this question tormented him deeply because he returned to it repeatedly in the author's correspondence. When the author named in this book the contributions of French mathematicians to the theory of the Volterra functions, his reaction was ..." je preférе "French Authors" á "French Mathematicians" ... Les bourbakistes dénient ce titre á ceux que vous mentionez.!! .." (from letter of June 16 1996). In his letter of August 10, 1998, this time in English, he wrote: "The results of the author investigations are lost....and after 1961 the author did not care about them, as the author was discouraged. Nobody was interested in the author's work except Pierre Barrucand and the author switched to other subjects: fluid mechanics, electromagnetic waves....Too much time has elapsed!... Erdélyi was somewhat interested. The author met him at Edinburgh in 1968 and the authors barely discussed the project." And also in December 23, 1997, in the same context ..."... the author remind that a paper the author submitted many years ago on Mittag-Leffler's general theorem was rejected on the ground that the subject had fallen in desuetude and the author's terminology unqualified. ...Nowadays Humbert-Agarwal work would not find a publisher in Europe. It is sad."

As a continuation to his stories about operational calculus, he sent the author copies of his correspondence during the 1951-1956 years with B. van der Pol (he valued him very highly) which dealt with the terminology and presentation of the δ-function in the well-known book of B. van der Pol and H. Bremmer "*Operational Calculus Based on the Two-Sided Laplace Integral*". As a result of Colombo comments and suggestions, some changes were introduced to the book, as is appreciated in the Preface to the second edition which appeared in 1954. Another item of interest is the copy of the lecture "*Le Calcul Symbolique d'Heaviside*" presented by P. Humbert at 19 November 1946, at the Ecole Polytechnique, Paris, where he gave the basic rules and a short history of operational calculus in the world, but especially in France mentioning amongst other names..."M. Serge COLOMBO, qui soutiendra dans quelques mois une Thèse de Doctorat sur le Calcul Symbolique et les fonctions discontinues." It is worthwhile also to mention here the short letter of Humbert to Colombo from 22 Mai 1953 "Cher ami, C'est simplement pour vous dire que les Indiens qui font du calcul symbolique à une et deux variables (Mitra, Agarwal, Chakra...), tous de l'Université de de Lucknow, désirent s'appeler "Humbert's Indian School". Qu'en dites vous? Serais-je célèbre jusqui'au 80° degré longitude Est? J'y suis évidemment plus prophète qu'après de Laurent Schwartz. Dire que c'est moi qui l'ai reçu à Normale (i.e. at prestigious École Normale) en 1934! Il est vrai qu'il avait fait une planche remarquable. Bien amicalement....P. Humbert". This letter was accompanied with Colombo's comments ..."un mathématicien inconnu" disait Henri Cartan who disliked Pierre Humbert, and perhaps finding "Modern Analysis" a "boring book" (i.e. about the textbook "A Course of Modern Analysis" by E.T. Whittaker and G.N. Watson) ...Yet, Pierre Humbert's works were much appreciated by Indian mathematicians...".

At 24 April 1976, Colombo itself presented, this time a popular lecture "Oliver Heaviside (1850-1925) et la théorie électromagnétique" at the Palais de la Découverte, Paris (Revue du Palais de la Découverte, vol. 5 – N^{0} 42, pp. 16 -26), which in the author's opinion, is probably the best short analysis of work and life of the actual initiator of the operational calculus. Probably, in his private archive, there are many interesting materials from the history of French mathematics in the last sixty years...

The authors' first meeting in 1994 marked the starting point to his renewed involvement with the Volterra functions. Initially, the authors considered to write this book together, but very quickly it was clear that his contribution will be limited and the major part of the efforts will be put on the author's shoulders. He liked the subject because as he said this returns him back to his younger years. He proceeded very slowly, absorbed also by other subjects and duties and evidently more and more by his health problems. The author is quoting here from a typical letter (22 December 1997) about his extraordinary vitality and he was then 87 year old! "the author must apologize. Now the author's health is better. But the author am asked to afford too many mundane activities. The author is member of the association "Les Amis de l'Université Catholique de Louvain" (alumni, emeritus professors) The French residents are numerous and a French section of this association was created few months ago. The "Recteur" (President) of the University visited Paris (Diner, lectures,...). The "Société d'Encouragement pour la Recherche et l'Invention" asked a complete account of the author's activities (research, teaching, publications, copies of documents, diplomas,...). the author must write letters, answer telephone calls, deliver speeches, (attend meetings at UNESCO...) the author find all this tiresome and cannot escape...".

Nevertheless, he was able to write the main part of Chapter 3 "Volterra Integral Equations" and Appendix B "Summation Formulas". They were initially written in French and I translated them into English. Based on these translations, Colombo rewrote them three-four times. The balance in his contribution, being professional mathematician, is evidently shifted more to mathematical rigorism than to a rather formalistic approach which prevails in this book. As a leading authority on the subject, it is of interest to mention from our correspondence what he said about his role and others who were involved with the Volterra functions. "Ramanujan's formula was also given by Touchard, independently... the author has no remembrance of Ramanujan's work dealing with $\mu(x,\alpha)$, $\mu(x,\alpha,\beta)$". "Volterra and Pèrés made little contributions to "Volterra functions". Indeed, they rather spoke of "the powers of the unit element of composition ring". In their book "*Théorie Générale du Fontionnelles*" they devote only 4 pages to Volterra's integral equations with logarithmic kernels"... " As a matter of fact μ and ν were not mentioned in Volterra's publications..." ... "Parodi - he never went deeply in the subject!" ... "It is evident that the author was the first to use the notations $\nu(z,n)$, $\mu(z,m,n)$ but be sure the author never claimed that these functions should be called "Colombo's functions" because the author has shown that they play an important part in the resolution of Volterra equations of the closed cycle with logarithmic kernels. For the time being (as well as in the future) it is impossible to dissociate μ, ν from Volterra's equations." ..."the author believe the author was the first to use the notations (1948 ?) of $\nu(t,n)$ (or $\nu_n(t)$), $\mu(t,m,n)$ (or $\mu_n(t,m)$), Erdélyi was pleased..." ..."Truly $\nu(e^z,n)$ are COLOMBO'S functions. But until now nobody knows anything about their zeros..." ..."At that time (1964) Barrucand wrote and spoke about "les fonctions de Colombo." Actually, this happened much earlier, because already in 1947 Maurice Parodi published a paper entitled "*Relations intégrales entre les fonctions* $\nu(t)$ et $\mu(t,x)$ *de Serge Colombo*" and in 1951 Pierre Barrucand "*Sur les fonctions de S. Colombo*".

The conjecture that the Volterra functions ν and μ are not solutions of linear, homogenous differential equations with polynomial coefficients absorbed Colombo for many years. However, he was unsatisfied with his proof considering that the conditions imposed are too stringent and finally he asked the author to remove it from this book.

Serge Colombo (12 June 1910 - 1 February 2000) passed away on his 89 year and as his daughter wrote me "His death happened during the night in his bed as he was reading a text about "time" as he could not sleep... He is burried in Normandy (Saint-Aubin-sur-Mer) near the sea facing England where he had pleasure to go. He had a lot of friends at his funeral and all of us keep a lot of things in the author's heart about him, of his human being and his thinks." This book is devoted to his memory.

The present book aims to present an up-to-date state of the Volterra functions as well as to serve as a reference book, being as complete as possible. For this purpose a comprehensive list of formulas and tables of the Volterra functions is given in this book. A special effort has been made to standardize the notation, to remove inconsistencies and to correct misprints and errors in the literature. Thus, this book includes a critically evaluated old and recent material on the Volterra functions, but also many new results that were obtained by the author during the preparation of this book and that are published here for the first time. These results include: the behavior of the Volterra functions as a function of parameters, the integral representations of the functions, many new Laplace and other one-dimensional and two-dimensional integral transforms, integrals and series and extensive numerical computations that are presented in graphical and tabular forms.

In Chapter 1 the appearance and historical development of the Volterra functions is described by presenting the most important investigations and results, but without going into specific details.

In Chapter 2 is devoted to the basic properties of the Volterra functions, their behaviour with regard to variables and parameters and integral representations.

Volterra integral equations and related topics as elements of the functional analysis, convolution products, fundamentals of the Laplace-Abel transformation and Laplace transforms of logarithmic functions are discussed in Chapter 3.

Various one- and two-dimensional direct and inverse integral transforms of the Volterra functions and the reciprocal gamma functions are considered in a detail in Chapter 4. The main part of this Chapter is dedicated to the Laplace transformation, but the Mellin, Stieltjes, Hankel, two-sided Laplace, factorial transformations and the Riemann-Liouville and Weyl fractional integrals are also presented together with corresponding two-dimensional integral transforms and with many interrelations between the Volterra functions and other special functions.

In Chapter 3 the Volterra functions were introduced as solutions of the Volterra integral equations of convolution type with the logarithmic kernels [3,26,27,30,108,114,115]. However, integral equations are considered here not only in the context of the Volterra functions but rather in general, including a number of related topics, elements of the functional analysis, convolution products and fundamentals of the Laplace-Abel transformation.

In Chapter 4 the occurence of the Volterra functions in the Laplace transformation is associated with the fact that on one hand their Laplace transforms are expressed in terms of the logarithmic functions, and are useful in a determination of many inverse transforms. On the other hand, the operational calculus is a powerful and an efficient method for the derivation of the Volterra function properties. Dealing with the Volterra functions, Humbert [61-67], Colombo [23-30], Parodi [89-95], Barrucand [12,13] and others [3-5,45] assumed that in most cases the functions under consideration are continous, differentiable and have the Laplace transforms and mathematical operations under the integral sign (e.g. differentiation with respect to variables and parameters, inversion of order of integration and etc.) when applying rules and theorems of operational calculus are permissible. Since our main interest lies in the application of the Volterra functions, the same formal approach will be followed here. The authors suppose that the readers are familiar with the basic properties and techniques of the Laplace transformation which can be easily found in the literature (e.g. [15,19,22,31,33,38,66,74,76,83,84,90,92,93,96,97,107,117] and many others). Occasionally however, in order to make the presented material self-contained, the related general formulas of operational calculus will also be presented using the common notation of the direct Laplace transform, $F(s) = L\{f(t)\}$ and the inverse Laplace transform, $f(t) = L^{-1}\{F(s)\}$ of the function $f(t)$.

Evidently, the main attention will be directed to the $\nu(t,\alpha)$ and less to the $\mu(t,\beta,\alpha)$ functions considering that many results can be extended by a simple generalization or by use of the recurrence and other relations which exist between different functions.

In Chapter 5 are discussed methods that can be used in the evaluation of series, integrals and limits associated with the Volterra functions.

There are four Appendixes dealing with a quite different subjects, but related to the material presented in this book. (A)-The Laplace transformation and the Mittag-Leffler type functions; (B)-Summation formulas; (C)-Application of the shifted Dirac delta function $\delta(x-1)$ to the evaluation of the asymptotic relations, limits of series, integrals and integral representations of elementary and special functions and (D)-Tables of Volterra functions and the list of symbols and definitions of special functions.

The book is mainly directed to researchers and graduate students of different disciplines of engineering and science, who encounter in their work problems associated with integral equations, integral transforms and special functions. The author hope that the appearance of this book will stimulate renewed research activity and interest in the Volterra functions.

There is a pleasure to express gratitude to the author's son Dr. Yoram Apelblat who read the text to help the author with English language and to the author's wife Ira for her patience, support and encouragement during so many years that this book was slowly written.

Alexander Apelblat
Chemical Engineering Department,
Ben Gurion University of the Negev, Beer Sheva, Israel.
August, 2001- August, 2006.

Chapter 1

1. Introduction

1.1. A Short Historical Survey

The Volterra functions which are defined by the following integrals

$$\nu(z) = \int_0^\infty \frac{z^u}{\Gamma(u+1)} du \tag{1.1.1}$$

$$\nu(z,\alpha) = \int_0^\infty \frac{z^{u+\alpha}}{\Gamma(u+\alpha+1)} du \tag{2.1.1}$$

$$\mu(z,\beta) = \int_0^\infty \frac{z^u u^\beta}{\Gamma(\beta+1)\Gamma(u+1)} du \quad \operatorname{Re}\beta > -1 \tag{3.1.1}$$

$$\mu(z,\beta,\alpha) = \int_0^\infty \frac{z^{u+\alpha} u^\beta}{\Gamma(\beta+1)\Gamma(u+\alpha+1)} du \quad \operatorname{Re}\beta > -1 \tag{4.1.1}$$

appeared for the first time in work of Ramanujan (1903-1914) [55], Touchard (1913) [110], Volterra (1916) [114] and Landau (1918) [73]. They introduced them in a context of different problems in the theory of prime numbers, definite integrals and integral equations. Actually, the last two functions are the generalizations of the Volterra functions and they were introduced by Colombo (1955) [28].

In Ramanujan's notebooks, a number of formulas can be found in the theory of definite integrals. One of them may be regarded as the Euler-Maclaurin summation formula [53-55]

$$\sum_{n=0}^{\infty}\Phi(n)=\int_0^{\infty}\Phi(u)du+\int_0^{\infty}\frac{\sum_{n=0}^{\infty}(-1)^n u^n\Phi(u)}{u[\pi^2+(\ln u)^2]}du \tag{5.1.1}$$

As pointed out by Hardy [55], this formula can be also deduced from the Plane formula (1820), which is discussed in a detail by Lindelöf (1904) in his *"Calcul des Résidus et Ses Applications"* [75]. In the special case of (5.1.1), $\Phi(u) = 1/\Gamma(u+1)$, the Volterra function $\nu(x)$ is encountered

$$\nu(x)=\int_0^{\infty}x^u\Phi(u)du=\int_0^{\infty}\frac{x^u}{\Gamma(u+1)}du \tag{6.1.1}$$

The Laplace transform of $\nu(x)$ is

$$L\{\nu(x)\}=\int_0^{\infty}e^{-sx}(\int_0^{\infty}\frac{x^u}{\Gamma(u+1)}du)dx=\int_0^{\infty}\frac{1}{\Gamma(u+1)}(\int_0^{\infty}e^{-sx}x^u\,dx)du=$$

$$\int_0^{\infty}(\frac{1}{s})^{u+1}du=\frac{1}{s\ln s} \tag{7.1.1}$$

$$\operatorname{Re} s>1$$

and from (7.1.1), the integral representation of $\nu(x)$ is reached by using the inverse formula of the Laplace transformation

$$\nu(x)=\frac{1}{2\pi i}\int_{c-i\infty}^{c+i\infty}\frac{e^{xs}}{s\ln s}ds \tag{8.1.1}$$

The result of the complex integration is [55]

$$\nu(x)=e^x-\int_0^{\infty}\frac{e^{-xu}}{u[\pi^2+(\ln u)^2]}du \tag{9.1.1}$$

which includes the famous *Ramanujan integral.* Ramanujan also proved a more general result [55]

$$\int_{-\xi}^{\infty}\frac{x^u}{\Gamma(u+1)}du+\int_0^{\infty}\frac{u^{\xi-1}e^{-xu}}{[\pi^2+(\ln u)^2]}(\cos(\pi\xi)-\frac{\sin(\pi\xi)}{\pi}\ln u)du=e^x \tag{10.1.1}$$

$$x\geq 0\quad,\quad \xi\geq 0$$

Hardy [55] showed that if $\Phi(u) = \Gamma(u+\alpha)/\Gamma(\alpha)\Gamma(u+1)$, the Ramanujan formula (5.1.1) gives

$$\frac{1}{(1-x)^{\alpha}} - \int_0^{\infty} \frac{x^u \Gamma(u+\alpha)}{\Gamma(\alpha)\Gamma(u+1)} du = \int_0^{\infty} \frac{(1+xu)^{-\alpha}}{u[\pi^2 + (\ln u)^2]} du \tag{11.1.1}$$

and if α is a positive integer the integrals in (11.1.1) can be evaluated in terms of elementary functions. The integral representation of $\nu(x)$ in the form given in (9.1.1) has been independently derived and used by Touchard [110] and Landau [73]. However, Touchard also gives two new integral representations of $\nu(x)$

$$\nu(z) = e^z - \frac{1}{2} - \frac{z}{\pi}\int_0^{\infty} e^{-zu} \tan^{-1}\left(\frac{\ln u}{\pi}\right) du \quad , \quad \operatorname{Re} z > 0 \tag{12.1.1}$$

$$\nu(z) = 2\sinh z + \frac{z}{\pi}\int_0^{\infty} e^{-zu} \tan^{-1}\left(\frac{\pi}{\ln u}\right) du \quad ; \quad \operatorname{Re} z > 0 \tag{13.1.1}$$

and

$$\nu\left(z, -\frac{1}{2}\right) = e^z + \frac{1}{\pi}\int_0^{\infty} \frac{e^{-zu} \ln u}{\sqrt{u}[\pi^2 + (\ln u)^2]} du \tag{14.1.1}$$

A special class of Volterra integral equations of the first kind is [116]

$$\int_0^x K(x,y)\Phi(y)dy = f(x) \tag{15.1.1}$$

where $\Phi(x)$ is an unknown function and the kernel $K(x,y)$ is the function $K(s)$ with $s = x-y$

$$\int_0^x K(x-y)\Phi(y)dy = f(x) \tag{16.1.1}$$

and therefore a certain function of one variable, have been called by Volterra [115,116] as *the integral equations of the closed cycle*. However, today they are rather called *the integral equations of the convolution type* because the convolution is a particular case of *the Volterra composition of the first kind* [115]

$$\overset{*}{g}\overset{*}{h} = \int_y^x g(x,\xi)h(\xi,y)d\xi \tag{17.1.1}$$

where $g(x,y) = g(x\text{-}y)$ and $h(x,y) = h(x\text{-}y)$. Introducing $u = \xi\text{-}y$ into (17.1.1) and setting $t = x\text{-}y$, we have *the convolution of two functions* $g(t)$ and $h(t)$

$$g(t) * h(t) = \int_0^t g(t-u)h(u)\,du \tag{18.1.1}$$

The detailed theory of composition and permutable functions is given in the Volterra and Pérès and Volterra books and in the Pérès thesis [95, 115,116].

The integral equations of the convolution type are usually solved using the Laplace transformation

$$L\{f(t)\} = \int_0^\infty e^{-st} f(t)\,dt = F(s) \tag{19.1.1}$$

because if $L\{f(t)\} = F(s)$ and $L\{g(t)\} = G(s)$, then

$$L\{f(t) * g(t)\} = F(s)G(s) \tag{20.1.1}$$

Thus, by introducing $L\{f(x)\} = F(s)$, $L\{K(x)\} = K(s)$ and $L\{\Phi(x)\} = \Phi(s)$, the Laplace transformed integral equation (16.1.1) becomes

$$F(s) = K(s)\Phi(s) \tag{21.1.1}$$

and the solution of integral equation is reduced to the inversion of the Laplace transform

$$\Phi(x) = L^{-1}\left\{\frac{F(s)}{K(s)}\right\} \tag{22.1.1}$$

The Volterra functions appear in the context of integral equations of convolution type in the case of logarithmic kernels. Volterra (1916) [114] and Volterra and Pérès (1924) [115] considered the solution of a general integral equation when the polynomial in $\ln(x\text{-}y)$ is the kernel

$$\int_0^x f(y)\left\{(x-y)^{\alpha-1}\sum_{k=0}^{n} a_k[\ln(x-y)]^{n-k} + (x-y)^{\alpha} F(y,x)\right\}dy = \Phi(x) \tag{23.1.1}$$

$$\alpha > 0$$

where a_k are constants, $F(y,x)$ and $\Phi(x)$ are finite, continuous and differentiable functions and $f(x)$ is an unknown function. In a particular case, of $n = 1$, $\alpha = 1$, $F(y,x) = 0$ with $a_0 = 1$ and $a_1 = \gamma = 0.57721...$ ($\gamma = -\psi(1)$ is the *Euler constant*)

$$\int_0^x f(y)\{\ln(x-y)+\gamma\}dy = \Phi(x) \tag{24.1.1}$$

they showed that the solution of (24.1.1) is given by

$$f(x) = -\frac{d^2}{dx^2}\left\{\int_0^x \Phi(y)[\int_0^\infty \frac{(x-y)^u}{\Gamma(u+1)}du]dy\right\} \tag{25.1.1}$$

where the inner integral denoted by them as $\lambda(x,y)$ is evidently the Volterra function $\nu(x\text{-}y)$. In the Volterra-Pérès notation, $\lambda(x,y;z)$ stands for $\nu(x\text{-}y,z)$ and they used this function to express the infinite integral which contains the ratio of psi (digamma) and gamma functions

$$\int_0^\infty \frac{(x-y)^{u+z}\,\psi(u+z+1)}{\Gamma(u+z+1)}du = \frac{(y-x)^z}{\Gamma(z+1)} + \ln(y-x)\,\nu(x-y,z) \tag{26.1.1}$$

The integral equations of the convolution type with logarithmic kernels continued sporadically to be considered, but mainly in relation to the Laplace transformation. The works of Fock (1924) [46,47], Poli (1935) [97], Colombo (1952,1953) [26,27], Butzer (1959) [17], Rühs (1962) [104] and of Rooney (1983) [103] belong to this group of studies. The use of the Laplace transform techniques (usually in the Carson-Laplace transformation version) to the kernels with logarithms is demonstrated by Fan (1942) [45], Parodi (1948-1950) [90-92], Colombo (1964) [30], Humbert and Colombo (1965) [66], Krasnov, Kissélev and Makarenko (1977) [71] and Strivastava and Buschman (1992) [108].

Solutions of the Volterra integral equations of the first kind (16.1.1) for special cases of the kernel

$$K(x-y) = \{a_0 + [a_1 + a_2\ln(x-y)]\,(x-y)^{\alpha-1}\} \tag{27.1.1}$$

have been given by Colombo (1952) [26] for $\alpha = 2$ and by Colombo (1953) [26], Krasnov, Kissélev and Makarenko (1977) [71] and Strivastava and Buschman (1992) [108] for $a_0 = a_1 = 0$ and $\alpha = 1$. Strivastava and Buschman gave also solutions for $a_0 = a_1 = 0$, $\alpha = 1/2$ and $a_0 = 0$, $a_1 = \psi(\alpha)$, $a_2 = -1$.

Besides logarithmic kernels, the Volterra functions are also associated with solutions of integral equations of the convolution type when kernels are themselves the Volterra functions. This case has been considered by Apelblat (1985) [3]. For the convolution kernels containing the exponential integral, Ei(-x), logarithmic integral, li(x), sine and cosine integral functions,

Si(x) and Ci(x), solutions of (16.1.1) that involve the Volterra functions are given by Strivastava and Buschman [108].

Another group of integral equations leads to numerous integral formulas of the type

$$\int_0^\infty K(x,y)\Phi(y)dy = f(x) \tag{28.1.1}$$

where both $\Phi(x)$ and $f(x)$ contain the Volterra functions. For example, in the simplest case

$$\int_0^\infty e^{-y^2/4x}\nu(y)dy = 2\sqrt{\pi x}\,\nu(x) \tag{29.1.1}$$

The kernels $K(x,y)$ include the heat transfer moments, $K(x,y) = y^n \exp(-y^2/4x)$, n = 0,1,2,3, ... , the error functions, Bessel and parabolic cylinder functions. The integral formulas have been derived by different methods, but if the Laplace transformation is considered, the investigations of Colombo (1943, 1955 and 1964) [23,24,28,30], Humbert (1944) [61], Humbert and Poli (1944) [67] and Parodi (1950 and 1957) [92,93] should be mentioned. Prior to 1953, the known integral formulas have been collected in *"Higher Transcendental Functions"* by Erdélyi, Magnus, Oberhettinger and Tricomi [43]. Some new results were reported by Apelblat [3,8].

At the same time, many basic properties of the Volterra functions have been investigated in conjunction with the development of operational calculus. This situation can be explained by the fact that on the one hand, the properties of Volterra functions can be determined immediately and directly by using the integral transform methods and on the other side, the Volterra functions are useful in inversion of some Laplace transforms. Specifically, the Volterra functions are the Laplace transforms of some expressions containing gamma and related functions, the Laplace transforms of Volterra functions are expressible in terms of the logarithmic functions and finally the Volterra functions are associated with the substitution formulas that include logarithms. For example, taking into account that the Laplace transform of function $f(t)$ is defined by

$$L\{f(t)\} = \int_0^\infty e^{-st} f(t)dt = F(s) \tag{30.1.1}$$

we have an illustration of this in few simple cases

$$L\{\frac{1}{\Gamma(t+1)}\} = \nu(e^{-s}) \tag{31.1.1}$$

$$L\{\nu(t)\} = \frac{1}{s\ln s} \quad ; \quad \operatorname{Re} s > 1 \tag{32.1.1}$$

$$L\{\int_0^\infty \frac{t^x}{\Gamma(x+1)} f(x)dx\} = \frac{1}{s} F(\ln s) \tag{33.1.1}$$

$$L\{\int_0^\infty \nu(t,x) f(x)dx\} = \frac{F(\ln s)}{s \ln s} \tag{34.1.1}$$

Corresponding formulas exist for other Volterra functions.

As a result of the extended activity observed in the field of operational calculus between the 1943-1953, the Volterra functions started to be treated more as an independent subject and less as solutions of the original Volterra integral equations. Significantly important contributions in this direction of research have been made by Humbert (1944, 1945 and 1950) [61-63], Humbert and Poli (1944) [67], Parodi (1950) [92], Colombo (1943, 1948, 1952, 1953 and 1955) [23-28], Barrucand and Colombo (1950) [14], Barrucand (1950 and 1951) [13,14] and Carstoiu (1950) [21]. These studies covered many aspects associated with the Volterra functions, namely the basic properties, recurrence, differential and integral relations [14,27,63,92], Laplace transforms and substitution formulas [14,21,26,27,61-63,92] and interrelations with other integral transforms (two-sided Laplace transformation, two-dimensional Laplace transformation, Mellin transformation and factorial transformation introduced by Fraser (1944, 1946) [48,49]) and special functions [14,25,27,63]. It is worthwhile to note that during the early stages of operational calculus, the Carson-Laplace transformation (p-multiplied Laplace transform) has been more frequently used than the ordinary one-sided Laplace transformation (30.1.1) and the same is true when the Volterra functions were considered. In part, the material dealing with the Volterra functions and covering this fruitful period has been collected in a small section 18.3, pp. 217-226 of *"Higher Transcendental Functions"* (1955) [43] under the title "*The function* $\nu(x)$ *and related functions*". A more complete description of the Volterra functions came only a decade later in 1965, in the second revised and enlarged edition (the first edition appeared in 1947) of the Humbert and Colombo book *"La Calcul Symbolique Et Ses Applications A La Physique Mathématique"* [66] and especially in the Studia Universitatis Lovanium No.18, (1964) [30] *"Étude des Transcendantes Intervenant Dans la Résolution des Équations Intégrales de Volterra a Noyaux Logarithmiques"*.

After the 1943-1953 period, with a gradual decline in the research activity in the field of operational calculus, the Volterra functions have been only sporadically mentioned in the literature. Dzrbashyan (1959 and 1960) [39,40] discussed properties of the integral transformation with the $\nu(ixy,1/2)$ kernel and the asymptotic behaviour and integral representations of $\nu(z)$ and $\nu(z,\alpha)$ functions. He considered the Volterra function $\nu(z,\alpha)$ as a limiting case of the *Mittag-Leffler type function* [43]

$$\nu(z,\alpha) = z^\alpha \lim_{\rho\to\infty} \{\frac{1}{\rho} E_{1/\rho,\alpha+1}(z^{1/\rho})\} \tag{35.1.1}$$

where $E_{\alpha,\beta}(z)$ is defined by [43]

$$E_{\alpha,\beta}(z) = \sum_{m=0}^{\infty} \frac{z^m}{\Gamma(\alpha m + \beta)} \tag{36.1.1}$$

Dorning, Nicolaenko and Thurber (1969) [37] analyzed the physical problem of time-dependent fast neutron field with simultaneous fission regeneration and slowing down. The asymptotic behaviour of the Ramanujan integral (9.1.1) represented the amplitude of a decaying fast-neutron field following a burst of neutrons at time zero. Such situation arises when the dimensions of the system supporting the neutron field are made sufficiently small so that pure exponential time-decay no longer exists. They derived the asymptotic behaviour of $\nu(t,\alpha)$ for real t

$$\nu(t,\alpha) = e^t - \frac{t^\alpha}{\Gamma(\alpha+1)\ln t} + O\left(\frac{t^\alpha \ln(\alpha+1)}{\Gamma(\alpha+1)(\ln t)^2}\right) \qquad \alpha > -1 \;,\; t \to \infty \tag{37.1.1}$$

and

$$\nu(t,\alpha) = \nu(t) - \frac{t^\alpha}{\ln t} \sum_{k=0}^{\infty} \frac{(-1)^k}{(\ln t)^k} \frac{d^k}{dz^k} \frac{1}{\Gamma(1+z)}\bigg|_{z=\alpha} + \frac{1}{\ln t} \sum_{k=0}^{\infty} \frac{(-1)^k}{(\ln t)^k} \frac{d^k}{dz^k} \frac{1}{\Gamma(z)}\bigg|_{z=1} \qquad \alpha > -1 \;,\; t > 1 \tag{38.1.1}$$

where the last term in (38.1.1) is the asymptotic expansion for the Ramanujan integral. The integral was numerically evaluated in the range $10^1 \leq t \leq 10^{12}$. The similar problem was also considered by Bouwkamp (1971) [16].

Only after about two decades, (Apelblat, 1985 and 1989 [3,6], Apelblat and Kravitsky, 1985 [4]) a revived interest in the Volterra functions was once again observed. This can be explained by the fact that certain infinite integrals of gamma and polygamma functions are expressible in terms of the Volterra functions. However, what is even more interesting, the infinite integrals with respect to the order of the Bessel and related functions are associated with the Volterra functions. A typical example includes the Struve $H_\mu(x)$ and $L_\mu(x)$ functions [6]

$$\int_0^\infty H_\mu(x)\,d\mu = \sqrt{\frac{2x}{\pi}} \int_0^1 \sin(xu)\,\nu\left[\frac{x}{2}(1-u^2)\right]du \tag{39.1.1}$$

$$\int_0^\infty L_\mu(x)\,d\mu = \sqrt{\frac{2x}{\pi}} \int_0^1 \sinh(xu)\,\nu\left[\frac{x}{2}(1-u^2)\right]du \tag{40.1.1}$$

and these integrals can be developed in the rapidly convergent series of the Volterra functions

$$\int_0^\infty H_\mu(x)d\mu = \sqrt{\frac{2x}{\pi}} \sum_{m=0}^{\infty} \frac{(-1)^m x^m \nu(\frac{x}{2}, m+\frac{1}{2})}{1\cdot 3\cdot 5\ldots(2m+1)} \tag{41.1.1}$$

$$\int_0^\infty L_\mu(x)d\mu = \sqrt{\frac{2x}{\pi}} \sum_{m=0}^{\infty} \frac{x^m \nu(\frac{x}{2}, m+\frac{1}{2})}{1\cdot 3\cdot 5\ldots(2m+1)} \tag{42.1.1}$$

In a context of this renewed research activity, a large number of infinite, indefinite and definite integrals of the Volterra functions was evaluated and for the first time, short tables of $\nu(x)$ function in the range $0.001 \leq x \leq 10$ [3] and of $\nu(x,-1/2)$ function in the range $0.0001 \leq x \leq 6$ [6] were presented. The derived asymptotic expressions of the Volterra functions for $x \to 0$ and $x \to \infty$, permitted an extension of the computed ranges.

From this short survey, it is clear that during about century the Volterra functions reached an independent, mature status. They passed from the position of being merely solutions of the convolution integral equations with the logarithmic kernels to a new one whereas the Volterra functions are involved in a large variety of quite different problems. The Volterra functions play an important role in several integral transforms, yet their position in the Laplace transformation is something unique. Finally, a gradual process is observed where the Volterra functions become more and more interrelated with other special functions.

Chapter 2

2. BASIC PROPERTIES OF VOLTERRA FUNCTIONS

2.1. DEFINITIONS AND NOTATIONS

As already pointed out earlier, the Volterra functions are defined by the improper integrals

$$\nu(z)=\int_0^{\infty}\frac{z^u}{\Gamma(u+1)}du \tag{1.2.1}$$

$$\nu(z,\alpha)=\int_0^{\infty}\frac{z^{u+\alpha}}{\Gamma(u+\alpha+1)}du \tag{2.2.1}$$

$$\mu(z,\beta)=\int_0^{\infty}\frac{z^u u^{\beta}}{\Gamma(\beta+1)\Gamma(u+1)}du \qquad \operatorname{Re}\beta>-1 \tag{3.2.1}$$

$$\mu(z,\beta,\alpha)=\int_0^{\infty}\frac{z^{u+\alpha}u^{\beta}}{\Gamma(\beta+1)\Gamma(u+\alpha+1)}du \qquad \operatorname{Re}\beta>-1 \tag{4.2.1}$$

The definition of $\mu(z,\beta,\alpha)$ and $\mu(z,\beta) = \mu(z,\beta, 0)$ can be extended to the entire β-plane by repeated integration by parts [43]

$$\mu(z,\beta,\alpha)=\frac{(-1)^n}{\Gamma(\beta+n+1)}\int_0^{\infty}u^{\beta+n}\frac{d^n}{du^n}\left\{\frac{z^{u+\alpha}}{\Gamma(u+\alpha+1)}\right\}du \qquad \operatorname{Re}\beta>-(n+1) \tag{5.2.1}$$

It follows from (5.2.1) that

$$\mu(z,-n,\alpha)=(-1)^{n-1}\frac{d^{n-1}}{d\alpha^{n-1}}\left(\frac{z^{\alpha}}{\Gamma(\alpha+1)}\right) \tag{6.2.1}$$

$$n=1,2,3,...$$

All Volterra functions are analytical functions of z with branch points at the origin and at infinity and they have no other singularities. The functions $\nu(z,\alpha)$ and $\mu(z,\beta,\alpha)$ are entire functions of α and $\mu(z,\beta)$ and $\mu(z,\beta,\alpha)$ are entire functions β. In older papers and some tables of integrals and integral transforms the factor $\Gamma(\beta+1)$ is omitted.

The functions $\nu(z)$ and $\nu(z,\alpha)$ have been explicitly introduced by Vito Volterra, but with a different notation. The last two functions $\mu(z,\beta)$ and $\mu(z,\beta,\alpha)$ appeared much later in the context of operational calculus [67]. The function $\nu(z,\alpha)$ is sometimes called *the incomplete Volterra function* $\nu(z)$, because it can be written as [67]

$$\nu(z,\alpha)=\int_{\alpha}^{\infty}\frac{z^{u}}{\Gamma(u+1)}du \tag{7.2.1}$$

If α is a positive integer, $\nu(z,\alpha)$ in (6.2.1) can be represented as repeated integration of $\nu(x)$ function [92]

$$\nu(z,n)=\int_{0}^{\infty}\frac{z^{u}}{\Gamma(u+1)}du=\int_{0}^{z}\dots\int_{0}^{z}\nu(x)(dx)^{n} \tag{8.2.1}$$

It seems convenient to name $\nu(z,\alpha)$ as *the Volterra function* considering that $\nu(z)$ is a particular case of it and similarly to call $\mu(z,\beta,\alpha)$ as *the generalized Volterra function,* while still preserving the special notation for $\nu(z)=\nu(z,0)$ and $\mu(z,\beta)=\mu(z,\beta,0)$.

In an analogy with the sine integral, *the Volterra integral function* is defined by [67]

$$\nu i(z,\alpha)=\int_{0}^{z}\frac{1}{u}\nu(u,\alpha)du \tag{9.2.1}$$

$$\alpha>0$$

and in a similar way it can be introduced

$$\mu i(z,\beta,\alpha)=\int_{0}^{z}\frac{1}{u}\mu(u,\beta,\alpha)du \tag{10.2.1}$$

but for $\alpha=0$ it follows that

$$\mu i(z,\beta)=\mu i(z,\beta,0)=\frac{1}{\beta}\mu(z,\beta-1) \tag{11.2.1}$$

2.2. Basic Properties and Interrelations between the Volterra Functions

There is a number of recurrence formulas which are valid for the Volterra functions

$$z\nu(z,\alpha-1)-\alpha\nu(z,\alpha)=\mu(z,1,\alpha) \tag{1.2.2}$$

$$z\mu(z,\beta,\alpha-1)-\alpha\mu(z,\beta,\alpha)=(\beta+1)\mu(z,\beta+1,\alpha) \tag{2.2.2}$$

$$\mu(z,\beta)=\mu(z,\beta,0)=\frac{z}{\beta}\mu(z,\beta-1,-1) \tag{3.2.2}$$

$$\nu(z,\alpha)=\mu(z,0,\alpha) \tag{4.2.2}$$

$$\nu(z)=\nu(z,0)=\mu(z,0)=\mu(z,0,0) \tag{5.2.2}$$

Introducing the complex variable $z = x + iy$ in the polar form $z = re^{i\theta}$ with $-\pi < \arg z < \pi$, *the real and imaginary parts of the generalized Volterra functions* are [28,30]

$$\operatorname{Re}\{\mu(re^{i\theta},\beta,\alpha)\}=\int_0^\infty\frac{r^{u+\alpha}u^\beta\cos[(u+\alpha)\theta]}{\Gamma(\beta+1)\Gamma(u+\alpha+1)}du \tag{6.2.2}$$

$$\operatorname{Im}\{\mu(re^{i\theta},\beta,\alpha)\}=\int_0^\infty\frac{r^{u+\alpha}u^\beta\sin[(u+\alpha)\theta]}{\Gamma(\beta+1)\Gamma(u+\alpha+1)}du \tag{7.2.2}$$

and similar formulas can be written for other Volterra functions using the reduction forms (3.2.2) - (5.2.2). For the purely imaginary arguments, $z = \pm iy$, we have [27,66]

$$\frac{\mu(iy,\beta,\alpha)+\mu(-iy,\beta,\alpha)}{2}=\int_\alpha^\infty\frac{y^u u^\beta\cos(\frac{\pi u}{2})}{\Gamma(\beta+1)\Gamma(u+1)}du \tag{8.2.2}$$

$$\frac{\mu(iy,\beta,\alpha)-\mu(-iy,\beta,\alpha)}{2}=\int_\alpha^\infty\frac{y^u u^\beta\sin(\frac{\pi u}{2})}{\Gamma(\beta+1)\Gamma(u+1)}du \tag{9.2.2}$$

The Volterra functions obey the following *symmetry relations*

$$\mu(-z,\beta,\alpha)=\mu(ze^{i\pi},\beta,\alpha) \tag{10.2.2}$$

$$\mu(\bar{z},\beta,\alpha)=\bar{\mu}(z,\beta,\alpha) \tag{11.2.2}$$

Barrucand [13] introduced the auxiliary function

$$Ts(z,\alpha)=\frac{1}{\pi\alpha}\int_0^\infty\frac{z^u\sin(\pi\alpha u)}{\Gamma(u+1)}du \tag{12.2.2}$$

and the second function with cosine can be denoted in a similar way as $Tc(z,\alpha)$. However, we propose to remove the normalization factor $1/\pi\alpha$, and define them as

$$Tc(z,\alpha)=\int_0^\infty\frac{z^u\cos(\pi\alpha u)}{\Gamma(u+1)}du \tag{13.2.2}$$

$$Ts(z,\alpha)=\int_0^\infty\frac{z^u\sin(\pi\alpha u)}{\Gamma(u+1)}du \tag{14.2.2}$$

because for $\alpha = 0$, $Tc(z,\alpha)$ is reduced to the Volterra function $\nu(z)$

$$Tc(z,0)=\nu(z) \tag{15.2.2}$$

$$Ts(z,0)=0 \tag{16.2.2}$$

when the Barrucand definition is less convenient because then $Ts(z,0) = \mu(z,1)$ and $Tc(z,0) \to \infty$. Introducing (13.2.2) and (14.2.2) into (6.2.2) - (9.2.2) we have

$$\mathrm{Re}\{\nu(re^{i\theta})\}=Tc(r,\frac{\theta}{\pi}) \tag{17.2.2}$$

$$\mathrm{Im}\{\nu(re^{i\theta})\}=Ts(r,\frac{\theta}{\pi}) \tag{18.2.2}$$

and

$$\frac{\nu(iy)+\nu(-iy)}{2}=Tc(y,\frac{1}{2}) \tag{19.2.2}$$

$$\frac{\nu(iy)-\nu(-iy)}{2}=Ts(y,\frac{1}{2}) \tag{20.2.2}$$

Derivatives with respect to the variable z follow from definitions of the Volterra functions are

$$\frac{d^n\nu(z)}{dz^n}=\nu(z,-n) \quad , \quad n=1,2,3,... \tag{21.2.2}$$

$$\frac{d^n\nu(z,\alpha)}{dz^n}=\nu(z,\alpha-n) \tag{22.2.2}$$

$$\frac{d^n\mu(z,\beta)}{dz^n}=\mu(z,\beta,-n) \tag{23.2.2.}$$

$$\frac{d^n\mu(z,\beta,\alpha)}{dz^n}=\mu(z,\beta,\alpha-n) \tag{24.2.2}$$

and from (22.2.2) and (24.2.2) it becomes

$$\frac{d^n\nu(z,2n)}{dz^n}=\nu(z,n) \quad , \quad n=1,2,3,... \tag{25.2.2}$$

$$\frac{d^n\mu(z,\beta,2n)}{dz^n}=\mu(z,\beta,n) \tag{26.2.2}$$

If in the expression

$$e^{-\alpha t}\nu(xe^t,\alpha)=\int_0^\infty \frac{u^{u+\alpha}e^{tu}}{\Gamma(u+\alpha+1)}du \tag{27.2.2}$$

the exponential function is expanded into the Taylor series

$$e^{-\alpha t}\nu(xe^t,\alpha)=\int_0^\infty \frac{u^{u+\alpha}}{\Gamma(u+\alpha+1)}[\sum_{n=0}^\infty \frac{(tu)^n}{\Gamma(n+1)}]du \tag{28.2.2}$$

and the order of integration and summation is interchanged then

$$\nu(xe^t,\alpha) = e^{\alpha t}\sum_{n=0}^{\infty} t^n \mu(x,n,\alpha) \tag{29.2.2}$$

Denoting $e^t = y$ or $e^t = 1/y$ we have

$$\nu(xy,\alpha) = y^{\alpha}\sum_{n=0}^{\infty} (\ln y)^n \mu(x,n,\alpha) \tag{30.2.2}$$

$$\nu(\frac{x}{y},\alpha) = y^{-\alpha}\sum_{n=0}^{\infty} (-1)^n (\ln y)^n \mu(x,n,\alpha) \tag{31.2.2}$$

For $\alpha = 0$, these series become

$$\nu(xe^t) = \sum_{n=0}^{\infty} t^n \mu(x,n) \tag{32.2.2}$$

$$\nu(xy) = \sum_{n=0}^{\infty} (\ln y)^n \mu(x,n) \tag{33.2.2}$$

$$\nu(\frac{x}{y}) = \sum_{n=0}^{\infty} (-1)^n (\ln y)^n \mu(x,n) \tag{34.2.2}$$

Applying the same procedure to the generalized Volterra function $\mu(x,\beta,\alpha)$, we have

$$\mu(xe^t,\beta,\alpha) = \frac{e^{\alpha t}}{\Gamma(\beta+1)}\sum_{n=0}^{\infty} \frac{\Gamma(\beta+n+1)}{\Gamma(n+1)} t^n \mu(x,\beta+n,\alpha) \tag{35.2.2}$$

which for integer values of β-parameter, can be written in the form

$$\mu(xe,m,\alpha) = \frac{e^{\alpha}}{m}\sum_{n=0}^{\infty} \frac{\mu(x,m+n,\alpha)}{B(m,n+1)} \quad , \quad m = 1,2,3,\ldots \tag{36.2.2}$$

2.3. Integral Representations of the Volterra Functions

The integral representation of $\nu(z)$ has been independently evaluated by Ramanujan [55] , Touchard [110] and Landau [73]

$$\nu(z) = e^z - \int_0^\infty \frac{e^{-zu}}{u[\pi^2 + (\ln u)^2]} du \tag{1.2.3}$$

$\operatorname{Re} z > 0$

Two other forms for ν(*z*) were given by Touchard [110]

$$\nu(z) = e^z - \frac{1}{2} - \frac{z}{\pi} \int_0^\infty e^{-zu} \tan^{-1}\left(\frac{\ln u}{\pi}\right) du \tag{2.2.3}$$

$$\nu(z) = 2\sinh z + \frac{z}{\pi} \int_0^\infty e^{-zu} \tan^{-1}\left(\frac{\pi}{\ln u}\right) du \tag{3.2.3}$$

$\operatorname{Re} z > 0$

who also presented the special case of the Volterra function ν(*z*,α)

$$\nu\left(z, -\frac{1}{2}\right) = e^z + \frac{1}{\pi} \int_0^\infty \frac{e^{-zu} \ln u}{\sqrt{u}[\pi^2 + (\ln u)^2]} du \tag{4.2.3}$$

For any value of α , the integral representation of ν(*z*,α) was given by Dzrbashyan [39,40]

$$\nu(z, \alpha) = e^z - \int_0^\infty \frac{e^{-zu}}{u[\pi^2 + (\ln u)^2]} du - \int_0^\alpha \frac{z^u}{\Gamma(u+1)} du \tag{5.2.3}$$

$\operatorname{Re} z \geq 0$

However, it is evident from (2.2.1), (6.2.1) and (1.2.3) that (5.2.3) expresses only the interrelation between the Volterra functions ν(*z*) and ν(*z*,α)

$$\nu(z, \alpha) = \nu(z) - \int_0^\alpha \frac{z^u}{\Gamma(u+1)} du \tag{6.2.3}$$

It is possible to present the Ramanujan result (10.1) in the following form

$$\nu(z, -\alpha) = e^z - \frac{1}{\pi} \int_0^\infty \frac{e^{-zu} \cos\left[\pi\alpha + \tan^{-1}\left(\frac{\ln u}{\pi}\right)\right]}{u^{1-\alpha}\sqrt{\pi^2 + (\ln u)^2}} du \tag{7.2.3}$$

$\alpha \geq 0 \quad , \quad z \geq 0$

In numerical computations, it is better to change the variable of integration in (7.2.3)

$$\nu(z,-\alpha) = e^{z} - \frac{1}{\pi}\int_{-\infty}^{\infty} \frac{e^{\alpha u - ze^{u}} \cos\,[\pi\alpha + \tan^{-1}(\frac{u}{\pi})]}{\sqrt{\pi^{2}+u^{2}}}du \tag{8.2.3}$$

$$\alpha \geq 0 \quad , \quad z \geq 0$$

Thus, (7.2.3) gives the integral representation of $\nu(z,\alpha)$ for negative values of α. For example, introducing $\alpha = 1/2$ into (7.2.3) and taking into account that

$$\cos\,[\frac{\pi}{2} + \tan^{-1}(\frac{\ln u}{\pi})] = -\frac{\ln u}{\sqrt{\pi^{2} + (\ln u)^{2}}} \tag{9.2.3}$$

the Touchard result (4.2.3) is recovered. For $\alpha = 0$, the integral representation (7.2.3) reduces to the Ramanujan integral (1.2.3) because

$$\cos\,[\tan^{-1}(\frac{\ln u}{\pi})] = \frac{\pi}{\sqrt{\pi^{2} + (\ln u)^{2}}} \tag{10.2.3}$$

In order to evaluate the integral representation of $\nu(z,\alpha)$ for positive values of α, the direct and inverse transforms of the Volterra function will be considered

$$L\{\nu(t,\alpha)\} = \int_{0}^{\infty} e^{-st}\nu(t,\alpha)\,dt = F(s,\alpha) \tag{11.2.3}$$

$$F(s,\alpha) = \frac{1}{s^{\alpha+1}\ln s} \quad , \quad \operatorname{Re}\alpha > -1 \quad , \quad \operatorname{Re}s > 1 \tag{12.2.3}$$

and

$$\nu(t,\alpha) = \frac{1}{2\pi i}\int_{c-i\infty}^{c+i\infty} \frac{e^{ts}}{s^{\alpha+1}\ln s}ds \tag{13.2.3}$$

where $\nu(t,\alpha)$ is a function of the real variable t and it equals zero for negative t.

The complex integral (13.2.3) can be evaluated by taking into account that the integrand has the branch point at $s = 0$ and the pole at $s = 1$. The Bromwich contour used in the integration is presented in figure 1. The contour consists of the straight line AB, a large semi-circle of radius R, a small circle about the origin and a cut along the negative axis. According to the Jordan lemma, the integrals over the large semi-circle parts BC and GA vanish as R →

∞. Thus contributions to the integral (13.2.3) come only from the CD, DEF and FG parts of the contour and from the pole at $s = 1$. The residue at this point is e^t and therefore

$$\begin{aligned} \nu(t,\alpha) &= e^t + \frac{1}{2\pi i}\int_0^\infty e^{-tu}[F(ue^{-i\pi},\alpha) - F(ue^{i\pi},\alpha)]du \\ &+ \frac{1}{2\pi i}\lim_{\varepsilon\to 0}\int_{-\pi}^{\pi} e^{t\varepsilon e^{i\theta}} F(\varepsilon e^{i\theta},\alpha)\varepsilon e^{i\theta} d\theta \end{aligned} \qquad (14.2.3)$$

However, it should be pointed out that as $\varepsilon \to 0$, the last integral in (14.2.3) converges only for $-1 < \alpha \le 0$ and for α values from this interval the integral over the small circle DEF vanishes. How to overcome this difficulty and extend the integration to positive values of α will be discussed later.

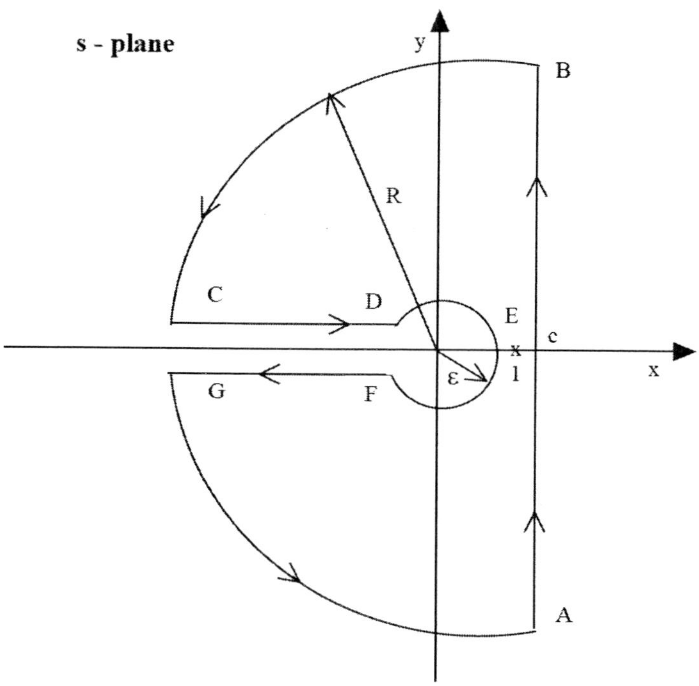

Figure 1. The Bromwich contour of integration of (13.2.3).

Introducing $F(s,\alpha)$ from (12.2.3) into (14.2.3), the final result of the complex integration is

$$v(t,\alpha)=e^{t}+\frac{1}{\pi}\int_{0}^{\infty}\frac{e^{-tu}\sin[\pi(\alpha+1)+\tan^{-1}(\frac{\pi}{\ln u})]}{u^{\alpha+1}\sqrt{\pi^{2}+(\ln u)^{2}}}du \tag{15.2.3}$$

$$-1<\alpha\leq 0 \quad , \quad t\geq 0$$

For α = 0, we have once again the Ramanujan integral

$$v(t)=e^{t}-\int_{0}^{\infty}\frac{e^{-tu}}{u[\pi^{2}+(\ln u)^{2}]}du \tag{16.2.3}$$

because

$$\sin[\pi+\tan^{-1}(\frac{\pi}{\ln u})]=-\sin[\tan^{-1}(\frac{\pi}{\ln u})]=-\frac{\pi}{\sqrt{\pi^{2}+(\ln u)^{2}}} \tag{17.2.3}$$

and since

$$\sin[\frac{\pi}{2}+\tan^{-1}(\frac{\pi}{\ln u})]=\cos[\tan^{-1}(\frac{\pi}{\ln u})]=\frac{\ln u}{\sqrt{\pi^{2}+(\ln u)^{2}}} \tag{18.2.3}$$

(15.2.3) reduces to the Touchard integral representation of $\nu(t,-1/2)$ given in (4.2.3).

There are additional cases when $\nu(t,\alpha)$ takes a rather simplified form

$$v(t,-\frac{1}{4})=e^{t}+\frac{1}{\sqrt{2}\pi}\int_{0}^{\infty}\frac{e^{-tu}(\ln u-\pi)}{u^{3/4}[\pi^{2}+(\ln u)^{2}]}du \tag{19.2.3}$$

$$v(t,-\frac{3}{4})=e^{t}+\frac{1}{\sqrt{2}\pi}\int_{0}^{\infty}\frac{e^{-tu}(\ln u+\pi)}{u^{1/4}[\pi^{2}+(\ln u)^{2}]}du \tag{20.2.3}$$

$$v(t,-\frac{1}{3})=e^{t}+\frac{1}{2\pi}\int_{0}^{\infty}\frac{e^{-tu}(\sqrt{3}\ln u-\pi)}{u^{2/3}[\pi^{2}+(\ln u)^{2}]}du \tag{21.2.3}$$

$$v(t,-\frac{2}{3})=e^{t}+\frac{1}{2\pi}\int_{0}^{\infty}\frac{e^{-tu}(\sqrt{3}\ln u+\pi)}{u^{1/3}[\pi^{2}+(\ln u)^{2}]}du \tag{22.2.3}$$

Now we will return to the problem of integral representation of $\nu(t,\alpha)$ but for positive values of α. If the Laplace transform of $f(t)$ is $F(s)$ and

$$f(0) = f'(0) = f''(0) = \ldots = f^{(n-1)}(0) = 0 \tag{23.2.3}$$

then the Laplace transform of the nth derivative of $f(t)$ is given by

$$L\{f^{(n)}(t)\} = s^n F(s) \tag{24.2.3}$$

Applying this rule to $\nu(t,\alpha)$ we have

$$L\{\nu^{(n)}(t,\alpha)\} = s^n F(s,\alpha) \quad , \quad n = 1,2,3,\ldots \tag{25.2.3}$$

by taking into consideration that for $\alpha > n - 1$

$$\begin{gathered} \nu^{(m)}(0,\alpha) = \nu(0,\alpha - m) = 0 \\ m = 1,2,3,\ldots(n-1) \end{gathered} \tag{26.2.3}$$

This follows from (2.2.1), because for any positive value of α, $\nu(0,\alpha)$ is equal zero. Thus, the inverse Laplace transform of (24.2.3) expresses the nth derivative of $\nu(t,\alpha)$

$$\begin{gathered} \nu^{(n)}(t,\alpha) = \frac{1}{2\pi i}\int_{c-i\infty}^{c+i\infty} \frac{e^{ts}}{s^{\alpha+1-n}\ln s} ds \\ n = 1,2,3\ldots \end{gathered} \tag{27.2.3}$$

Using in the complex integration the same Bromwich contour (figure 1) and considering that the convergence of the last integral in (14.2.3) imposes conditions on the values of α, the inverse of (27.2.3) is

$$\begin{gathered} \nu^{(n)}(t,\alpha) = e^t + \frac{1}{\pi}\int_0^\infty \frac{e^{-tu}\sin\left[\pi(\alpha+1-n) + \tan^{-1}\left(\frac{\pi}{\ln u}\right)\right]}{u^{\alpha+1-n}\sqrt{\pi^2 + (\ln u)^2}} du \\ n = 1,2,3,\ldots \quad , \quad n-1 < \alpha < n \end{gathered} \tag{28.2.3}$$

Evidently, the function $\nu(t,\alpha)$ can be evaluated from (28.2.3) with use of (26.2.3) by repeated integration with respect to the variable t. Let us denote the n-fold iterated integrals as

$$g_n(\tau) = \frac{1}{(n-1)!}\int_0^\tau \xi^{n-1} e^{(\tau-\xi)} d\xi \tag{29.2.3}$$

$$h_n(\tau) = \frac{1}{(n-1)!}\int_0^{\tau} \xi^{n-1} e^{-(\tau-\xi)} d\xi \qquad (30.2.3)$$
$$n = 1, 2, 3\ldots$$

then the integral representation of $\nu(t,\alpha)$ for positive values of α is given by

$$\nu(t,\alpha) = g_n(t) + \frac{1}{\pi}\int_0^{\infty} \frac{h_n(tu)\sin\left[\pi(\alpha+1-n) + \tan^{-1}\left(\frac{\pi}{\ln u}\right)\right]}{u^{\alpha+1-n}\sqrt{\pi^2 + (\ln u)^2}} du \qquad (31.2.3)$$
$$n = 1, 2, 3, \ldots \quad , \quad n-1 \le \alpha < n$$

The first three pairs of integrals from (29.2.3) and (30.2.3) are

$$g_1 = e^{\tau} - 1 \quad , \quad h_1 = 1 - e^{-\tau} \qquad (32.2.3)$$

$$g_2 = e^{\tau} - \tau - 1 \quad , \quad h_2 = e^{-\tau} + \tau - 1 \qquad (33.2.3)$$

$$g_3 = e^{\tau} - \frac{\tau^2}{2} - \tau - 1 \quad , \quad h_3 = 1 - \tau + \frac{\tau^2}{2} - e^{-\tau} \qquad (34.2.3)$$

For example, from (31.2.3) and (32.2.3) for $\alpha = 1/2$ we have

$$\nu(t,\frac{1}{2}) = e^t - 1 + \frac{1}{\pi}\int_0^{\infty} \frac{(1-e^{-tu})\ln u}{u^{3/2}[\pi^2 + (\ln u)^2]} du \qquad (35.2.3)$$

or generalizing for half-odd integers

$$\nu(t,n-\frac{1}{2}) = g_n(t) + \frac{1}{\pi}\int_0^{\infty} \frac{h_n(tu)\ln u}{u^{n+1/2}[\pi^2 + (\ln u)^2]} du \qquad (36.2.3)$$
$$n = 1, 2, 3, \ldots$$

In a similar way it is possible to obtain

$$\nu(t,n-\frac{3}{4}) = g_n(t) + \frac{1}{\sqrt{2}\pi}\int_0^{\infty} \frac{h_n(tu)[\ln u + \pi]}{u^{n+1/4}[\pi^2 + (\ln u)^2]} du \qquad (37.2.3)$$

$$\nu(t,n-\frac{1}{4}) = g_n(t) + \frac{1}{\sqrt{2}\pi}\int_0^\infty \frac{h_n(tu)[\ln u - \pi]}{u^{n+3/4}[\pi^2 + (\ln u)^2]}du \qquad (38.2.3)$$

$n = 1,2,3,...$

and

$$\nu(t,n-\frac{2}{3}) = g_n(t) + \frac{1}{2\pi}\int_0^\infty \frac{h_n(tu)[\sqrt{3}\ln u + \pi]}{u^{n+1/3}[\pi^2 + (\ln u)^2]}du \qquad (39.2.3)$$

$$\nu(t,n-\frac{1}{3}) = g_n(t) + \frac{1}{2\pi}\int_0^\infty \frac{h_n(tu)[\sqrt{3}\ln u - \pi]}{u^{n+2/3}[\pi^2 + (\ln u)^2]}du \qquad (40.2.3)$$

$n = 1,2,3,...$

For positive integer values of α, (31.2.3) takes the form

$$\nu(t,n) = g_{n+1}(t) + \int_0^\infty \frac{h_{n+1}(tu)}{u^{n+1}[\pi^2 + (\ln u)^2]}du \quad , \quad n = 1,2,3,... \qquad (41.2.3)$$

Using the recurrence formulas (1.2.2) and (2.2.2) it is possible to extend the integral representations (7.2.3) and (31.2.3) of $\nu(t,\alpha)$ to the generalized Volterra functions, but only with the β-parameter being a positive integer. They can be consecutively evaluated and the result is a combination of the Volterra functions with different α. For example, for $\beta = 1$ and $\beta = 2$

$$\mu(t,1,\alpha) = t\nu(t,\alpha-1) - \alpha\nu(t,\alpha) \qquad (42.2.3)$$

$$\mu(t,2,\alpha) = \frac{1}{2}\{t^2\nu(t,\alpha-2) - (2\alpha-1)t\nu(t,\alpha-1) + \alpha^2\nu(t,\alpha)\} \qquad (43.2.3)$$

However, the integral representation of the generalized Volterra functions with $\beta = 1,2,3,...n$ can be evaluated directly from

$$\mu(t,n,\alpha) = \frac{1}{2\pi i}\int_{c-i\infty}^{c+i\infty} \frac{e^{ts}ds}{s^{\alpha+1}(\ln s)^{n+1}} \qquad (44.2.3)$$

$\mathrm{Re}\,\alpha > -1 \quad , \quad \mathrm{Re}\,s > 1 \quad , \quad n = 1,2,3,...$

in a similar way to that given in (14.2.3) by using the Bromwich contour of integration (figure 1). For negative values of α, the result of the complex integration is

$$\mu(t,n,\alpha)=\frac{1}{n!}\frac{d^n}{ds^n}\left\{\frac{e^{ts}}{s^{\alpha+1}}\left[\frac{s-1}{\ln s}\right]^{n+1}\right\}\Bigg|_{s=1}+$$

$$\frac{1}{\pi}\int_0^\infty \frac{e^{-tu}\sin\,[\pi(\alpha+1)+(n+1)\tan^{-1}(\frac{\pi}{\ln u})]}{u^{\alpha+1}[\pi^2+(\ln u)^2]^{(n+1)/2}}du \tag{45.2.3}$$

$$-1<\alpha\leq 0\quad,\quad t\geq 0\quad,\quad n=1,2,3,\ldots$$

where the first term in (45.2.3) is the residue at $s = 1$.

For $\beta = 1$, equation (45.2.3) gives

$$\mu(t,1,\alpha)=(t-\alpha)e^t+\frac{1}{\pi}\int_0^\infty \frac{e^{-tu}\sin\,[\pi(\alpha+1)+2\tan^{-1}(\frac{\pi}{\ln u})]}{u^{\alpha+1}[\pi^2+(\ln u)^2]}du \tag{46.2.3}$$

$$-1<\alpha\leq 0$$

The integral representation of the generalized Volterra functions $\mu(t,1,\alpha)$ can be simplified for $\alpha = 0$ and $\alpha = -1/2$ taking into account that

$$\sin[2\tan^{-1}(\frac{\pi}{\ln u})]=\frac{2\pi\ln u}{[\pi^2+(\ln u)^2]} \tag{47.2.3}$$

$$\cos[2\tan^{-1}(\frac{\pi}{\ln u})]=-\frac{\pi^2-(\ln u)^2}{[\pi^2+(\ln u)^2]} \tag{48.2.3}$$

From (46.2.3), (47.2.3) and (48.2.3) we have

$$\mu(t,1)=te^t-2\int_0^\infty \frac{e^{-tu}\ln u}{u[\pi^2+(\ln u)^2]^2}du \tag{49.2.3}$$

$$\mu(t,1,-\frac{1}{2})=(t+\frac{1}{2})e^t-\frac{1}{\pi}\int_0^\infty \frac{e^{-tu}[\pi^2-(\ln u)^2]}{u^{1/2}[\pi^2+(\ln u)^2]^2}du \tag{50.2.3}$$

It follows from (45.2.3) that for $\beta = 2$

$$\mu(t,2,\alpha)=\frac{1}{2}[t^2-(2\alpha-1)t+\alpha^2]e^t+$$

$$\frac{1}{\pi}\int_0^\infty \frac{e^{-tu}\sin\,[\pi(\alpha+1)+3\tan^{-1}(\frac{\pi}{\ln u})]}{u^{\alpha+1}[\pi^2+(\ln u)^2]^{3/2}}du \tag{51.2.3}$$

$$-1<\alpha\leq 0$$

and using

$$\sin[3\tan^{-1}(\frac{\pi}{\ln u})]=\frac{\pi[3(\ln u)^2-\pi^2]}{[\pi^2+(\ln u)^2]^{3/2}} \tag{52.2.3}$$

$$\cos[3\tan^{-1}(\frac{\pi}{\ln u})]=\frac{\ln u[(\ln u)^2-3\pi^2]}{[\pi^2+(\ln u)^2]^{3/2}} \tag{53.2.3}$$

we have

$$\mu(t,2)=\frac{1}{2}(t+1)te^t-\int_0^\infty\frac{e^{-tu}[3(\ln u)^2-\pi^2]}{u[\pi^2+(\ln u)^2]^3}du \tag{54.2.3}$$

$$\mu(t,2,-\frac{1}{2})=[(t+1)^2-\frac{3}{4}]e^t-\frac{1}{\pi}\int_0^\infty\frac{e^{-tu}\ln u[3\pi^2-(\ln u)^2]}{u^{1/2}[\pi^2+(\ln u)^2]^3}du \tag{55.2.3}$$

An extension to positive values of α can be performed in a similar way as in (23.2.3) - (27.2.3), by considering that for $\alpha > m-1$

$$\begin{gathered}\mu^{(m)}(0,n,\alpha)=\mu(0,n,\alpha-m)=0\\ m=1,2,3,\ldots\quad,\quad n=1,2,3\ldots\end{gathered} \tag{56.2.3}$$

The mth derivative of the generalized Volterra function is therefore represented by

$$\mu^{(m)}(t,n,\alpha) = \frac{1}{n!}\frac{d^n}{ds^n}\left\{\frac{e^{ts}}{s^{\alpha+1-m}}\left[\frac{s-1}{\ln s}\right]^{n+1}\right\}\Bigg|_{s=1} +$$

$$\frac{1}{\pi}\int_0^\infty \frac{e^{-tu}\sin\,[\pi(\alpha+1-m)+(n+1)\tan^{-1}(\frac{\pi}{\ln u})]}{u^{\alpha+1-m}[\pi^2+(\ln u)^2]^{(n+1)/2}}du \tag{57.2.3}$$

$$m-1\le\alpha\le m \quad , \quad m=1,2,3,\ldots \quad , \quad n=1,2,3,\ldots$$

and $\mu(t,n,\alpha)$ can be evaluated by repeated integration of (57.2.3) with respect to the variable t. For example, if $0 \le \alpha < 1$

$$\mu(t,1,\alpha) = \alpha + (t-\alpha)e^t + \frac{1}{\pi}\int_0^\infty \frac{(1-e^{-tu})\sin\,[\pi(\alpha+1)+2\tan^{-1}(\frac{\pi}{\ln u})]}{u^{\alpha+1}[\pi^2+(\ln u)^2]}du \tag{58.2.3}$$

and using (47.2.3)

$$\mu(t,1,\frac{1}{2}) = \frac{1}{2} + (t-\frac{1}{2})e^t - \frac{1}{\pi}\int_0^\infty \frac{(1-e^{-tu})[\pi^2-(\ln u)^2]}{u^{3/2}[\pi^2+(\ln u)^2]^2}du \tag{59.2.3}$$

2.4. Asymptotic Behavior, Zeros and Extrema of the Volterra Functions

For real positive t, it follows from (7.2.3) and (31.2.3) that

$$\nu(t,\alpha) \sim e^t \quad \text{as} \quad t\to\infty \tag{1.2.4}$$

and from (42.2.3) and (43.2.3) that

$$\mu(t,n,\alpha) \sim t^n e^t \quad \text{as} \quad t\to\infty \tag{2.2.4}$$
$$n = 1,2,3,\ldots$$

For complex variable z, the asymptotic behavior of $\nu(z,\alpha)$ has been established by Dzrbashyan [39,40]. If $|z| \to \infty$ and $\pi/2 < \theta < \pi$, then

$$\nu(z,\alpha) = e^z + O(\frac{|z|}{\ln|z|}^{\alpha}) \quad , \quad \arg|z| \le \theta \tag{3.2.4}$$

and

$$\nu(z,\alpha) = O\left(\frac{|z|^{\alpha}}{\ln|z|}\right) \quad , \quad \arg|z| > \theta \tag{4.2.4}$$

The asymptotic behavior of $\nu(t,\alpha)$ for $t \to 0$ can be determined taking into account that for the Laplace integrals [41]

$$\int_0^{\infty} e^{-x\xi} f(\xi) d\xi \sim \sum_{n=0}^{\infty} \frac{1}{x^{n+1}} f^{(n)}(0) \quad \text{as} \quad x \to \infty \tag{5.2.4}$$

and by presenting $\nu(t,\alpha)$ as

$$\nu(t,\alpha) = \int_0^{\infty} \frac{t^{u+\alpha}}{\Gamma(u+\alpha+1)} du = \int_0^{\infty} \frac{t^{\alpha} e^{-u\ln(1/t)}}{\Gamma(u+\alpha+1)} du \tag{6.2.4}$$

Thus, from (5.2.4), the asymptotic expansion of $\nu(t,\alpha)$ in descending powers of $\ln(1/t)$, is [6,43]

$$\begin{aligned} &\nu(t \to 0, \alpha) \sim \frac{t^{\alpha}}{\Gamma(\alpha+1)} \sum_{n=0}^{\infty} \frac{a_n}{[\ln(1/t)]^{n+1}} \\ &a_n = \Gamma(\alpha+1) \frac{d^n}{d\xi^n} \left\{ \frac{1}{\Gamma(\xi+\alpha+1)} \right\} \Bigg|_{\xi=0} \\ &\alpha \neq -1, -2, -3, \ldots \end{aligned} \tag{7.2.4}$$

where the first four a_n coefficients are

$$\begin{aligned} a_0 &= 1 \\ a_1 &= -\psi(\alpha+1) \\ a_2 &= \psi(\alpha+1)^2 - \psi^{(1)}(\alpha+1) \\ a_3 &= 3\psi(\alpha+1)\,\psi^{(1)}(\alpha+1) - \psi^{(2)}(\alpha+1) - \psi(\alpha+1)^3 \end{aligned} \tag{8.2.4}$$

In the case of α being a negative integer, the corresponding expression can be derived starting from [28,30]

$$t^{-\alpha}\nu(t,\alpha)(\ln t)^2 = \int_0^{\infty} \frac{t^u (\ln t)^2}{\Gamma(u+\alpha+1)} du = \int_0^{\infty} \frac{\ln t}{\Gamma(u+\alpha+1)} d(t^u) \tag{9.2.4}$$

and performing twice the integration by parts

$$t^{-\alpha}\nu(t,\alpha)(\ln t)^2 = -\frac{\psi(\alpha+1)}{\Gamma(\alpha+1)} - \int_0^{\infty} t^u \frac{d}{du}\left\{\frac{\psi(u+\alpha+1)}{\Gamma(u+\alpha+1)}\right\} du \tag{10.2.4}$$

Since the quotient $\psi(\alpha+1)/\Gamma(\alpha+1)$ remains finite for α = -1,-2,-3, ... and the integral in (10.2.4) tends to zero with t, the final result is therefore

$$\nu(t,-n) \sim \frac{(-1)^{n+1}(n-1)!}{t^n(\ln t)^2} \quad \text{as} \quad t \to 0 \tag{11.2.4}$$

$$n = 1,2,3,...$$

For $t > 0$, to the great extent, the behavior of $\nu(t,\alpha)$ as a function of t, depends on the value of the parameter α.

If $\alpha \geq 1$, the first and second derivatives of $\nu(t,\alpha)$ are

$$\nu'(t,\alpha) = \nu(t,\alpha-1) = \int_0^{\infty} \frac{(u+\alpha)t^{u+\alpha-1}}{\Gamma(u+\alpha+1)} du \tag{12.2.4}$$

$$\nu''(t,\alpha) = \nu(t,\alpha-2) = \int_0^{\infty} \frac{(u+\alpha)(u+\alpha-1)t^{u+\alpha-2}}{\Gamma(u+\alpha+1)} du \tag{13.2.4}$$

and they are always positive because the integrands of (12.2.4) and (13.2.4) are always positive. Considering also that in this case $\nu(0,\alpha) = 0$, the Volterra function $\nu(t,\alpha)$ with $\alpha \geq 1$ increases continuously with t having no zeros or extrema.

However, for $0 \leq \alpha < 1$, the first derivative continues to be positive, but the integrand of (13.2.4) changes sign in the interval of integration and therefore the second derivative $\nu''(t,\alpha) = \nu(t,\alpha-2)$, vanishes at some value of t. Since the third derivative $\nu'''(t,\alpha) = \nu(t,\alpha-3)$ differs from zero and is positive at this point, the Volterra function $\nu(t,\alpha)$ has an inflection point and the function $\nu(t,\alpha-1)$ has a minimum. For instance, for $\alpha = 0$, this occurs at t = 0.241835521...where $\nu(t)$ has the inflection point, $\nu(t,-1)$ has the minimal value of 1.6242599... and this is the only root of $\nu(t,-2)$. The behavior of these three Volterra functions is shown in figure 2.

From (6.2.3)

$$\nu(z,\alpha) = \nu(z) - \int_0^{\infty} \frac{z^u}{\Gamma(u+1)} du \tag{14.2.4}$$

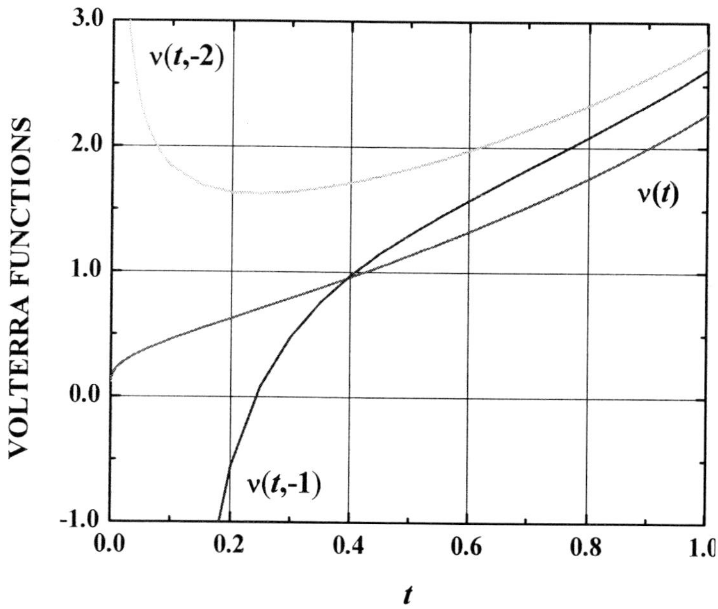

Figure 2. Volterra functions $\nu(t)$, $\nu(t,-1)$ and $\nu(t,-2)$.

it follows that for $t > 0$ and $\alpha > 0$

$$\nu(t) > \nu(t,\alpha_1) > \nu(t,\alpha_2) \quad \text{if} \quad \alpha_2 > \alpha_1 > 0 \tag{15.2.4}$$

Using the Ramanujan integral

$$e^t = \nu(t) + \int_0^\infty \frac{e^{-tu}}{u[\pi^2 + (\ln u)^2]} du \tag{16.2.4}$$

and

$$0 < \int_0^\infty \frac{e^{-tu}}{u[\pi^2 + (\ln u)^2]} du < \int_0^\infty \frac{1}{u[\pi^2 + (\ln u)^2]} du = 1 \tag{17.2.4}$$

it is evident that for $t \geq 0$, the Volterra function $\nu(t)$ is bounded by

$$e^t - 1 < \nu(t) < e^t \tag{18.2.4}$$

The case of negative values of α, can be analyzed by considering the asymptotic behavior of the Volterra functions for $t \to 0$ and $t \to \infty$.

From (7.2.4)

$$\nu(t \to 0,\alpha) \sim \frac{t^{\alpha}}{\Gamma(\alpha+1)\ln(1/t)} \tag{19.2.4}$$

it follows that at the origin the Volterra function approaches an infinite value and the sign of it depends on the gamma factor. The gamma function changes the sign when passing through negative integers. $\Gamma(\alpha+1)$ is positive for $-1 \leq \alpha \leq 0$ and therefore $\nu(t \to 0) \to +\infty$, but for $-2 \leq \alpha \leq -1$, the gamma function $\Gamma(\alpha+1)$ is negative and therefore $\nu(t \to 0) \to -\infty$. This can be generalized to

$$\lim_{t\to 0} \nu(t,\alpha) = -\infty \quad \text{for} \quad -2n \leq \alpha \leq 1-2n \tag{20.2.4}$$

$$\lim_{t\to 0} \nu(t,\alpha) = +\infty \quad \text{for} \quad 1-2n \leq \alpha \leq 2-2n \qquad n = 1,2,3\ldots \tag{21.2.4}$$

This result was already derived by Colombo [30], but in a quite different way. On the other hand, for any positive or negative α, we have from (1.2.4) that

$$\lim_{t\to \infty} \nu(t,\alpha) = +\infty \tag{22.2.4}$$

Therefore, by combining for example (20.2.4) with (22.2.4) it is clear that at least one value of t exists for which $\nu(t,\alpha) = 0$. Thus, if α belongs to one of the intervals (20.2.24), that can be denoted as (-,+) intervals, the Volterra function is zero (see for example behavior of the function $\nu(t,-2)$ in figure 2). However, $\nu(t,\alpha)$ is the first derivative of the Volterra function $\nu(t,\alpha+1)$, and therefore simultaneously $\nu'(t,\alpha+1) = 0$. Thus, at the same value of t, the Volterra function $\nu(t,\alpha+1)$ has an extremum, but the change from α to $\alpha+1$ is accompanied by a different behavior of the Volterra function. Now, $\alpha+1$ belongs to one of the intervals from (21.2.4), that will be denoted as (+,+) intervals. By combining (21.2.4) with (22.2.4) the extremum of $\nu(t,\alpha+1)$ is a minimum as shown for example, in the behavior of the function $\nu(t,-1)$ in figure 2. Gradual passing of the Volterra functions from the interval $0 \leq \alpha \leq 1$ to the (+,+) interval, $-1 \leq \alpha \leq 0$, and shifting the minimum points in a positive direction with the increasing of α is illustrated in figure 3.

Thus, the root of $\nu(t,\alpha)$ is the point at which $\nu(t,\alpha+1)$ has a minimum and the Volterra function $\nu(t,\alpha+2)$ has an inflection point because simultaneously $\nu(t,\alpha)$ and $\nu''(t,\alpha+2)$ equal to zero. Since $\alpha+2$ belongs to the (-,+) interval, the function $\nu(t,\alpha+2)$ possesses at least one zero but at different t. Such behavior is expected only in the simplest case when $\nu(t,\alpha+1)$ from the (+,+) interval has one minimum and the function is always positive. If $\nu(t,\alpha+1)$ is negative for some range of t, $t_1 < t < t_2$, having in addition to the minimum also two zeros, at $\nu(t_1,\alpha+1) = 0$ and at $\nu(t_2,\alpha+1) = 0$, then, $\nu(t,\alpha+2)$ from the (-,+) interval will have one root and two extrema, a maximum at t_1 and a minimum at t_2

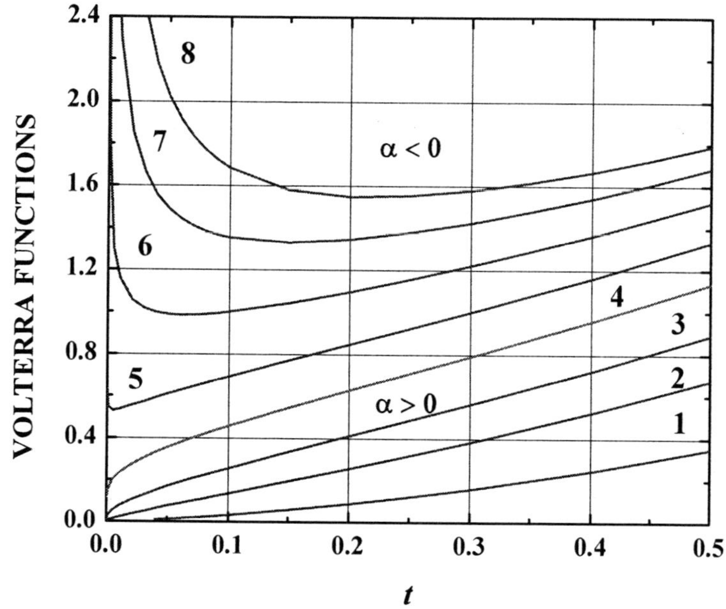

Figure 3. Behavior of the Volterra functions ν(*t*,α) with following α–parameters: **1** α = 1.0 ; **2** α = 0.50 ; **3** α = 0.25 ; **4** α = 0.0 ; **5** α = - 0.20 ; **6** α = - 0.40 ; **7** α = - 0.60 ; **8** α = - 0.80.

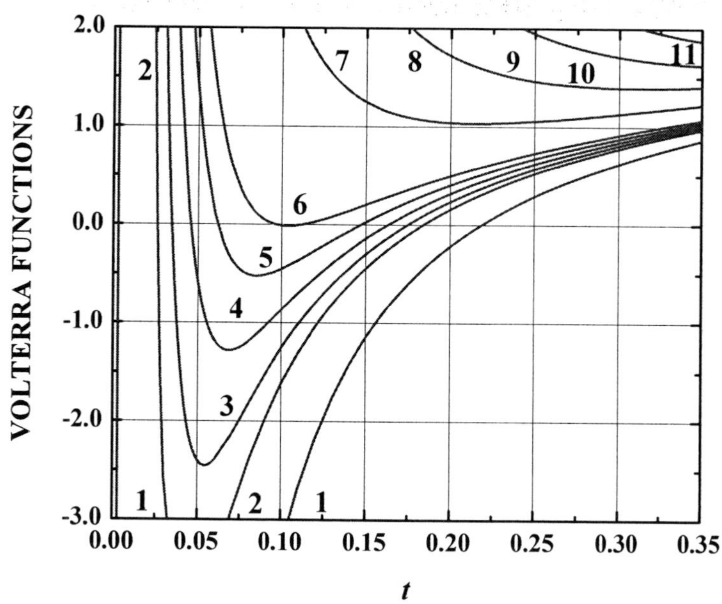

Figure 4. Behavior of the Volterra functions ν(*t*,α) with following α–parameters: **1** α = - 2.15 ; **2** α = - 2.21 ; **3** α = - 2.22 ; **4** α = - 2.23 ; **5** α = - 2.24 ; **6** α = - 2.25 ; **7** α = - 2.30 ; **8** α = - 2.35 ; **9** α = - 2.40 ; **10** α = - 2.45 ; **11** α = - 2.50.

In figures 4 and 5, a number of the Volterra functions from the consecutive $-3 < \alpha < -2$, (+,+) and $-4 < \alpha < -3$, (-,+) intervals are plotted to demonstrate this behavior. For example, the function ν(*t*,-2.24) has two zeros (see figure 4, curve 5) and therefore ν(*t*,-1.24) will have two extrema, a maximum and a minimum at the same values of *t* that ν(*t*,-2.24) has two roots. Similarly, function ν(*t*,-3.27) (see figure 5, curve 3) possesses two extrema and this means that ν(*t*,-4.27) will have two roots.

As can be seen, the behavior of the Volterra functions for $t > 0$ is rather complex with an exception of positive values of α when ν(*t*,α) has no zeros or extrema and only inflection points for $0 \leq \alpha < 1$. If α is negative, the Volterra functions will have zeros, extrema and inflection points, primarily depending on the interrelation between the Volterra functions with α, α+1 and α+2 parameters and to which group of intervals α belongs, (+,+) or (-,+).

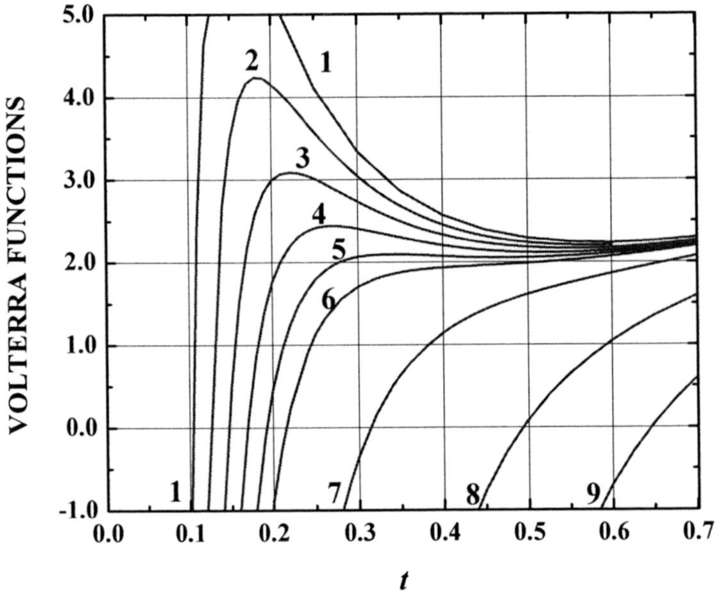

Figure 5. Behavior of the Volterra functions ν(*t*,α) with following α–parameters: **1** α = - 3.25 ; **2** α = - 3.26 ; **3** α = - 3.27 ; **4** α = - 3.28 ; **5** α = - 3.29 ; **6** α = - 3.30 ; **7** α = - 3.35 ; **8** α = - 3.50 ; **9** α = - 3.75.

2.5. Volterra Functions as a Function of the α and β-Parameters

Now, the properties of ν(*t*,α) will be discussed considering the parameter α as a variable. Applying the Leibnitz rule for differentiation of an integral to (7.2.1) we have

$$\frac{\partial \nu(t,\alpha)}{\partial \alpha} = -\frac{t^{\alpha}}{\Gamma(\alpha+1)} \tag{1.2.5}$$

It follows immediately from (1.2.5) that for a given t, the Volterra function $\nu(t,\alpha)$ passes through infinite number of extrema, which are all located at the negative axis, at $\alpha = -1, -2, -3, \ldots$, and from (6.2.1) we also have that

$$\lim_{\alpha\to\infty} \nu(t,\alpha) = 0 \tag{2.2.5}$$

The second derivative of $\nu(t,\alpha)$ with respect to α is

$$\frac{\partial^2 \nu(t,\alpha)}{\partial \alpha^2} = \frac{t^\alpha[\psi(\alpha+1) - \ln t]}{\Gamma(\alpha+1)} \tag{3.2.5}$$

and for a given t, it vanishes if

$$\psi(\alpha+1) = \ln t \tag{4.2.5}$$

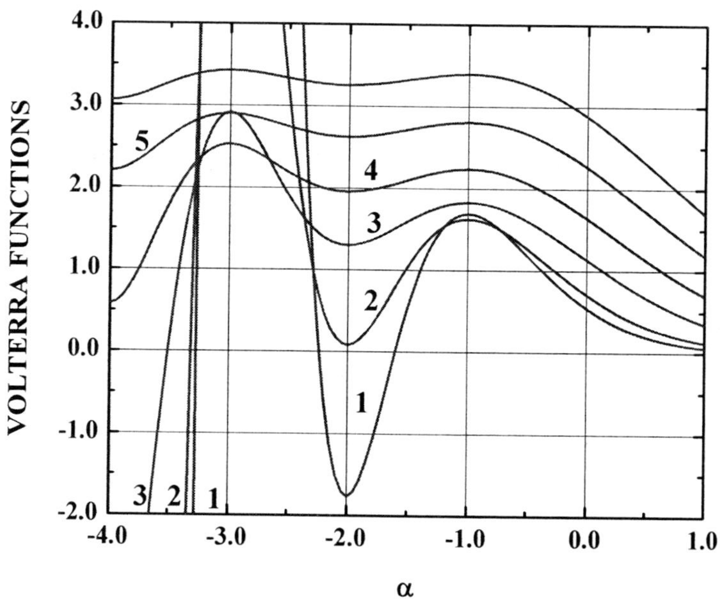

Figure 6. Volterra functions $\nu(t,\alpha)$ as a function of parameter α with: **1** $t = 0.15$; **2** $t = 0.25$; **3** $t = 0.50$; **4** $t = 0.75$; **5** $t = 1.00$; **6** $t = 1.20$.

Considering that the psi (digamma) function $\psi(\alpha+1)$ has one branch for $\alpha+1 > 0$ and an infinite number of branches for $\alpha+1 < 0$, equation (4.2.5) has one positive root and an infinite number of negative roots. Together with the fact that the third derivative with respect to α differs from zero, this means that the Volterra function has one inflection point for positive values of α and an infinite number of inflection points for negative values of α that are located between the extrema of $\nu(t,\alpha)$.

For example, if $t = 1$, the first three roots of $\psi(\alpha+1) = 0$ are $\alpha_1 = 0.463121..$, $\alpha_2 = -1.496623...$ and $\alpha_3 = -2.431334...$ Typical behavior of $\nu(t,\alpha)$ as a function of α-parameter for a number of variables t is illustrated in figure 6.

Since for a given t

$$\Delta\nu(t,\alpha) = \nu(t,\alpha+\Delta\alpha) - \nu(t,\alpha) \approx \frac{\partial\nu(t,\alpha)}{\partial\alpha}\Delta\alpha \tag{5.2.5}$$

it is possible to obtain from (1.2.5) and (5.2.5) the inequalities which are valid for the Volterra functions with negative values of α. For positive values of α, the inequalities are given in (15.2.4).

If α_1 and α_2 are from the same (+,+) interval, then

$$\nu(t,\alpha_2) > \nu(t,\alpha_1) \quad \text{if} \quad |\alpha_2| > |\alpha_1| \tag{6.2.5}$$

and similarly if α_1 and α_2 are from the same (-,+) interval, then

$$\nu(t,\alpha_2) < \nu(t,\alpha_1) \quad \text{if} \quad |\alpha_2| > |\alpha_1| \tag{7.2.5}$$

The surface $\nu(t,\alpha)$ for $0 \le t \le 2.0$ and $-4.0 \le \alpha \le 1$ is given in figure 7.

Successive differentiation of (1.2.5) yields the Taylor expansion of $\nu(t,\alpha)$

$$\nu(t,\alpha) = \nu(t,0) + \sum_{n=0}^{\infty} \frac{\alpha^n}{n!} \frac{\partial^n}{\partial\alpha^n} \left\{ -\frac{t^\alpha}{\Gamma(\alpha+1)} \right\} \tag{8.2.5}$$

and the first three terms are

$$\nu(t,\alpha) = \nu(t,0) - \frac{t^\alpha}{\Gamma(\alpha)}\{1 - \frac{\alpha[\psi(\alpha+1) - \ln t]}{2!} + \frac{\alpha^2[(\ln t)^2 + \psi(\alpha+1)^2 - \psi'(\alpha+1)]}{3!} - ...\} \tag{9.2.5}$$

By comparison of (9.2.5) with (6.2.3)

$$\int_0^\alpha \frac{t^u}{\Gamma(u+1)}du = \frac{t^\alpha}{\Gamma(\alpha)}\{1-\frac{\alpha[\psi(\alpha+1)-\ln t]}{2!}+ \frac{\alpha^2[(\ln t)^2+\psi(\alpha+1)^2-\psi'(\alpha+1)]}{3!}-...\} \tag{10.2.5}$$

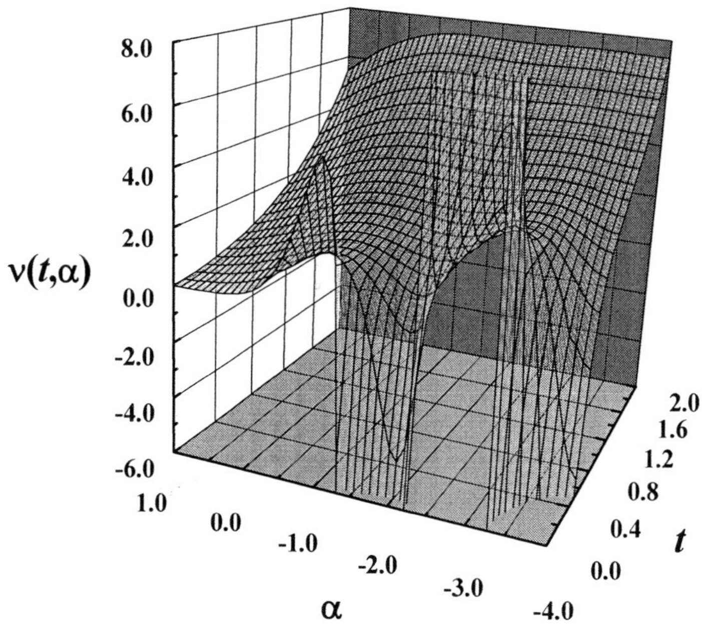

Figure 7. Volterra functions $\nu(t,\alpha)$ as a function of variable t and parameter α.

The series (10.2.5) takes a simple form for $t = 1$

$$\int_0^\alpha \frac{1}{\Gamma(u+1)}du = \frac{1}{\Gamma(\alpha)}\sum_{n=0}^{\infty}\frac{a_n\alpha^n}{(n+1)!} \tag{11.2.5}$$

where the first four a_n coefficients have been already given in (8.2.4). If also $\alpha = 1$, then

$$\int_0^1 \frac{1}{\Gamma(u+1)}du = \sum_{n=0}^{\infty}\frac{a_n}{(n+1)!} = 1.08514266... \tag{12.2.5}$$

where

$$\begin{aligned} a_0 &= 1 \\ a_1 &= \gamma - 1 \\ a_2 &= [(1-\gamma)^2 - (\frac{\pi^2}{6} - 1)] \\ a_3 &= [3(1-\gamma)(\frac{\pi^2}{6} - 1) - \psi^{(2)}(2) - (1-\gamma)^3] \end{aligned} \tag{13.2.5}$$

Direct differentiation of the integral (2.2.1) gives

$$\frac{\partial}{\partial \alpha} \int_0^\infty \frac{t^{u+\alpha}}{\Gamma(u+\alpha+1)} du = \int_0^\infty \frac{t^{u+\alpha}[\ln t - \psi(u+\alpha+1)]}{\Gamma(u+\alpha+1)} du \tag{14.2.5}$$

and by comparison with (1.2.5), yields as a by-product

$$\int_0^\infty \frac{t^{u+\alpha}[\ln t - \psi(u+\alpha+1)]}{\Gamma(u+\alpha+1)} du = -\frac{t^\alpha}{\Gamma(\alpha+1)} \tag{15.2.5}$$

which becomes for $t = 1$

$$\int_0^\infty \frac{\psi(u+\alpha+1)}{\Gamma(u+\alpha+1)} du = \frac{1}{\Gamma(\alpha+1)} \tag{16.2.5}$$

This result was already derived by Apelblat [3], but in a different way. Letting $\alpha = 0$ in (16.2.5) we have

$$\int_0^\infty \frac{\psi(u+1)}{\Gamma(u+1)} du = 1 \tag{17.2.5}$$

In a similar manner it is possible to obtain

$$\int_0^\infty \frac{t^{u+\alpha}[(\ln t)^2 - 2\ln t\,\psi(u+\alpha+1) + \psi(u+\alpha+1)^2 - \psi'(u+\alpha+1)]}{\Gamma(u+\alpha+1)} du = \frac{\psi(\alpha+1) - \ln t}{\Gamma(\alpha+1)} \tag{18.2.5}$$

which reduces for $t = 1$ to [3]

$$\int_0^\infty \frac{[\psi(u+\alpha+1)^2 - \psi'(u+\alpha+1)]}{\Gamma(u+\alpha+1)} du = \frac{\psi(\alpha+1)}{\Gamma(\alpha+1)} \tag{19.2.5}$$

and if also $\alpha = 0$, the integral is equal to the Euler constant

$$\int_0^\infty \frac{[\psi(u+1)^2 - \psi'(u+1)]}{\Gamma(u+1)} du = \gamma \tag{20.2.5}$$

These integrals, especially (15.2.5) and (19.2.5), will be considered later when the integrals associated with the Volterra functions will be discussed.

In an analogy with the logarithmic derivative of the gamma function, we introduce a similar function that will be denoted as $\lambda(t,\alpha)$

$$\lambda(t,\alpha) = \frac{d \ln \nu(t,\alpha)}{dt} \tag{21.2.5}$$

From (22.2.2) and (6.2.3)

$$\lambda(t,\alpha) = \frac{\nu(t,\alpha-1)}{\nu(t,\alpha)} = 1 + \frac{1}{\nu(t,\alpha)} \int_{\alpha-1}^{\alpha} \frac{t^u}{\Gamma(u+1)} du \tag{22.2.5}$$

Thus, the function $\lambda(t,\alpha)$ expresses the ratio of the Volterra functions with the α - 1 and α parameters. Evidently, the ratio of the Volterra functions with the α - n and α parameters, where n is a positive integer is given by the product of the λ-functions

$$\frac{\nu(t,\alpha-n)}{\nu(t,\alpha)} = \lambda(t,\alpha-n+1)\lambda(t,\alpha-n+2)\ldots\lambda(t,\alpha) \tag{23.2.5}$$

The asymptotic behavior of $\lambda(t,\alpha)$ follows from (1.2.4) and (7.2.4)

$$\lim_{t\to\infty} \lambda(t,\alpha) = 1 \tag{24.2.5}$$

$$\lim_{t\to 0} t\lambda(t,\alpha) = \alpha \tag{25.2.5}$$

In terms of the λ-functions, the first and second derivatives of $\lambda(t,\alpha)$ are

$$\frac{d\lambda(t,\alpha)}{dt} = \lambda(t,\alpha)[\lambda(t,\alpha-1) - \lambda(t,\alpha)] \tag{26.2.5}$$

$$\frac{d^2\lambda(t,\alpha)}{dt^2} = \lambda(t,\alpha)[\lambda(t,\alpha-2)\lambda(t,\alpha-1) - 3\lambda(t,\alpha-1)\lambda(t,\alpha) + 2\lambda(t,\alpha)^2] \tag{27.2.5}$$

Differentiation of $\lambda(t,\alpha)$ with respect to α gives

$$\frac{d\lambda(t,\alpha)}{d\alpha} = \sigma(t,\alpha)[\frac{\alpha}{t} - \lambda(t,\alpha)] \tag{28.2.5}$$

where

$$\sigma(t,\alpha) = \frac{\partial \ln \nu(t,\alpha)}{\partial \alpha} = -\frac{t^\alpha}{\Gamma(\alpha+1)\nu(t,\alpha)} \tag{29.2.5}$$

For α equal to zero or a positive integer, (26.2.5) becomes

$$\left.\frac{\partial\lambda(t,\alpha)}{\partial\alpha}\right|_{\alpha=n} = \frac{t^n}{n!\,\nu(t,n)}\left\{\frac{\nu(t,n-1)}{\nu(t,n)} - \frac{n}{t}\right\} \qquad n = 0,1,2,3\ldots \tag{30.2.5}$$

and the derivative vanishes when α is a negative integer.

The second derivative of $\lambda(t,\alpha)$ with respect to α is

$$\frac{\partial^2\lambda(t,\alpha)}{\partial\alpha^2} = 2\sigma(t,\alpha)^2[\lambda(t,\alpha) - \frac{\alpha}{t}] + \sigma(t,\alpha)\{\lambda(t,\alpha)[\psi(\alpha+1) - \ln t] - \frac{\alpha}{t}[\psi(\alpha) - \ln t]\} \tag{31.2.5}$$

and it takes a simple form only for $\alpha = 0$

$$\left.\frac{\partial^2\lambda(t,\alpha)}{\partial\alpha^2}\right|_{\alpha=0} = \frac{\nu(t,-1)}{\nu(t,n)^2}[\frac{2}{\nu(t)} + \gamma + \ln t] \tag{32.2.5}$$

Behavior of the generalized Volterra functions $\mu(t,\beta,\alpha)$ which is virtually unknown, is evidently more complex, resulting from the introduction of the additional parameter β. However, these functions can be analyzed in the same way as $\nu(t,\alpha)$ functions. Some general results are presented, but only for generalized Volterra functions that have the Laplace transforms [43]

$$F(s,\beta,\alpha)=L\{\mu(t,\beta,\alpha)\}=\frac{1}{s^{\alpha+1}(\ln s)^{\beta+1}}$$
$$\operatorname{Re}\alpha>-1 \quad , \quad \operatorname{Re}\beta>-1 \quad , \quad \operatorname{Re}s>1 \tag{33.2.5}$$

when t is the real variable and $\mu(t,\beta,\alpha)$ is identically zero for $t < 0$, and

$$\mu(t,\beta,\alpha)=\int_0^\infty \frac{t^{u+\alpha}u^\beta}{\Gamma(\beta+1)\Gamma(u+\alpha+1)}du=\int_\alpha^\infty \frac{t^x(x-\alpha)^\beta}{\Gamma(\beta+1)\Gamma(x+1)}dx \tag{34.2.5}$$

From (2.2.4) and (34.2.5) or using the properties of the Laplace transforms

$$\lim_{s\to 0} sF(s,\beta,\alpha)=\mu(t\to\infty,\beta,\alpha) \tag{35.2.5}$$

$$\lim_{s\to 0} sF(s,\beta,\alpha)=\mu(t\to 0,\beta,\alpha) \tag{36.2.5}$$

then for positive values of α, it follows that

$$\lim_{t\to\infty}\mu(t,\beta,\alpha)=+\infty$$
$$\alpha>-1 \quad , \quad \beta>-1 \tag{37.2.5}$$

and

$$\lim_{t\to 0}\mu(t,\beta,\alpha)=0$$
$$\alpha\geq 0 \quad , \quad \beta>-1 \tag{38.2.5}$$

In order to obtain the asymptotic behavior of $\mu(t,\beta,\alpha)$ for $t \to 0$ when $-1 < \alpha < 0$, the generalized Volterra functions from (34.2.5) are presented as a sum of two integrals

$$\mu(t,\beta,\alpha)=\int_0^\infty \frac{t^u(u+|\alpha|)^\beta}{\Gamma(\beta+1)\Gamma(u+1)}du+\int_{|\alpha|}^0 \frac{t^u(u+|\alpha|)^\beta}{\Gamma(\beta+1)\Gamma(u+1)}du \tag{39.2.5}$$

However, for $t \to 0$, the first integral approaches zero with t and therefore only a contribution from the second integral in (39.2.5) is of importance. By applying the mean value theorem of integrals, it follows that

$$\mu(t \to 0, \beta, \alpha) \sim \frac{|\alpha|^{\beta+1}(1-\xi)^{\beta} t^{-\xi|\alpha|}}{\Gamma(\beta+1)\Gamma(1-\xi|\alpha|)} \quad (40.2.5)$$

$$0 < \xi < 1$$

and therefore

$$\lim_{t \to 0} \mu(t, \beta, \alpha) = +\infty \quad (41.2.5)$$

$$-1 < \alpha < 0 \quad , \quad \beta > -1$$

The expected behavior of the generalized Volterra functions coming from (37.2.5), (38.2.5) and (41.2.5) is demonstrated in figures 8 and 9 in which $\mu(t, 1,\alpha)$ and $\mu(t,-0.25,\alpha)$ are plotted with positive and negative values of α. The surface of the generalized Volterra functions $\mu(t,-0.25,\alpha)$ for $0 \le t \le 2.0$ and $-1.0 \le \alpha \le 1$ is given in figure 10.

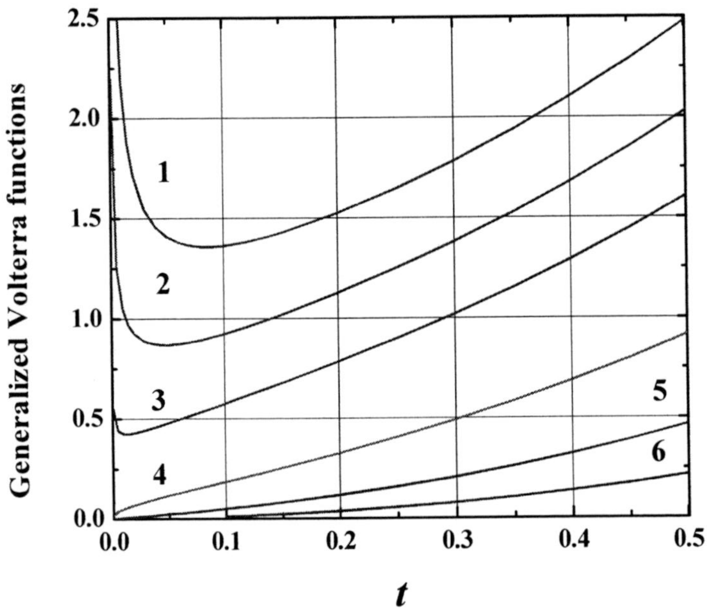

Figure 8. Generalized Volterra functions $\mu(t,1,\alpha)$ as a function of the variable t and the parameter α. **1** α = - 1.00 ; **2** α = - 0.75 ; **3** α = -0.50 ; **4** α = 0.0 ; **5** α = 0.50 ; **6** α = 1.00.

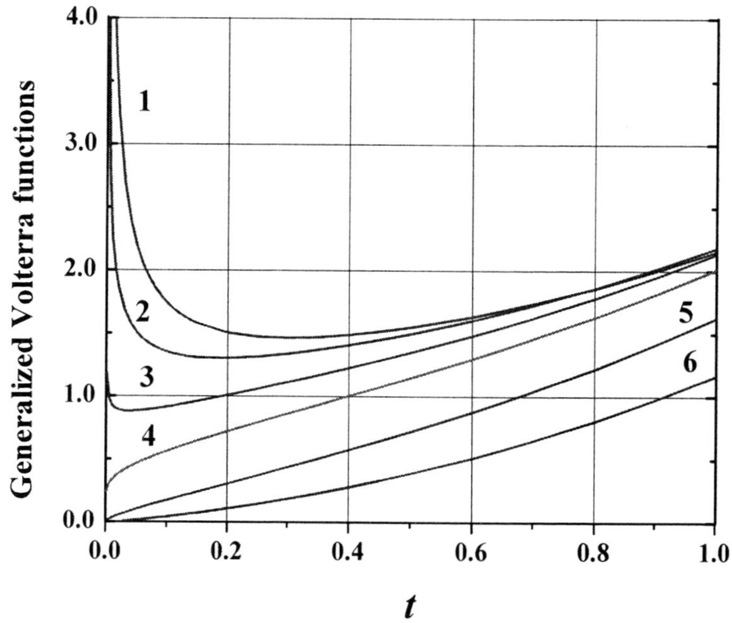

Figure 9. Generalized Volterra functions $\mu(t,-0.25,\alpha)$ as a function of the variable t and the parameter α. **1** $\alpha = -0.75$; **2** $\alpha = -0.50$; **3** $\alpha = -0.25$; **4** $\alpha = 0.0$; **5** $\alpha = 0.50$; **6** $\alpha = 1.00$.

Derivatives of the generalized Volterra functions with respect to variable t have been already given in (24.2.2). The first derivative of $\mu(t,\beta,\alpha)$ as a function of α can be obtained by applying the Leibnitz rule for differentiation of an integral to (34.2.5)

$$\frac{\partial}{\partial\alpha}\int_0^\infty \frac{t^{u+\alpha}u^\beta}{\Gamma(\beta+1)\Gamma(u+\alpha+1)}du = \int_0^\infty \frac{t^{u+\alpha}[\ln t - \psi(u+\alpha+1)]}{\Gamma(\beta+1)\Gamma(u+\alpha+1)}du \tag{42.2.5}$$

$$\frac{\partial}{\partial\alpha}\int_\alpha^\infty \frac{t^x(x-\alpha)^\beta}{\Gamma(\beta+1)\Gamma(x+1)}dx = -\beta\int_\alpha^\infty \frac{t^x(x-\alpha)^{\beta-1}}{\Gamma(\beta+1)\Gamma(x+1)}dx = -\mu(t,\beta-1,\alpha) \tag{43.2.5}$$

and therefore from (42.2.5) and (43.2.5)

$$\int_0^\infty \frac{t^{u+\alpha}u^\beta\psi(u+\alpha+1)}{\Gamma(\beta+1)\Gamma(u+\alpha+1)}du = \mu(t,\beta-1,\alpha) + \ln t\,\mu(t,\beta,\alpha) \tag{44.2.5}$$

This result was already differently derived by Apelblat [3].

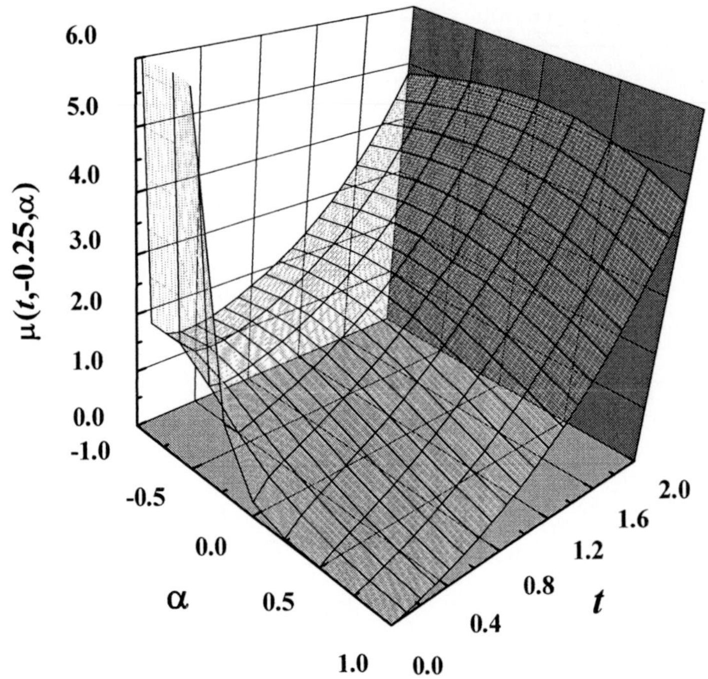

Figure 10. Generalized Volterra functions μ(*t*,-0.25,α) as a function of variable *t* and parameter α.

For $t = 1$, (42.2.5) becomes

$$\int_0^\infty \frac{u^\beta \psi(u+\alpha+1)}{\Gamma(u+\alpha+1)} du = \Gamma(\beta+1)\mu(1,\beta-1,\alpha) \tag{45.2.5}$$

which expresses the equality of integrals

$$\int_0^\infty \frac{u^\beta \psi(u+\alpha+1)}{\Gamma(u+\alpha+1)} du = \beta \int_0^\infty \frac{u^{\beta-1}}{\Gamma(u+\alpha+1)} du \qquad \beta > 0 \tag{46.2.5}$$

For β = 1, (46.2.5) reduces to

$$\int_0^\infty \frac{u\psi(u+\alpha+1)}{\Gamma(u+\alpha+1)} du = \int_0^\infty \frac{1}{\Gamma(u+\alpha+1)} du = \nu(1,\alpha) \tag{47.2.5}$$

If β is a positive integer, then the sum of the geometric and the arithmetic-geometric progressions as applied to (46.2.5) gives

$$\int_0^\infty \frac{[u^{n+1}-1]\,\psi(u+\alpha+1)}{(u-1)\Gamma(u+\alpha+1)}du = \frac{1}{\Gamma(\alpha+1)} + \int_0^\infty \frac{\{u^n[n(u-1)-1]+1\}}{(u-1)^2\Gamma(u+\alpha+1)}du \qquad (48.2.5)$$

$n = 1,2,3,\ldots$

The next differentiation of (34.2.5) with respect to α produces

$$\int_0^\infty \frac{t^{u+\alpha}u^\beta[(\ln t)^2 - 2\ln t\psi(u+\alpha+1)+\psi(u+\alpha+1)^2-\psi'(u+\alpha+1)]}{\Gamma(\beta+1)\Gamma(u+\alpha+1)}du \qquad (49.2.5)$$

$$= \mu(t,\beta-2,\alpha)$$

But using (44.2.5) it follows that

$$\int_0^\infty \frac{t^{u+\alpha}u^\beta[\psi(u+\alpha+1)^2-\psi'(u+\alpha+1)]}{\Gamma(\beta+1)\Gamma(u+\alpha+1)}du = \mu(t,\beta-2,\alpha)+$$

$$2\ln t\mu(t,\beta-1,\alpha)+(\ln t)^2\mu(t,\beta,\alpha) \qquad (50.2.5)$$

$\beta > 1$

For $t = 1$, (50.2.5) expresses the equality of integrals

$$\int_0^\infty \frac{u^\beta[\psi(u+\alpha+1)^2-\psi'(u+\alpha+1)]}{\Gamma(u+\alpha+1)}du = \beta(\beta-1)\int_0^\infty \frac{u^{\beta-2}}{\Gamma(u+\alpha+1)}du \qquad (51.2.5)$$

$\beta > 1$

The left-hand side integral in (51.2.5) has been already evaluated for β = 0 in (19.2.5)

$$\int_0^\infty \frac{[\psi(u+\alpha+1)^2-\psi'(u+\alpha+1)]}{\Gamma(u+\alpha+1)}du = \frac{\psi(\alpha+1)}{\Gamma(\alpha+1)} \qquad (52.2.5)$$

Considering that for β = 1, in the limit from (43.2.5) we have

$$\lim_{\beta\to 1}\mu(t,\beta-2,\alpha) = -\lim_{\beta\to 1}\left(\frac{\partial\mu(t,\beta-1,\alpha)}{\partial\alpha}\right) = -\frac{\partial\nu(t,\alpha)}{\partial\alpha} \qquad (53.2.5)$$

and using equation (15.2.5) with $t = 1$

$$\int_0^\infty \frac{u[\psi(u+\alpha+1)^2-\psi'(u+\alpha+1)]}{\Gamma(u+\alpha+1)}du=\frac{1}{\Gamma(\alpha+1)} \tag{54.2.5}$$

the following equality of integrals is reached

$$\int_0^\infty \frac{u[\psi(u+\alpha+1)^2-\psi'(u+\alpha+1)]}{\Gamma(u+\alpha+1)}du=\int_0^\infty \frac{\psi(u+\alpha+1)}{\Gamma(u+\alpha+1)}du \tag{55.2.5}$$

For β = 2, from (51.2.5) it follows that

$$\int_0^\infty \frac{u^2[\psi(u+\alpha+1)^2-\psi'(u+\alpha+1)]}{\Gamma(u+\alpha+1)}du=2\nu(1,\alpha) \tag{56.2.5}$$

and by comparison with (47.2.5)

$$\int_0^\infty \frac{u^2[\psi(u+\alpha+1)^2-\psi'(u+\alpha+1)]}{\Gamma(u+\alpha+1)}du=2\int_0^\infty \frac{u\psi(u+\alpha+1)}{\Gamma(u+\alpha+1)}du \tag{57.2.5}$$

Combining (52.2.5), (54.2.5) and (56.2.5) and putting $\alpha = \mathbf{0}$, the following result is obtained

$$\int_1^\infty \frac{u^2[\psi(u)^2-\psi'(u)]}{\Gamma(u)}du=2-\gamma+2\nu(1) \tag{58.2.5}$$

Repeated differentiation of μ(t,β,α) with respect to α gives

$$\frac{\partial^n \mu(t,\beta,\alpha)}{\partial\alpha^n}=(-1)^n\mu(t,\beta-n,\alpha) \tag{59.2.5}$$

and using (59.2.5) the Taylor expansion of the generalized Volterra function is

$$\mu(t,\beta,\alpha)=\sum_{n=0}^{\infty}\frac{(-1)^n\alpha^n\mu(t,\beta-n,\alpha)}{n!} \tag{60.2.5}$$

Finally, the first two derivatives of μ(t,β,α) with respect to the parameter β are

$$\frac{\partial \mu(t,\beta,\alpha)}{\partial \beta} = \int_0^\infty \frac{t^{u+\alpha} u^\beta \ln u}{\Gamma(\beta+1)\Gamma(u+\alpha+1)} du - \psi(\beta+1)\mu(t,\beta,\alpha) \tag{61.2.5}$$

and

$$\frac{\partial^2 \mu(t,\beta,\alpha)}{\partial \beta^2} = \int_0^\infty \frac{t^{u+\alpha} u^\beta [(\ln u)^2 - 2\psi(\beta+1)\ln u]}{\Gamma(\beta+1)\Gamma(u+\alpha+1)} du + [\psi(\beta+1)^2 - \psi'(\beta+1)]\mu(t,\beta,\alpha) \tag{62.2.5}$$

Chapter 3

3. VOLTERRA INTEGRAL EQUATIONS

As mentioned earlier, the Volterra functions were introduced as solutions of the Volterra integral equations of convolution type with the logarithmic kernels [3,26,27,30,108,114,115]. However, integral equations are considered here not only in the context of the Volterra functions but rather in general, including a number of related topics, elements of the functional analysis, convolution products and fundamentals of the Laplace-Abel transformation.

3.1. PRELIMINARIES ON THE VOLTERRA INTEGRAL EQUATIONS

An integral equation

$$f(x) = \int_0^x K(x,y)\phi(y)dy \tag{1.3.1}$$

where $f(x)$ and $K(x,y)$ are given functions and $\phi(y)$ is a function to be determined, is the *Volterra integral equation of the first kind.* The function $K(x,y)$ is called the *kernel of the integral equation.*

Similarly, an integral equation

$$\phi(x) = f(x) + \int_0^x K(x,y)\phi(y)dy \tag{2.3.1}$$

is the *Volterra integral equation of the second kind.* If $K(x,x) \neq 0$, the problem of solving the integral equation of the first kind can be reduced to that of the integral equation of the second kind. In fact, differentiation of (1.3.1) with respect to x converts (1.3.1) into (2.3.1)

$$\phi(x)=\frac{f'(x)}{K(x,x)}+\int_0^x L(x,y)\phi(y)dy \tag{3.3.1}$$

where

$$L(x,y)=-\frac{1}{K(x,x)}\frac{\partial}{\partial x}K(x,y) \tag{4.3.1}$$

Historically, prior to Volterra, Abel in 1826 [111] considered the following integral equation of the first kind

$$f(x)=\int_0^x \frac{\phi(y)}{(x-y)^\alpha}dy \qquad 0<\alpha<1 \tag{5.3.1}$$

which has the solution [120]

$$\phi(y)=\frac{\sin(\pi\alpha)}{\pi}\frac{d}{dy}\int_0^y \frac{\phi(z)}{(y-z)^{1-\alpha}}dz \tag{6.3.1}$$

The case α = 1/2 deals with a classical problem from mechanics. Starting from rest, a particle slides without friction down the curve to the origin in the xy-plane. If the time of descent is independent of the starting point, the curve is called the *tautochronous curve.* In this case, the solution of the integral equation leads to the cycloid. In 1673, Huygens showed that the tautochronous curve is an arch of an inverted cycloid and constructed a *cycloidal tautochronous pendulum* by considering that the evolute of a cycloid is a cycloid whose generating circle has the same diameter. Later on, in 1694 Johann Bernoulli and Leibnitz showed that the time of descent fulfills an extremum condition, i.e. the time is minimal with respect to any other curve. The cycloid is also encountered in hydrodynamics. Gerstner showed in 1802 that the profile of a sea wave is an elongated cycloid.

Integral equation (5.3.1) is also involved in the problem of the *fractional derivatives and integrals* [79,101].

3.2. Solution of the Volterra Integral Equations of the Second Kind

The classical *method of successive approximations* preconized by Emile Picard (1856-1941) promptly leads to the solution of the Volterra integral equations of the second kind (2.3.1), showing also the availability of a *resolvent kernel*

$$H(x,y)=\sum_{n=1}^{\infty}K_n(x,y) \tag{1.3.2}$$

for a suitable expression of the unknown $\phi(x)$ where $\{K_n(x,y)\}$ is the sequence of *iterated kernels.* The starting point of Picard's method consists in the consideration of the following equations:

$$\begin{aligned}
&\phi_0(x)=f(x)\\
&\phi_1(x)=f(x)+\int_0^x K(x,y)f(y)dy\\
&\phi_2(x)=f(x)+\int_0^x K(x,y)\phi_1(x)dy\\
&\cdots\cdots\cdots\cdots\cdots\cdots\cdots\cdots\cdots\cdots\cdots\cdots\\
&\phi_n(x)=f(x)+\int_0^x K(x,y)\phi_{n-1}(x)dy
\end{aligned} \tag{2.3.2}$$

By introducing

$$\phi_n(x)-\phi_{n-1}(x)=\psi_n(x) \tag{3.3.2}$$

it follows that

$$\begin{aligned}
&\psi_n(x)=\int_0^x K(x,y)\psi_{n-1}(y)dy\\
&n=1,2,3,...
\end{aligned} \tag{4.3.2}$$

and

$$\begin{aligned}
&\phi_0(x)=f(x)\\
&\phi_n(x)=\sum_{k=0}^{n}\psi_k(x)
\end{aligned} \tag{5.3.2}$$

Thus, it remains to verify that

$$\lim_{n\to\infty}\phi_n(x)=\phi(x)=\sum_{k=0}^{\infty}\psi_k(x) \tag{6.3.2}$$

is the solution of the integral equation (2.3.1).

The first term of the sequence $\{\psi_n(x)\}$ is

$$\psi_1(x) = \int_0^x K(x,y)f(y)dy \tag{7.3.2}$$

and the second term is

$$\psi_2(x) = \int_{z=0}^{z=x} K(x,z)dz \int_{y=0}^{y=z} K(z,y)f(y)dy \tag{8.3.2}$$

This iterated integral is equal to a double integral over a domain (D) in the yz-plane formed by the triangle with sides $y = 0$, $z = x$, $y = z$ and the element of area is the rectangle $dydz$ (see Figure 11)

$$\psi_2(x) = \iint_{(D)} K(x,z)K(z,y)dydz \tag{9.3.2}$$

Interchanging the order of integration, the integral (9.3.2) becomes

$$\psi_2(x) = \int_{y=0}^{z=x} K(x,z)K(z,y)f(y)dydz = \int_0^x K_2(x,y)f(y)dy \tag{10.3.2}$$

where

$$K_2(x,y) = \int_y^x K(x,z)K(z,y)dz \tag{11.3.2}$$

Continuing this procedure

$$\begin{aligned} &K_n(x,y) = \int_y^x K(x,z)K_{n-1}(z,y)dz \\ &K_1(x,y) = K(x,y) \\ &n = 2,3,4,... \end{aligned} \tag{12.3.2}$$

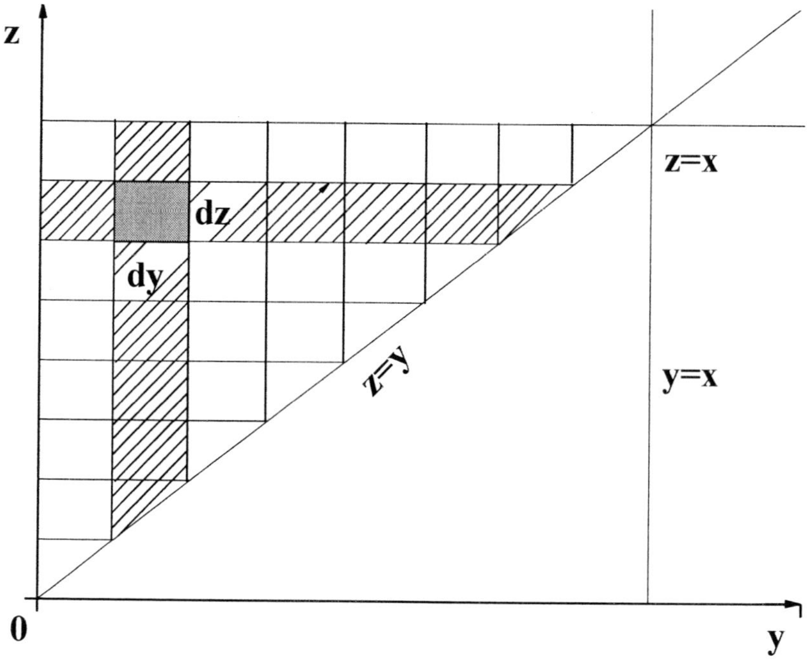

Figure 11. Domain (D) in the yz-plane.

we obtain the sequence of the iterated kernels $\{K_n(x,y)\}$. Thus, the Picard method of successive approximations leads to the sequence

$$\phi_n(x) = f(x) + \int_0^x [\sum_{m=1}^{n} K_m(x,y)] f(y)dy \tag{13.3.2}$$

and finally to the solution

$$\phi(x) = f(x) - \int_0^x H(x,y) f(y)dy \tag{14.3.2}$$

where the resolvent kernel is given in terms of the iterated kernels

$$H(x,y) = -\sum_{n=0}^{\infty} K_{n+1}(x,y) \tag{15.3.2}$$

It remains to determine conditions under which the series (15.3.2) is convergent. If $K(x,y)$ is continuous in the square $0 \le x \le h$, $0 \le y \le h$ and $f(x)$ is continuous for $0 \le x \le h$, then it is possible to find a *majorant series* for the resolvent kernel $H(x,y)$ and to ensure that the solution is unique [122].

3.3. THE L_2 SPACE

Since in some applications it is necessary to consider unbounded kernels and discontinuities of function $f(x)$, discontinuities of the first and second kind, we will extend our discussion to a more general and rigorous analysis. The theory of integration of Lebesgue, published already after appearance of the Volterra works, is a suitable framework to study integral equations.

The readers are supposed to be familiar with basic concepts of Lebesgue (1875-1941) theory of measure and integration [18]. In order to avoid too frequent references to classical treatises on the subject, some definitions will be briefly recalled and the properties of the complete metric space L_2 will be retraced.

Suppose that the following are given:

1. a measurable set E of points $(x_1,x_2, \dots ,x_n) = x$ in $\mathbf{R}^n$ (n-dimensional Euclidean space).
2. a real function $f(x)$ from $x \in E$ to $\mathbf{R}$.
3. two sets, $\overset{+}{F}$ and $\overset{-}{F}$ of points $(y_1,y_2, \dots ,y_{n+1}) = y$ in R^{n+1} defined as follows: $y_k = x_k$, $k = 1,2, \dots , n$ and $y_{n+1} = f(x_1,x_2, \dots ,x_n)$

then:

$$
\begin{aligned}
F^+ &= \{(y_1, y_2, \dots, y_n) \in E \subset R^{n+1} \quad , \quad 0 \le y_{n+1} \le f(x)\} \\
F^- &= \{(y_1, y_2, \dots, y_n) \in E \subset R^{n+1} \quad , \quad 0 \ge y_{n+1} \ge f(x)\}
\end{aligned}
\tag{1.3.3}
$$

The function $f(x)$ is said to be measurable on E if both sets $\overset{+}{F}$ and F^- are measurable in $\mathbf{R}^{n+1}$ (as by definition $f(x)$ is measurable on a set E *if and only if* for each real number α , the set of values $x \in E$ for which $f(x) > \alpha$ is measurable).

Let μ_1 and μ_2 be the measures of $\overset{+}{F}$ and F^- respectively, then f(x) is said to be *integrable over E* if at least one of the measures is finite. The Lebesgue integral of function $f(x)$ over the set E is defined by

$$
\int_E f(x)dx = \mu_1 - \mu_2 \tag{2.3.3}
$$

The value of (2.3.3) is finite if the measures μ_1 and μ_2 of the sets $\overset{+}{F}$ and F^- in $\mathbf{R}^{n+1}$ are finite. The Lebesgue integral is equal to + ∞ if only μ_2 is finite and is equal to - ∞ if only μ_1 is finite.

The function $f(x)$ measurable on E, is said to be *summable on E*, if $f(x)$ is integrable over E and if both measures μ_1 and μ_2 are finite. In other words, the necessary and sufficient condition for $f(x)$ to be summable on E is that the integral $\int_E |f(x)|dx$ is finite.

Two measurable functions f and g in $\mathbf{R}^n$ are said to be *equivalent in a set E of this space* if the set of points, $Q \in E$ where $f(Q) \neq g(Q)$, is of a measure zero. This can also be expressed by saying that the equality $f(Q) = g(Q)$ is satisfied *almost everywhere.* If two measurable functions f and g are continuous in a compact set and if a point Q of this compact set exists for which $f(Q) \neq g(Q)$, then these functions can not be equivalent. A set of equivalent functions constitutes an infinite class of functions which contains at least only one continuous function. Hereafter, the phrase *almost everywhere* is indicating that some proposition or equality is true, except perhaps on a set of zero measure.

If $f(x)$ is summable on a set E, the value of its Lebesgue integral remains unchanged whenever $f(x)$ is replaced by an equivalent function, in the sense already mentioned above.

The set of *square-summable functions on a set E*, is the set of functions which satisfy the condition

$$\int_E |f(x)|^2 dx < +\infty \tag{3.3.3}$$

It defines a *complete metric space*: every Cauchy sequence of square-summable functions admits a limit which is a square-summable function and the involved integral is a Lebesgue integral. This space is called the L_2 *space*. Hereafter in this section, *integral* and *integrable* stand for *Lebesgue integral* and *summable* respectively.

So far, x represented a point in the space $\mathbf{R}^n$. Now, we will consider functions of the real variable x which are square-integrable over $0 \leq x \leq h$ as well as functions of two real variable x and y which are square-integrable over $0 \leq x \leq h, 0 \leq y \leq h$. Functions of two real variables are associated with the kernels, iterative kernels and resolvent kernels of integral equations. The field of values of these functions is situated over the set of real numbers $\mathbf{R}$, but sometimes in physical applications, also over its complex counterpart $\mathbf{C}$. In the latter case, $f(x) = f_1(x) + i f_2(x)$ and it is supposed that $f_1(x)$ and $f_2(x)$ are measurable, i.e. $f_1(x) = \mathrm{Re}[f(x)]$ and $f_2(x) = \mathrm{Im}[f(x)]$. In this case one has

$$\begin{gathered} \int_a^b f(x)dx = \int_a^b f_1(x)dx + i\int_a^b f_2(x)dx \\ |f(x)| = \sqrt{[f_1(x)]^2 + f_2[(x)]^2} \end{gathered} \tag{4.3.3}$$

and if $f_1(x)$ and $f_2(x)$ are almost everywhere zero, then

$$\int_a^b |f(x)|dx = 0 \tag{5.3.3}$$

Evidently, the concept of functions which are integrable in the Lebesgue sense can easily be extended to functions of two variables x and y, defined in some domain of the xy-plane.

The *inner or scalar product,* (f,g) of two functions f and g which are integrable in the Lebesgue sense is defined by

$$(f,g)=\int_a^b f(x)\overline{g(x)}dx \tag{6.3.3}$$

where $\overline{g(x)}$ denotes the complex conjugate to $g(x)$. The inner product of two functions is characterized by the following properties

$$\overline{(f,g)}=(g,f) \tag{7.3.3}$$

$$(f_1+f_2,g_1+g_2)=(f_1,g_1)+(f_2,g_1)+(f_1,g_2)+(f_2,g_2) \tag{8.3.3}$$

$$(\alpha f,\beta g)=\alpha\overline{\beta}(f,g) \tag{9.3.3}$$

where α and β are constants.

Two functions f and g are called *orthogonal* if their inner product is zero, $(f,g) = 0$. A function which is almost everywhere zero over [a,b] is orthogonal to any function which is integrable in [a,b].

The *norm* $\|f\|$ of a function f, is defined by

$$\|f\|^2=(f,f)=\int_a^b |f(x)^2|dx\geq 0 \tag{10.3.3}$$

and it satisfies the following, easily verified relations

$$\|\alpha f\|=|\alpha|\|f\| \tag{11.3.3}$$

$$|(f,g)|\leq\|f\|\|g\| \tag{12.3.3}$$

$$\|f+g\|\leq\|f\|+\|g\| \tag{13.3.3}$$

where α is a constant. The inequality (12.3.3) is called the *Schwarz inequality* while (13.3.3) is the *Minkowski* or the *triangle inequality*.

The *linear space* L_2 is the space of functions with a *finite norm*, in other words, this is the linear space of functions which are square-integrable over the closed interval [a,b]. The *vectorial* space L_2 is a *metric space*. The *distance* (metric) between two elements f and g is defined by

$$d(f,g)=\|f-g\| \tag{14.3.3}$$

If m elements $\phi_1,\phi_2,\ldots,\phi_m$, are orthogonal with each other, then

$$\left\|\sum_{k=1}^{m}\phi_k\right\|^2 = \sum_{k=1}^{m}\|\phi_k\|^2 \tag{15.3.3}$$

which is the *Pythagorean relation.* A sequence $\{\phi_n\}$ of square-integrable functions is said to converge to $\phi \in L_2$ if

$$\lim_{n\to\infty}\|\phi-\phi_n\| = \lim_{n\to\infty}[\int_a^b|\phi(x)-\phi_n(x)|^2dx]^{1/2} \equiv 0 \tag{16.3.3}$$

Since for the space L_2, the Cauchy convergence criterion

$$\lim_{m,n\to\infty}\|\phi_m-\phi_n\| = 0 \tag{17.3.3}$$

remains valid, this space is also said to be *complete.* A complete space endowed with an *inner product* is called a *Hilbert space* [57-59,109].

If a sequence $\{\phi_n\}$ is given of L_2-functions such that for any $n \geq 2$, $\phi_1, \phi_2, \ldots, \phi_n$, are linearly independent, then following the Schmidt procedure, it is possible to construct an *orthonormal basis* $\psi_1, \psi_2, \ldots, \psi_n$, $((\psi_j, \psi_k) = \delta_{jk}$ where δ_{jk} denotes the Kronecker delta). In this context, the Riesz-Fisher theorem (1907) should be mentioned. If the series $\sum_{n=1}^{\infty} a_n^2$ converges, then there exists $f \in L_2$ such that $a_n = (f, \psi_n)$.

At this point, it is worthwhile to remark on the statement made at the beginning of this section. As reminded at section 3.2, the existence and uniqueness of solution for the Volterra equation are granted whenever the kernel $K(x,y)$ and $f(x)$ are continuous. The formally obtained solution (14.3.2) is altogether justified as a consequence of the series (15.3.2) being uniformly convergent [122].

In the space L_2, a more general assumption is possible. The norms of $K(x,y)$ and $f(x)$ are finite

$$\|K\| = \{\int_0^h\int_0^h|K(x,y)|^2\,dxdy\}^{1/2} < \infty$$
$$\|f\| = \{\int_0^h[f(x)]^2dx\}^{1/2} < \infty \tag{18.3.3}$$

i.e. $\|K\| < +\infty$ and similarly for the function f, that $\|f\| < +\infty$. The proof is based on the *Schwartz inequality* and the *Fubini theorem* [111].

Let

$$A(x) = \{ \int_0^h [K(x,y)]^2 dy \}^{1/2}$$
$$B(y) = \{ \int_0^h [K(x,y)]^2 dx \}^{1/2} \tag{19.3.3}$$

where $A(x)$ is defined *almost everywhere* in $0 \le x \le h$ and $B(y)$ is defined *almost everywhere* in $0 \le y \le h$.

Then

$$\|K\|^2 = \int_0^h [A(x)]^2 dx = \int_0^h [B(y)]^2 dy \tag{20.3.3}$$

and (10.3.2) implies successively

$$[K_2(x,y)]^2 = [\int_y^x K(x,z)K(z,y)dz]^2$$
$$\le \int_y^x [K(x,z)]^2 dz \int_y^x [K(z,y)]^2 dz \tag{21.3.3}$$
$$\le \int_0^h [K(x,z)]^2 dz \int_0^h [K(z,y)]^2 dz$$
$$= [A(x)]^2 [B(y)]^2$$

and

$$[K_3(x,y)]^2 \le \int_y^x [K(x,z)]^2 dz \int_y^x [K_2(z,y)]^2 dz$$
$$\le \int_0^h [K(x,z)]^2 dz \int_0^h [K_2(z,y)]^2 dz \tag{22.3.3}$$
$$= [A(x)]^2 [B(y)]^2 \int_y^x [A(z)]^2 dz$$

we have in the general case

$$[K_{n+2}(x,y)]^2 \le [A(x)]^2 [B(y)]^2 F_n(x,y) \quad , \quad n = 1,2,3,... \tag{23.3.3}$$

where $F_n(x,y)$ are defined by

$$F_1(x,y)=\int_y^x [A(z)]^2 dz$$
$$F_n(x,y)=\int_y^x [A(z)]^2 F_{n-1}(z,y)dz \quad , \quad n=2,3,4,... \tag{24.3.3}$$

In order to justify the solution of integral equation (13.3.2) and the resolvent kernel (14.3.2) for $K \in L_2$ and $f \in L_2$, it is sufficient to show that

$$F_n(x,y)=\frac{1}{n!}[F_1(x,y)]^n \tag{25.3.3}$$

This can be done by applying the method of mathematical induction. Evidently, the relation (25.3.3) is valid for n = 1 and by assuming that this also happens for n, the expected result for n +1 is

$$F_{n+1}(x,y)=\int_y^x \frac{1}{n!}[F_1(z,y)]^n[A(z)]^2 dz=\frac{1}{n!}\int_y^x [F_1(z,y)]^n \frac{\partial}{\partial z}F_1(z,y)dz$$
$$=\frac{1}{n!}\{\frac{1}{n+1}[F_1(z,y)]^{n+1}\}\Bigg|_{z=y}^{z=x}=\frac{1}{(n+1)!}[F_1(x,y)]^{n+1} \tag{26.3.3}$$

Denoting

$$a^2=\int_0^h [A(z)]^2 dz \tag{27.3.3}$$

we finally have

$$0 \le F_1(x,y) \le \int_0^h [A(x,y)]^2 dx \le a^2$$
$$0 \le F_n(x,y) \le \frac{1}{n!}a^{2n} \tag{28.3.3}$$
$$0 \le x \le h \quad , \quad 0 \le y \le h \quad , \quad n=1,2,3,...$$

From (23.3.3) and (28.3.3) it follows that

$$\left|K_{n+1}(x,y)\right| \leq A(x)B(y)\frac{a^{n}}{\sqrt{n!}} \tag{29.3.3}$$

$$n = 0,1,2,3,...$$

The power series with the general term $u_n(z) = z^n / \sqrt{n!}$ has an infinite radius of convergence. However, the series with the general term $u_n(x,y) = A(x)B(y)x^n / \sqrt{n!}$ can not be considered as the majorant series for (14.3.2) because the functions $A(x)$ and $B(y)$ may be unbounded. Despite this, the resolvent kernel $H(x,y)$ and the solution given in (13.3.2) remain valid. This is the consequence of the *dominated convergence theorem* [18].

Suppose that $\{u_n(x)\}$ *is a sequence of measurable functions on the set E and a function M(x) is measurable on E.* If for every $|s_n(x)|$,

$$|s_n(x)| = \left|\sum_{k=1}^{n} u_k(x)\right| \leq M(x) \tag{30.3.3}$$

and if almost everywhere

$$\lim_{n\to\infty} s_n(x) = s(x) \tag{31.3.3}$$

then

$$\int_E s(x)dx = \sum_{n=1}^{\infty} \int_E u_n(x)dx \tag{32.3.3}$$

The uniqueness of solution is immediately established in the context of the Hilbert space L_2. If ϕ_1 and ϕ_2 are two different solutions (on a set of non-zero measure) then $\psi(x) = \phi_1(x) - \phi_2(x)$ satisfies

$$\psi(x) = \int_0^x K(x,y)\psi(x)dx \tag{33.3.3}$$

and therefore $\psi(x) = 0$ almost everywhere in the interval $0 \leq x \leq h$. This defines a sequence of solutions which establishes a unique solution in the sense adopted in this section.

3.4. SOLUTION OF ORDINARY DIFFERENTIAL EQUATIONS

Solving of linear, non-homogeneous differential equations is immediately reduced to a solution of the Volterra integral equations [111]. Consider the function

$$F(x) = \int_0^x \frac{(x-y)^n}{(n-1)!}\phi(y)dy \tag{1.3.4}$$

which is equal to zero together with its n-1 first derivatives at x = 0 and $F^{(n)}(x) = \phi(x)$. This is a consequence of the Leibniz rule for differentiation of integrals

$$\frac{d}{d\lambda}\int_{a(\lambda)}^{b(\lambda)} f(x,\lambda)dx = f[b(\lambda)]\frac{db}{d\lambda} - f[a(\lambda)]\frac{da}{d\lambda} + \int_{a(\lambda)}^{b(\lambda)} \frac{\partial f}{\partial\lambda}dx \tag{2.3.4}$$

which holds if f is a continuous function of both variables x and λ [120] and in the particular case of $a(\lambda) = 0$ and $b(\lambda) = \lambda$..

If the linear non-homogeneous differential equation

$$w^{(n)} + a_1(x)w^{(n-1)} + \ \ldots \ + a_n(x)w = \Phi(x) \tag{3.3.4}$$

with the initial conditions

$$w(0) = c_0 \ , \quad w'(0) = c_1 \ , \quad w''(0) = c_2 \ , \quad \ldots w^{(n-1)}(0) = c_{n-1} \tag{4.3.4}$$

admits a solution $\phi(x)$, this satisfies the Volterra integral equation of the second kind (2.3.1) for which

$$K(x,y) = -\sum_{m=1}^{n} a_m(x)\frac{(x-y)^{m-1}}{(m-1)!} \tag{5.3.4}$$

and

$$f(x) = \Phi(x) - c_{n-1}a_1(x) - (c_{n-1}x + c_{n-2})a_2(x) - \ \ldots$$
$$\ldots - (c_{n-1}\frac{x^{n-1}}{(n-1)!} + \ \ldots \ + c_1x + c_0)a_n(x) \tag{6.3.4}$$

Moreover, a different method for solving linear homogeneous differential equations was proposed by Fubini. This method is illustrated by considering the second order differential equation, but it can be extended without difficulty to higher order differential equations.

If the second order linear differential homogeneous equation can be written in the form [68]

$$y'' + p(x)y' + q(x)y = \alpha(x)y'' + \beta(x)y' + \gamma(x)y \tag{7.3.4}$$

and if $F_1(x)$ and $F_2(x)$ are two linearly independent solutions of

$$y'' + p(x)y' + q(x)y = 0 \tag{8.3.4}$$

then the Cauchy problem associated with (7.3.4) can be reduced to solving the system of two Volterra integral equations [111].

In fact, following the *Lagrange method of variation of constants*, the solution of linear non-homogeneous differential equation is expected to be in the form

$$y(x) = C_1(x)F_1(x) + C_2(x)F_2(x) \tag{9.3.4}$$

where $C_1(x)$ and $C_2(x)$ satisfy the condition

$$C_1'(x)F_1(x) + C_2'(x)F_2(x) = 0 \tag{10.3.4}$$

It follows from (9.3.4) and (10.3.4) that

$$y'(x) = C_1(x)F_1'(x) + C_2(x)F_2'(x) \tag{11.3.4}$$

and

$$y''(x) = C_1(x)F_1''(x) + C_2(x)F_2''(x) + C_1'(x)F_1'(x) + C_2'(x)F_2'(x) \tag{12.3.4}$$

Introducing these expressions into (7.3.4) one has

$$C_1'F_1' + C_2'F_2' = \frac{\alpha F_1'' + \beta F_1' + \gamma F_1}{1-\alpha} C_1 + \frac{\alpha F_2'' + \beta F_2' + \gamma F_2}{1-\alpha} C_2 \tag{13.3.4}$$

and

$$\begin{aligned} C_1' &= -(C_1\Phi_1 + C_2\Phi_2)F_2 \\ C_2' &= (C_1\Phi_1 + C_2\Phi_2)F_1 \end{aligned} \tag{14.3.4}$$

where

$$\Phi_m = \frac{\alpha F_m'' + \beta F_m' + \gamma F_m}{(1-\alpha)W(x)} \qquad (15.3.4)$$

$$m = 1,2, \quad \alpha \neq 1$$

and it is assumed that the *Wronskian determinant*, $W(x)$,

$$W(x) \equiv \begin{vmatrix} F_1(x) & F_2(x) \\ F_1'(x) & F_2'(x) \end{vmatrix} \qquad (16.3.4)$$

does not vanish in the interval $0 \le x \le h$.

The system of two differential equations in (14.3.4) can also be written as

$$\begin{aligned} C_1(x) &= c_1 - \int_0^x [C_1(\xi)\Phi_1(\xi) + C_2(\xi)\Phi_2(\xi)]F_2(\xi)d\xi \\ C_2(x) &= c_2 - \int_0^x [C_1(\xi)\Phi_1(\xi) + C_2(\xi)\Phi_2(\xi)]F_1(\xi)d\xi \end{aligned} \qquad (17.3.4)$$

where $c_1 = C_1(0)$ and $c_2 = C_2(0)$ are prescribed values of the particular solution $y(x)$ and its first derivative $y'(x)$. Thus, in this way, the problem of solving linear differential equations is easily reduced to a solution of the Volterra integral equation of the second kind (2.3.1) with

$$\phi(x) = C_1(x)\Phi_1(x) + C_2(x)\Phi_2(x) \qquad (18.3.4)$$

In the framework of the Fubini method, Tricomi [111] considered stable solutions in which $F_1(x)$ and $F_2(x)$ remain bounded for $x \to +\infty$ and an extension of the procedure to higher-order differential equations and a choice of different forms of in the right-hand side of (7.3.4).

3.5. Special Types of Kernels

A kernel $K(x,y)$ is called *separable* or *degenerate kernel* if it can be expressed as the product of a function of x only and a function of y only, i.e.

$$K(x, y) = L(x)M(y) \qquad (1.3.5)$$

In this case the Volterra integral equation of the second kind (2.3.1) becomes

$$\frac{\phi(x)}{L(x)} = \frac{f(x)}{L(x)} + \int_0^x M(y)\phi(y)dy \tag{2.3.5}$$

and this is equivalent to the linear differential equation

$$\Phi'(x) - M(x)L(x)\Phi(x) = F(x) \tag{3.3.5}$$

where

$$\Phi(x) = \frac{\phi(x)}{L(x)}$$
$$F(x) = \frac{d}{dx}\left[\frac{f(x)}{L(x)}\right] \tag{4.3.5}$$

The *composition product* of two functions $L(x,y)$ and $M(x,y)$ of two variables x,y is defined by

$$L(x,y) * M(x,y) = \int_y^x L(x,s)M(s,y)ds \tag{5.3.5}$$

Such product leads to the sequence of iterated kernels $\{K_n(x,y)\}$ which were used to construct the resolvent kernel $H(x,y)$ in (14.3.2). In the same manner, the kernel $K_n(x,y)$ can be expressed as the sequence of the *nth power of compositions* $\{(\overset{*}{K})^n\}$

$$\{(\overset{*}{K})^n\} = \int_y^x K_{n-1}(x,s)K(s,y)ds$$
$$n \geq 2 \tag{6.3.5}$$

The composition product is evidently *distributive with respect to addition*

$$L(x,y) * [M_1(x,y) + M_2(x,y)] = L(x,y) * M_1(x,y) + L(x,y) * M_2(x,y) \tag{7.3.5}$$

and *associative with regard to multiplication*

$$L(x,y) * [M(x,y) * N(x,y)] = [L(x,y) * M(x,y)] * N(x,y) \tag{8.3.5}$$

In fact, denoting the interior of the isosceles triangle (Figure 12) as a domain (D), it is possible to write

$$\int_y^x L(x,s)ds \int_s^y M(s,t)N(t,s)dt = \iint_{(D)} L(x,s)M(s,t)N(t,y)dsdt =$$
$$\int_{s=y}^{s=t} L(x,s)M(s,t)ds \int_{t=y}^{t=x} N(y,t)dt \qquad (9.3.5)$$

The composition product does not obey the *commutative law*. In general $L * M$ differs from $M * L$. However, if $L(x,y) = L(y,x)$ and $M(x,y) = M(y,x)$ then also $L * M \equiv M * L$.

A problem which is frequently encountered, is to determine a class of *commutable* (also called *permutable*) functions that are permutable with a given function $f(x,y)$ [95].

These functions are permutable with one other and with their powers of composition $\overset{*}{f}^2, \overset{*}{f}^3, \ldots \overset{*}{f}^n$; thus every polynomial $a_1 f + a_2 \overset{*}{f}^2 + a_3 \overset{*}{f}^3 + \ldots a_n \overset{*}{f}^n$ where $a_1, a_2, \ldots, a_n$, are constants, is permutable with f.

A kernel $K(x,y)$ belongs to a *group of closed cycle* if it is a function of the difference $s = x-y$, i.e. $K(x,y) = K(s) = K(x-y)$. This case is often encountered in general theory of linear systems.

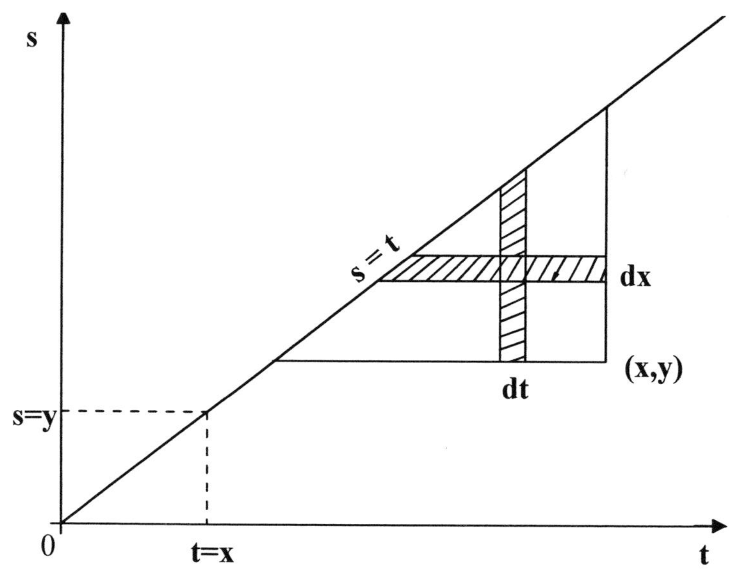

Figure 12. Domain (D) in the ts-plane

A physical system is said to obey the *condition of closed cycle* if its evolution during time, $g(t)$, under action of a given cause $f(t)$ remains independent of a chosen initial moment of time t_0. In other words, whenever the cause $f(t-t_0)$ produces the resulting effect $g(t-t_0)$. A particular case which illustrates this concept is a four-terminal electrical network. The network, has two entrance terminals (inputs) and two exit terminals (outputs) which are connected by an electrical circuit made from resistors, inductors and capacitors. The "cause" $f(t)$ is the voltage drop acting on the entrance terminals and the "effect" $g(t)$ is the voltage across the exit terminals. Examples for four-terminal networks are electrical transformers and transmission lines. Systems of this type cease to satisfy the condition of closed cycle if the circuit elements (resistances, inductances and capacitances) depend on the time.

A study of systems which obey the condition of closed cycle is facilitated by considering the functional transformation of Laplace and the *Horn-Borel theorem* associated with the *convolution product.*

3.6. Convolution Products

The *convolution product* of two functions F and G, defined for $t > 0$, is the function $H(t)$ given by

$$H(t) = \int_0^t F(t-\xi)G(\xi)d\xi \qquad t > 0 \tag{1.3.6}$$

The functions F and G are supposed to be *locally summable* i.e. summable in the Lebesgue sense on every measurable set bounded on a real positive semi-axis. Yet it does not imply summability on a measurable unbounded set. However, in the applications that will follow, it is possible to suppose that F and G are *sectionally (piecewise) continuous*. In other words, these functions may have a finite number of discontinuities in every bounded interval. Furthermore, if in a neighborhood of discontinuity point, $t = c$, the function under consideration is unbounded, its integral exists as a *generalized Riemann integral*, i.e.

$$\lim_{\varepsilon_1,\varepsilon_2\to 0}\{\int_a^{c-\varepsilon_1} F(t)dt + \int_{c+\varepsilon_2}^{b} F(t)dt\} \qquad a < c < b \tag{2.3.6}$$

where the limit is finite and independent of the mode in which ε_1 and ε_2 tend to zero.

The convolution product (3.6.1) is a particular case of the convolution product

$$H(t) = \int_{-\infty}^{+\infty} F(t-\xi)G(\xi)d\xi \tag{3.3.6}$$

which is encountered in a study of the Fourier integrals [50,51]. It is evident that this convolution product reduces to (1.3.6) when F and G identically vanish for $t < 0$. In the general theory of linear systems, this case appears frequently when functions responsible for the "causes" start to act after an arbitrary moment of time. Finally, when $F(t)$ and $G(t)$ are unbounded in the vicinity of the origin $t = 0$, it is appropriate to suppose that the integral $\int_0^{\varepsilon} F(t)dt$, $\varepsilon > 0$, exists as a generalized Riemann integral.

Sometimes, it is preferable to denote the convolution product as $F * G(t)$ rather than commonly used symbol $F(t) * G(t)$ or $F * G$, because the same notation represent also the composition product. Usually, the confusion is easily avoided considering in what context the products were used or by denoting the composition product with two variables and the convolution product with only one variable. Obviously, it is possible to emphasize the difference by using the symbol ♦ for the composition product and the symbol $*$ for the convolution product (hereafter only used in this book).

The principal properties of the convolution product are:

Commutative law of multiplication

$$F(t) * G(t) = G(t) * F(t) \tag{4.3.6}$$

which immediately can be verified by introducing in the right-hand side of (1.3.6) a new variable $u = t - \xi$.

Distributive law with respect to addition

$$F * \sum_{j=1}^{n} G_j(t) = \sum_{j=1}^{n} F * G_j(t) \tag{5.3.6}$$

Associative law of multiplication

$$(F * G) * L = F * (G * L) \tag{6.3.6}$$

which results from the possibility to change the order of integration in double integral. From

$$F * G = \int_0^t F(t-\xi)G(\xi)d\xi = H(t) \tag{7.3.6}$$

it follows that

$$\begin{gathered}(F * G) * L = H * L = \int_0^t H(u)L(t-u)du = \\ \int_0^t L(t-u)\int_0^u F(\xi)G(u-\xi)d\xi du = \iint_{(D)} F(\xi)G(u-\xi)L(t-u)d\xi du\end{gathered} \tag{8.3.6}$$

where the domain of integration (D) is the area bounded by sides $\xi = 0$, $u = t$, $u = \xi$ of triangle in the ξu-plane.

The inversion of the order of integration leads to

$$(F * G) * L = \int_0^t F(\xi) d\xi \int_\xi^t G(u-\xi) L(t-u) du \tag{9.3.6}$$

and by introducing $u = w+\xi$ in the last integral

$$(F * G) * L = \int_0^t F(\xi) d\xi \int_0^{t-\xi} G(w) L[(t-\xi) - w)] dw = \int_0^t F(\xi)[G * L(t-\xi)] d\xi = F * (G * L) \tag{10.3.6}$$

In virtue of these three properties, the ordinary addition and the convolution product are two binary operations that permit the construction of a *commutative ring* which is called the *convolution ring*. This ring does not have a *unit element*. However, it is possible to endow the convolution ring with a unit element provided that this ring is extended to the *distributions* or *generalized* functions. The *Dirac distribution* $\delta(t)$ satisfies the requirement:

$$\int_0^t F(\xi)\,\delta(t-\xi) d\xi = \int_0^t F(t-\xi)\,\delta(\xi) d\xi = F(t) \tag{11.3.6}$$

It is worthwhile to mention two properties of the convolution product.

a) The convolution product of function $F(t)$ with a constant $C = G(t)$ is given by

$$F(t) * C = C * F(t) = C \int_0^t F(\xi) d\xi \tag{12.3.6}$$

b) Introduction of the exponential factor, e^{at}, where a is constant, into the convolution product leads to

$$e^{at} F(t) * e^{at} G(t) = e^{at} F * G(t) \tag{13.3.6}$$

3.7. The Laplace Transformation and Convolution Products

In the Laplace method for solving linear, homogeneous differential equations, the solution $w(z)$ is supposed to be in the form

$$w(z) = \int_C e^{z\zeta} v(\zeta) d\zeta \tag{1.3.7}$$

This integral often called the *Laplace integral*, is taken along a contour C in the plane of the complex variable ζ. In particular, the Laplace method is useful in the case of linear differential equations with polynomial coefficients. If a linear differential equation admitting the solution $v(\zeta)$ is of order n and its coefficients are polynomials $P_k(\zeta)$, k = 0, 1, 2, ..., n of degree m, then $w(z)$ satisfies a linear differential equation of order m with polynomial coefficients $Q_k(z)$, k = 0, 1, 2, ..., m of degree n. Hence the differential equation with m < n and especially for m = 1, is usually easier than that of the original differential equation [68]. Integrals of the type

$$\int_0^\infty e^{-st} f(t) dt = \Phi(s) \tag{2.3.7}$$

were considered by Abel. Nowadays, they are known as the *Laplace-Abel integrals*. $\Phi(s)$ is called the *Laplace-Abel transform* of $f(t)$ and is denoted $\Phi(s) = L\{f(t)\}$. Some aspects of this integral transformation were already considered in Chapter 2. The basic properties associated with the theory of the Volterra integral equations are presented here (see also S. Colombo papers).

A rigorous description of Laplace-Abel transformation should be based on the Laplace-Abel integral (2.3.7) taken in the Lebesgue sense and $f(t)$ *locally summable* on the real positive semi-axis (A function is called locally summable on unbounded set E if it is summable in the Lebesgue sense on every measurable bounded subset included in E). However, in applications considered here, it is possible, without disadvantage, to suppose that the Laplace-Abel integral is the generalized Riemann integral

$$\Phi(s) = \lim_{\substack{a \to +0 \\ b \to +\infty}} \int_a^b e^{-st} f(t) dt \tag{3.3.7}$$

Other assumptions on $f(t)$ should be recalled: In particular, the points at which $f(t)$ is discontinuous must be the *isolated points*. Let $t = \tau$ be the only point of discontinuity on the closed interval a $\le t \le$ b. If $f(t)$ is unbounded in a neighborhood of τ, then $|f(t)|$ must be integrable as a generalized Riemann integral

$$\lim_{\varepsilon_1,\varepsilon_2\to 0}\{\int_a^{\tau-\varepsilon_1}|f(t)|dt+\int_{\tau+\varepsilon_2}^{b}|f(t)|dt\} \qquad (4.3.7)$$

$$\varepsilon_1>0 \quad , \quad \varepsilon_2>0$$

Yet one may not assert that in (3.3.7), a limit exists when $b\to+\infty$.

Convergence of the Laplace-Abel integral at $s = s_0$, imposes its convergence for every value of s such that Res > Res_0. Thus, the Laplace-Abel transform $\Phi(s)$ is holomorphic in the half plane Res > α, where α is called *abscissa of convergence.* On the straight line Res = α the Laplace-Abel integral may or may not converge. Similarly, one may define for $\int_0^\infty|e^{-st}f(t)|dt$ an *abscissa of absolute convergence* Res = β. [33,50,66,105]. But here convergence at a point s_0 implies convergence on the whole straight line Res = Res_0 and evidently $\alpha \le \beta$.

At this point it seems appropriate to point out that many problems in mathematical analysis lead to functions which are holomorphic in a half-plane. Since such functions can be put in a one to one correspondence, making use of a Möbius mapping with the functions holomorphic in the unit disc, the extension to the half-plane is easy. Yet, one meets representations which are not easily studied by mapping on the unit disc and at the beginning of the twentieth century several mathematicians discussed this problem [58].

The reader may recall that the Laplace transforms of non-analytic functions are analytic functions and that some transcendental functions can be represented by Laplace integrals. For example, denoting the largest integer less than or equal to the real positive number x, as $[x]$ it is possible to write

$$L\{[e^t]\}=\sum_{n=1}^{\infty}n\int_{\ln n}^{\ln(n+1)}e^{-st}dt=\frac{1}{s}\sum_{n=1}^{\infty}n[\frac{1}{n^s}-\frac{1}{(n+1)^s}]= \qquad (5.3.7)$$
$$\frac{1}{s}\sum_{n=1}^{\infty}\frac{1}{n^s}=\frac{\zeta(s)}{s}$$

where $\zeta(s)$ is the *Riemann zeta function.*

$$\zeta(s)=\{\prod_p(1-\frac{1}{p^s})\}^{-1} \qquad (6.3.7)$$

and the infinite product is taken over the sequence $\{p\}$ of all primes.

The representations (5.3.7) and (6.3.7) are valid for Res > 1, but analytic continuation is possible in these cases. The famous conjecture of Riemann that all zeros of $\zeta(s)$ situated in the strip $0\le$ Res ≤ 1, lie on the line Res = 1/2, has not yet been proved [120].

The result given in (5.3.7) demonstrates the procedure which permits to generalize the Laplace-Abel integral by considering the *Dirichlet series*, $\sum_{n=1}^{\infty} a_n e^{-\lambda_n s}$, as Laplace transforms; here $\{\lambda_n\}$ is a monotonous increasing sequence, $0 \le \lambda_1 < \lambda_2 < \lambda_3 \dots$, with $\lim_{n\to\infty} \lambda_n = +\infty$ while $\{a_n\}$ denotes a sequence of complex numbers.

On the other hand, the integral taken in the Stieltjes sense

$$\int_0^{+\infty} e^{-st} dg(t) = \Phi(s) \tag{7.3.7}$$

is called the *Laplace-Stieltjes transform* of function $g(t)$ of bounded variation [57-59,121]. Introducing $g(0) = 0$ and $g(t) = \sum_{n<t} a_n$, one has

$$\Phi(s) = \sum_{n=1}^{\infty} a_n e^{-ns} \tag{8.3.7}$$

In the case considered in (5.3.7), $g(0) = 0$, $g(t) = \sum_{\ln n<t} a_n$, and therefore the Dirichlet series becomes

$$\Phi(s) = \sum_{n=1}^{\infty} a_n n^{-s} \tag{9.3.7}$$

This useful generalization is mentioned here because if $g(t)$ is a differentiable function then the Laplace-Stieltjes integral reduces to the *Carson-Laplace integral* (p-multiplied Laplace transformation).

Operational rules of the Laplace-Abel transformation, predominantly called to-day the Laplace transformation, are used throughout this book. These rules are involved not only in solution of the Volterra integral equations with kernels of the closed cycle type, but they are also helpful in the establishment of many relations between the Volterra functions and with other special functions. Operational rules as applied to $f(t)$ and $\Phi(s)$, considerably enlarge a number of available function-transform pairs [1,44,50,51,66,80,88,96,102] However, a validity of these rules implies conditions on the original function $f(t)$ which are more restrictive than to assure exclusively the existence of transform $\Phi(s)$. Usually, it is supposed that $f(t)$ satisfies two conditions:

(a) the original function $f(t)$ is of *exponential order*, $f(t) = O(e^{at})$, i.e. if exists $a > 0$ such that the limit $\lim_{t\to\infty} e^{-at} f(t)$ is finite.

(b) the function $f(t)$ is continuous for $t > 0$ and $\lim_{\varepsilon \to +0} \int_{\varepsilon}^{a} |f(t)| dt$, $a > 0$, exists.

The principal operational rules of the Laplace transformation are:

(I) If $a > 0$ and $u(t)$ is the *unit step function* ($u(t) = 0$ for $t < 0$ and $u(t) = 1$ for $t \geq 0$), then

$$L\{u(t-a)f(t-a)\} = e^{-as}\Phi(s) \tag{10.3.7}$$

(II) If $f(t)$ and $f'(t)$ fulfill conditions (a) and (b) then

$$L\{f'(t)\} = s\Phi(s) - f(+0) \tag{11.3.7}$$

This operational rule remains still valid if $f'(t)$ has discontinuities of the first kind, i.e. if at every point of discontinuity of $f'(t)$, $t = \tau$, both limits $\lim_{\varepsilon \to +0} f'(\tau - \varepsilon)$ and $\lim_{\varepsilon \to +0} f'(\tau + \varepsilon)$ exist and are finite. In particular, this is the case when $f'(t)$ is a piecewise continuous function bounded in a neighborhood of every point of discontinuity [66]. The existence of $\lim_{\varepsilon \to +0} f(t) = f(+0)$ is assured by the conditions (a) and (b). The operational rule (11.3.7) can be generalized if $f(t)$, $f'(t)$, $f''(t)$, ... , $f^{(n-1)}(t)$ are continuous for $t > 0$

$$L\{f^{(n)}(t)\} = s^n\Phi(s) - \sum_{k=1}^{n} s^{n-k} f^{(n-k)}(+0) \tag{12.3.7}$$

where $f^{(0)} = f(t)$. If $f(t)$ and its derivatives have isolated discontinuities in the interval $0 < t < +\infty$, relations (11.3.7) and (12.3.7) are preserved in the sense used in the theory of distributions [70,106] or generalized functions [42,81,82]. For example, the Dirac delta function $\delta(t)$ from (11.3.6) is nothing else than the generalized derivative (in the distributions sense) of the unit step function $u(t)$ [31]. Tables of transforms of distributions (pseudo-functions) are available in the literature [31].

The following corollary comes directly from (11.3.7)

$$L\{\int_0^t f(u)du\} = \frac{\Phi(s)}{s} \tag{13.3.7}$$

(III) If $L\{f(t)\} = \Phi(s)$ and $L\{g(t)\} = \Psi(s)$ then the Laplace transform of the convolution product is

$$L\{f(t) * g(t)\} = \Phi(s)\Psi(s) \tag{14.3.7}$$

This operational rule is valid in some half plane where both functions $\Phi(s)$ and $\Psi(s)$ are holomorphic.

More precisely, if the rule (14.3.7) is valid for $s = s_0$ and if both integrals $\int_0^\infty e^{-s_0 t} f(t)dt$, $\int_0^\infty e^{-s_0 t} g(t)dt$ $\int_0^\infty e^{-s_0 t} g(t)dt$ are absolutely convergent, the Laplace integral $\int_0^\infty e^{-s_0 t}[f(t) * g(t)]dt$ is absolutely convergent.

On the other hand, using (13.3.6) and denoting $F(t) = e^{-st}f(t)$ and $G(t) = e^{-st}g(t)$ we have

$$\int_0^\infty [F(t) * G(t)]dt = [\int_0^\infty F(t)dt][\int_0^\infty G(t)dt] \qquad (15.3.7)$$

This relation exhibits an isomorphism between rules of convergence for series and those of convergence for Laplace transforms of convolution products. In fact, considering the product of two series

$$[\sum_{n=0}^\infty a_n][\sum_{n=0}^\infty b_n] = \sum_{n=0}^\infty c_n \qquad (16.3.7)$$

where

$$c_n = \sum_{k=0}^n a_{n-k} b_k \qquad (17.3.7)$$

the relation (16.3.7) is valid in the following three cases:

1) series $\sum a_n$ and $\sum b_n$ are absolutely convergent and then the series $\sum c_n$ is absolutely convergent (the Cauchy theorem).

2) at least one of the series $\sum a_n$ and $\sum b_n$ is absolutely convergent then the series $\sum c_n$ is convergent and its sum is AB where A = $\sum a_n$ and B = $\sum b_n$ (Mertens theorem).

3) the series $\sum a_n$, $\sum b_n$ and $\sum c_n$ are convergent then $\sum c_n$ = AB (the Abel theorem).

Similarly, the operational rule for convolution products (14.3.7) holds in the following three cases:

a) both Laplace integrals $L\{f(t)\}$ and $L\{g(t)\}$ are absolutely convergent.

b) if one of the Laplace integrals is convergent and the other integral is absolutely convergent then $L\{f(t) * g(t)\}$ is convergent.

c) the operational rule is valid if all three Laplace integrals are convergent.

(IV) The *complex inversion formula,* sometimes called the *Bromwich-Wagner formula* provides the original function $f(t)$ from its transform $\Phi(s)$

$$f(t) = \frac{1}{2\pi i} \lim_{\lambda\to\infty} \int_{c-i\lambda}^{c+i\lambda} e^{ts}\Phi(s)ds = L^{-1}\{\Phi(s)\} \qquad (18.3.7)$$

with $c > \alpha$ where α is the abscissa of convergence. Formulas (3.3.7) and (18.3.7) are associated with a particular aspect of the *double Fourier integral* [84, 96]

$$\lim_{\lambda\to\infty} \int_{-\lambda}^{\lambda} e^{ix\omega} d\omega \int_{-\infty}^{\infty} h(\xi) e^{-i\omega\xi} d\xi = \pi\{h(x+0) + h(x-0)\} \qquad (19.3.7)$$

which leads also to the *reciprocal Fourier integrals* (the *exponential Fourier transformation*)

$$\phi(x) = \frac{1}{\sqrt{2\pi}} \int_{-\infty}^{\infty} e^{-ix\omega}\Phi(\omega)d\omega$$

$$\Phi(\omega) = \frac{1}{\sqrt{2\pi}} \int_{-\infty}^{\infty} e^{-i\omega x}\phi(x)dx \qquad (20.3.7)$$

In practice, usage of the Bromwich-Wagner formula is reduced to the evaluation of the complex integral $\frac{1}{2\pi i}\int_{(C)} e^{st}\Phi(s)ds$ taken along a closed contour (C) consisting of a vertical line Re$s = c$, from c-iλ to c+iλ and the semi-circle of radius λ with center at point $s = c$. The real number c is arbitrary but chosen in such way that the line Re$s = c$ lies to the right of all singularities of $\Phi(s)$. The evaluation of $f(t)$ from $\Phi(s)$ in this procedure involves the usage of the Cauchy residue theorem and a modification of the contour (C) in case of branch points [76,77]. The Bromwich-Wagner inversion formula automatically gives $f(0) = 0$ for $t < 0$. The uniqueness of the inverse Laplace transforms is assured by the *Lerch theorem* [50].

(V) The operational rule associated with derivatives of the Laplace transforms

$$(-1)^n \Phi^{(n)}(s) = L\{t^n f(t)\} \qquad (21.3.7)$$

is often useful.

As was already pointed out, the interest in operational rules lies because they help to enrich tables of integral transforms. Laplace transforms of functions of exponential order such as polynomials, trigonometric and hyperbolic functions and the *interval functions* (functions which assign values to intervals, e.g. the "staircase" function defined by $f(t) = 0$ for $t < 0$ and $f(t) = [t]+1$ for $t \geq 0$) are easily available but they are suitable to rather few applications. However, they were sufficient to solve numerous problems associated with dynamic systems and with transient phenomena occuring in electrical circuits. These problems are represented by systems of linear, homogeneous or non-homogeneous differential equations with constant coefficients [19,20,50,76,77,96].

Since then, two procedures have found a broad usage, especially in the study of special functions by integral transform methods. The first makes use of expansions into series of functions. The second uses differentiation and integration with respect to a parameter appearing in the original function and its integral transform.

In all cases involved with an expansion into series of functions, it is enough to mention that the convergence of $\int_a^{\infty}[\sum_{n=1}^{\infty}|u_n(x)|dx]$ or of $\sum_{n=1}^{\infty}[\int_a^{\infty}|u_n(x)|dx]$ is sufficient to justify the equality

$$\int_a^{\infty}[\sum_{n=1}^{\infty}u_n(x)dx]=\sum_{n=1}^{\infty}[\int_a^{\infty}u_n(x)dx] \tag{22.3.7}$$

Integration with respect to the parameter λ is permissible if the integral $\int_a^{\infty}f(t,\lambda)dt$ is uniformly convergent for $\alpha \leq \lambda \leq \beta$ and if

$$\int_a^b dt\int_{\alpha}^{\beta}f(t,\lambda)d\lambda=\int_{\alpha}^{\beta}d\lambda\int_a^b f(t,\lambda)dt \tag{23.3.7}$$

It should be noted that uniform convergence of the integral $\int_a^{\infty}f(t,\lambda)dt$ over every closed interval [α,β] where β > α , can not necessarily assure its uniform convergence on the unbounded interval $[\alpha,+\infty)$.

3.8. The Laplace Transformation and the Volterra Integral Equations with a Kernel of Closed Cycle

Consider an integral equation of the first kind

$$F(t) = \int_0^t K(t-\xi)\Phi(\xi)d\xi \tag{1.3.8}$$

and three Laplace transforms $f(s) = L\{F(t)\}$, $K(s) = L\{K(t)\}$ and $\phi(s) = L\{\Phi(t)\}$ which satisfy the condition $f(s) = K(s)\cdot\phi(s)$, then the function to be determined $\Phi(t)$, is given by the complex inversion formula (18.3.7)

$$\Phi(t) = \frac{1}{2\pi i}\int_{c-i\infty}^{c+i\infty} e^{ts}\frac{f(s)}{K(s)}ds \tag{2.3.8}$$

In the case of an integral equation of the second kind

$$F(t) = \Phi(t) + \int_0^t K(t-\xi)\Phi(\xi)d\xi \tag{3.3.8}$$

we have

$$\Phi(t) = L^{-1}\{\frac{f(s)}{1+K(s)}\} = \frac{1}{2\pi i}\int_{c-i\infty}^{c+i\infty} e^{ts}\frac{f(s)}{1+K(s)}ds \tag{4.3.8}$$

Denoting

$$L\{\Phi(t)\} = \phi(s) = f(s)[1 - \frac{K(s)}{1+K(s)}] \tag{5.3.8}$$

and

$$L^{-1}\{\frac{K(s)}{1+K(s)}\} = R(t) \tag{6.3.8}$$

where $R(t)$ is the resolvent kernel (see section 3.3), then

$$R(t) = K(t) - [\overset{*}{K(t)}]^2 + [\overset{*}{K(t)}]^3 - [\overset{*}{K(t)}]^4 + \cdots \tag{7.3.8}$$

and

$$\Phi(t) = F(t) - \int_0^t R(t-\xi)F(\xi)d\xi \tag{8.3.8}$$

Among kernels of the closed cycle type, certain logarithmic kernels of the polynomial form

$$K(t-\xi) = \sum_{j=1}^{m} a_j [\ln(t-\xi)]^j \tag{9.3.8}$$

lead to solutions of integral equations which are easily expressed in terms of new transcendental functions. Laplace transforms of these functions are of the type $\Phi(s,\omega)$ where Φ is an algebraic function of s and $\omega = 1/(\ln s)$. Thus, it is necessary to consider the determination of the Laplace transforms of functions in the form $f(t) = g(\ln t)$. The new transcendental functions depend on one or two parameters, $\nu(t,\alpha)$ and $\mu(t,\beta,\alpha)$, and they are nowadays called the *Volterra functions* [43].

3.9. Laplace Transforms of Logarithmic Functions

In this section some Laplace transforms of logarithmic functions will be discussed. From

$$L\{t^\lambda\} = s^{-\lambda-1}\Gamma(\lambda+1) \qquad \operatorname{Re}\lambda > -1 \tag{1.3.9}$$

differentiation with respect to λ of both sides of (1.3.9) gives

$$L\{t^\lambda \ln t\} = s^{-\lambda-1}[\Gamma(\lambda+1)\ln(\frac{1}{s}) + \Gamma'(\lambda+1)] \tag{2.3.9}$$

Since the integral defining the Laplace transform of $t^\lambda \ln t$ is uniformly convergent if $\lambda \geq 0$,

$$L\{\ln t\} = -\frac{1}{s}(\ln s + \gamma) \tag{3.3.9}$$

where γ is the Euler constant

$$\gamma = -\Gamma'(1) = \lim_{n\to\infty}(\sum_{k=1}^{n}\frac{1}{k} - \ln n) = 0.57721566... \tag{4.3.9}$$

The second differentiation of (1.3.9) leads to

$$L\{t^{\lambda}(\ln t)^2\} = s^{-\lambda-1}[\Gamma(\lambda+1)[\ln(\frac{1}{s})]^2 + 2\Gamma'(\lambda+1)]\ln(\frac{1}{s}) + \Gamma''(\lambda+1)] \tag{5.3.9}$$

and

$$L\{(\ln t)^2\} = \frac{1}{s}\{[\ln(\frac{1}{s})]^2 - 2\gamma\ln(\frac{1}{s}) + (\gamma^2 + \frac{\pi^2}{6})\} \tag{6.3.9}$$

In general, after m successive differentiations of (1.3.9) we have

$$\begin{aligned} L\{t^{\lambda}(\ln t)^m\} = s^{-\lambda-1}\{\Gamma(\lambda+1)[\ln(\frac{1}{s})]^m + \binom{m}{1}\Gamma'(\lambda+1)[\ln(\frac{1}{s})]^{m-1} + \\ \dots + \binom{m}{k}\Gamma^{(k)}(\lambda+1)[\ln(\frac{1}{s})]^{m-k} + \dots + \Gamma^{(m)}(\lambda+1)\} \end{aligned} \tag{7.3.9}$$

Thus, this general and important result shows that the Laplace transform of the product of polynomial of ln t and t^{λ} is a polynomial of ln s of the same degree multiplied by $s^{-\lambda-1}$. However, in solving of the Volterra integral equation with the kernel given in (9.3.8) it is better to start with

$$L^{-1}\{s^{-\lambda-1}\} = \frac{t^{\lambda}}{\Gamma(\lambda+1)} \tag{8.3.9}$$

Differentiation of this relation with respect to λ gives

$$L^{-1}\{s^{-\lambda-1}\ln(\frac{1}{s})\} = \frac{t^{\lambda}}{\Gamma(\lambda+1)}[\ln t - \psi(\lambda+1)] \tag{9.3.9}$$

and the second differentiation leads to

$$L^{-1}\{s^{-\lambda-1}[\ln(\frac{1}{s})]^2\} = \frac{t^{\lambda}}{\Gamma(\lambda+1)}\{[\ln t - \psi(\lambda+1)]^2 - \psi'(\lambda+1)\} \tag{10.3.9}$$

where $\psi(z)$ denotes the *psi-function,* the logarithmic derivative of $\Gamma(z)$. For $\lambda = 0$, the relations (9.3.9) and (10.3.9) become

$$L^{-1}\{s^{-1}\ln(\frac{1}{s})\} = \ln t - \psi(1) \tag{11.3.9}$$

and

$$L^{-1}\{s^{-1}[\ln(\frac{1}{s})]^2\} = [\ln t - \psi(1)]^2 - \psi'(1) \tag{12.3.9}$$

As expected, these formulas are equivalent to those in (9.3.3) and (6.3.9).

Using the Leibnitz rule in successive differentiation of the product $t^\lambda\,[\Gamma(\lambda+1)]^{-1}$

$$\begin{aligned}\frac{\partial^m}{\partial\lambda^m}\{t^\lambda[\Gamma(\lambda+1)]^{-1}\} = t^\lambda\frac{\partial^m}{\partial\lambda^m}\frac{1}{\Gamma(\lambda+1)} + \binom{m}{1}(\frac{\partial}{\partial\lambda}t^\lambda)\frac{\partial^{m-1}}{\partial\lambda^{m-1}}\frac{1}{\Gamma(\lambda+1)} + \\ \dots + \binom{m}{k}(\frac{\partial^k}{\partial\lambda^k}t^\lambda)\frac{\partial^{m-k}}{\partial\lambda^{m-k}}\frac{1}{\Gamma(\lambda+1)} + \dots + (\frac{\partial^m}{\partial\lambda^m}t^\lambda)\frac{1}{\Gamma(\lambda+1)}\end{aligned} \tag{13.3.9}$$

and introducing the series expansion of $1/\,\Gamma(\lambda+1)$

$$\frac{1}{\Gamma(z+1)} = \sum_{k=1}^{\infty}\alpha_k\frac{z^k}{k!} \tag{14.3.9}$$

where the coefficients α_k are called the *Bourguet numbers,* it is possible to obtain

$$\begin{aligned}L^{-1}\{(-1)^m s^{-1}(\ln s)^m\} = (\ln t)^m + \alpha_1\binom{m}{1}(\ln t)^{m-1} + \alpha_2\binom{m}{2}(\ln t)^{m-2} + \\ \dots + \alpha_k\binom{m}{k}(\ln t)^{m-k} + \dots + \alpha_m\end{aligned} \tag{15.3.9}$$

In the Appendix B "Summation Formulas", the first few coefficients α_k are presented. The series expansion (14.3.9) up to k = 26 is also given in the literature [1]. In this way, using relations (10.3.9) and (15.3.9), solutions of the Volterra integral equations with kernels of the closed cycle (9.3.9) can be systematically evaluated using the Bromwich-Wagner formula.

$$\Phi(t) = \frac{1}{2\pi i}\int_{c-i\infty}^{c+i\infty} e^{ts}\frac{F(s)}{K(s)}\,ds \tag{16.3.9}$$

where the Laplace transforms $F(s)$ and $K(s)$ are

$$\begin{aligned}F(s) &= L\{f(t)\} \\ K(s) &= L\{\sum_{j=1}^{m} a_j[\ln(t-\xi)]^j\}\end{aligned} \tag{17.3.9}$$

Alternatively, new transcendental functions, i.e. the Volterra functions, express the solution in a more direct and elegant manner. This happens because the integrand of the Bromwich-Wagner complex integral (16.3.9) includes $[K(s)]^{-1}$ which contents the factor $[P(\ln s)]^{-1}$ where $P(u)$ denotes a polynomial of u with real coefficients. This factor can be decomposed into partial fractions and therefore the solution of the Volterra integral equation of the first kind with the closed cycle kernel (9.3.8) can be expressed in terms of convolution products. One element in the convolution product should be chosen among functions that have the following Laplace transforms

$$
\begin{aligned}
&(\ln s+a)^{-1}=[\ln(se^{a})]^{-1}\\
&(\ln s+a)^{-m}=[\ln(se^{a})]^{-m}\\
&[(\ln s+a)^{2}+b^{2}]^{-1}=\{[\ln(se^{a})]^{2}+b^{2}\}^{-1}\\
&[(\ln s+a)^{2}+b^{2}]^{-m}=\{[\ln(se^{a})]^{2}+b^{2}\}^{-m}
\end{aligned}
\tag{18.3.9}
$$

where a and b are real constants and m is a positive integer.

Applying the operational rule

$$
\begin{aligned}
&L\{f(\frac{t}{a})\}=a\Phi(as)\\
&a>0
\end{aligned}
\tag{19.3.9}
$$

it is possible to express the solution under consideration, as a sum of convolution products. Every convolution product contains a function, such that its Laplace transform belongs to the following group of four transforms

$$
\begin{aligned}
&s^{-\alpha-1}(\ln s)^{-1} \quad , \quad s^{-\alpha-1}(\ln s)^{-m-1}\\
&s^{-\alpha-1}[(\ln s)^{2}+b^{2}]^{-1} \quad , \quad s^{-\alpha-1}[(\ln s)^{2}+b^{2}]^{-m-1}\\
&\alpha>-1 \quad , \quad m=0,1,2,3,...
\end{aligned}
\tag{20.3.9}
$$

The first two expressions in (20.3.9) are the Laplace transforms of the Volterra functions

$$
\nu(t,\alpha)=L^{-1}\{s^{-\alpha-1}(\ln s)^{-1}\}
\tag{21.3.9}
$$

and

$$
\begin{aligned}
&\mu(t,m,\alpha)=L^{-1}\{s^{-\alpha-1}(\ln s)^{-m-1}\}\\
&t>0 \quad , \quad \alpha>-1 \quad , \quad m=0,1,2,...
\end{aligned}
\tag{22.3.9}
$$

where the real variable t can be analytically extended to the plane of the complex variable z. The remaining expressions in (20.3.9) represent linear combinations of the real and imaginary parts of the Volterra functions (see (6.2.2) and (7.2.2)).

Replacement of s by ln s in $\phi(s) = L\{f(t)\}$ results in the operational rule which is especially suitable for a study of the Volterra functions

$$s^{-1}\phi(\ln s) = \int_0^\infty s^{-1}e^{-u\ln s}f(u)du = \int_0^\infty s^{-u-1}f(u)du \tag{23.3.9}$$

which is valid for Re(ln s) > α, where α is the abscissa of convergence of $\phi(s)$.

Since

$$s^{-u-1} = \int_0^\infty \frac{e^{-st}t^u}{\Gamma(u+1)}dt \tag{24.3.9}$$

equation (23.3.9) takes the form

$$s^{-1}\phi(\ln s) = \int_0^\infty \frac{f(u)}{\Gamma(u+1)}du\int_0^\infty e^{-st}t^u dt \tag{25.3.9}$$

and when it is possible to invert the order of integration, we have

$$s^{-1}\phi(\ln s) = L\{\int_0^\infty \frac{t^u}{\Gamma(u+1)}f(u)du\} \tag{26.3.9}$$

This operational rule allows for a new definition of the Volterra function $\nu(t,\alpha)$ and $\mu(t,m,\alpha)$ with the aid of integrals taken over the real positive semi-axis, where until now in this Chapter, the Volterra functions were defined in terms of the Bromwich-Wagner complex integrals.

Indeed, by integrating both sides of the relation

$$\int_0^\infty e^{-st}\frac{t^\lambda}{\Gamma(\lambda+1)}dt = s^{-\lambda-1} \qquad \lambda > -1 \tag{27.3.9}$$

with respect to parameter λ, between λ = α and λ = ∞ and assuming that Res > 1, we have

$$\int_\alpha^\infty d\lambda\int_0^\infty e^{-st}\frac{t^\lambda}{\Gamma(\lambda+1)}dt = \int_\alpha^\infty s^{-\lambda-1}d\lambda = s^{-\alpha-1}(\ln s)^{-1} \tag{28.3.9}$$

It is possible to invert the order of integration in the left-hand side integral

$$L\{\int_{\alpha}^{\infty}\frac{t^{\lambda}}{\Gamma(\lambda+1)}d\lambda\}=L\{\int_{0}^{\infty}\frac{t^{u+\alpha}}{\Gamma(\lambda+1)}du\}=s^{-\alpha-1}(\ln s)^{-1} \tag{29.3.9}$$

$$\alpha>-1 \quad , \quad \mathrm{Re}\, s>1$$

Finally, the inversion of (29.3.9) with the help of the Bromwich-Wagner complex integral gives

$$\nu(t,\alpha)=\int_{0}^{\infty}\frac{t^{u+\alpha}}{\Gamma(u+\alpha+1)}du \tag{30.3.9}$$

An analogous formula for $\mu(t,m,\alpha)$ can be determined starting from

$$f(t)=\frac{(t-\alpha)^{m}}{\Gamma(m+1)}u(t-\alpha) \tag{31.3.9}$$

and by applying the operational rule (10.3.7) to (31.3.9)

$$\phi(s)=L\{f(t)\}=s^{-m-1}e^{-\alpha s} \tag{32.3.9}$$

Replacing s by $\ln s$ in (32.3.9) finally leads to

$$\mu(t,m,\alpha)=\int_{0}^{\infty}\frac{t^{u+\alpha}u^{m}}{\Gamma(m+1)\Gamma(u+\alpha+1)}du \tag{33.3.9}$$

Substitution of the real positive variable t by the complex variable $z = x+iy = \rho e^{i\theta}$ gives

$$\nu(z,\alpha)=\int_{0}^{\infty}\frac{\rho^{u+\alpha}e^{i(u+\alpha)}}{\Gamma(u+\alpha+1)}du= \\ \int_{0}^{\infty}\frac{\rho^{u+\alpha}\cos[(u+\alpha)\theta]}{\Gamma(u+\alpha+1)}du+i\int_{0}^{\infty}\frac{\rho^{u+\alpha}\sin[(u+\alpha)\theta]}{\Gamma(u+\alpha+1)}du \tag{34.3.9}$$

and

$$\mu(z,m,\alpha)=\int_0^\infty \frac{\rho^{u+\alpha}e^{i(u+\alpha)}u^m}{\Gamma(u+\alpha+1)}du=$$
$$\int_0^\infty \frac{\rho^{u+\alpha}u^m\cos[(u+\alpha)\theta]}{\Gamma(m+1)\Gamma(u+\alpha+1)}du+i\int_0^\infty \frac{\rho^{u+\alpha}u^m\sin[(u+\alpha)\theta]}{\Gamma(m+1)\Gamma(u+\alpha+1)}du \tag{35.3.9}$$

3.10. Examples

This section deals with examples of Volterra integral equations with kernels of the closed cycle group involving logarithmic functions. The Horn-Borel formula for Laplace transforms of convolution products offers a direct method for resolving the integral equation

$$\int_0^t \phi(u)K(t-u)du=g(t) \tag{1.3.10}$$

where $g(t)$ is a given differentiable function.

We start with the simplest case

$$\int_0^t \phi(u)\ln(t-u)du=g(t) \tag{2.3.10}$$
$$g(0)=g'(0)=0$$

Let $G(s)=L\{g(t)\}$, $\Phi(s)=L\{\phi(t)\}$ then

$$L\{\phi(t)\}=-\frac{sG(s)}{\ln s+\gamma}=-[s\ln(se^{\gamma})]^{-1}s^2G(s) \tag{3.3.10}$$

If $g(t)$ is twice differentiable and since $\nu(0)=0$, we obtain

$$\phi(t)=-\int_0^t g''(u)\nu[(t-u)e^{-\gamma}]du \tag{4.3.10}$$

This result was derived by Volterra and Pérès in1936.

If $g(0)=0$ but $g'(0)\neq 0$ then (4.3.10) takes the form

$$\phi(t) = -\int_0^t g''(u)\nu[(t-u)e^{-\gamma}]du - g'(0)\int_0^\infty \frac{t^u e^{-\gamma u}}{\Gamma(u+1)}du \tag{5.3.10}$$

and in the particular case of $g(t) = t$ is

$$\phi(t) = -\int_0^\infty \frac{t^u e^{-\gamma u}}{\Gamma(u+1)}du = -\nu(te^{-\gamma}) \tag{6.3.10}$$

As a direct consequence of (2.3.9), the Volterra integral equation

$$\begin{gathered}\int_0^t \phi(u)(t-u)\ln(t-u)du = g(t) \\ g(0) = g'(0) = 0\end{gathered} \tag{7.3.10}$$

admits a solution whose Laplace transform is

$$\Phi(s) = -\frac{s^2 G(s)}{\Gamma(2)[\ln s - \Gamma'(2)]} \tag{8.3.10}$$

The Bromwich-Wagner formula (18.3.7) provides the solution

$$\phi(t) = \frac{1}{2\pi i}\int_{c-i\infty}^{c+i\infty} \frac{e^{ts}s^2 G(s)}{[\ln\frac{1}{s} + \Gamma'(2)]}ds \tag{9.3.10}$$

This solution admits also a representation as a convolution product of $g''(t)$ and the derivative of $\nu(t,0) = \nu(t)$. The conditions of differentiability for $g(t)$ may be discarded if solutions of distributional type are considered.

Since equation (6.3.9) may be rewritten as

$$L\{(\ln t)^2\} = \frac{1}{s}\{[\ln \gamma s]^2 + \frac{\pi^2}{6}\} = \frac{1}{s}[\ln(\ln \gamma s) + i\frac{\pi}{\sqrt{6}}][\ln(\ln \gamma s) - i\frac{\pi}{\sqrt{6}}] \tag{10.3.10}$$

the solution of

$$\int_0^t \phi(u)[\ln(t-u)]^2 du = g(t) \tag{11.3.10}$$

also admits two representations

$$\phi(t) = \frac{1}{2\pi i} \int_{c-i\infty}^{c+i\infty} \frac{e^{ts} sG(s)}{[(\ln \gamma s)^2 + \frac{\pi^2}{6}]} ds \tag{12.3.10}$$

Here the computation of the Bromwich-Wagner integral compels the choice of an unbounded, yet closed contour of integration, never crossing the negative real axis at finite distance, the later acting as a 'cut' restoring an uniform character for the multivalued integrand.

The Laplace transform (10.3.10) involves the second representation by a convolution product of three factors

a) a function $g(t)$ or one of its derivatives

b) a function $f_1(t)$ satisfying condition $L\{f_1(t)\} = s^{-\alpha}[\ln(\gamma e^{i\pi/\sqrt{6}} s)]^{-1}$, $\alpha > 0$

c) a function $f_2(t)$ satisfying condition $L\{f_2(t)\} = s^{-\beta}[\ln(\gamma e^{i\pi\sqrt{6}} s]^{-1}$

where the sum $\alpha + \beta$ determines the order of the derivative of $g(t)$.

It seems wise to restrain of giving more details on all these computations. To obtain a general formula one is liable to make use of the real part and of the imaginary part of the $\nu(t,\alpha)$ function, as given by (6.2.2) and (7.2.2)

$$\begin{aligned} \operatorname{Re}\{\nu(re^{i\theta},\alpha)\} &= \int_0^\infty \frac{r^{u+\alpha}\cos(u+\alpha)}{\Gamma(u+\alpha+1)} du \\ \operatorname{Im}\{\nu(re^{i\theta},\alpha)\} &= \int_0^\infty \frac{r^{u+\alpha}\sin(u+\alpha)}{\Gamma(u+\alpha+1)} du \end{aligned} \tag{11.3.10}$$

All these examples illustrate how the Volterra functions play an important part in the resolution of the Volterra integral equations with logarithmic kernels.

Few more solutions of integral equations which are related in some way to the Volterra functions and numerical methods can be found in the books of Srivastava and Buschman [108], Linz [78] and Polyanin and Manzhirov [99].

Chapter 4

4. Integral Transforms and Volterra Functions

As already mentioned, the occurence of the Volterra functions in the Laplace transformation is associated with the fact that on one hand their Laplace transforms are expressed in terms of the logarithmic functions, and are useful in a determination of many inverse transforms. On the other hand, the operational calculus is a powerful and an efficient method for the derivation of the Volterra function properties. Dealing with the Volterra functions, Humbert [61-67], Colombo [23-30], Parodi [89-95], Barrucand [12,13] and others [3-5,45] assumed that in most cases the functions under consideration are continous, differentiable and have the Laplace transforms and mathematical operations under the integral sign (e.g. differentiation with respect to variables and parameters, inversion of order of integration and etc.) when applying rules and theorems of operational calculus are permissible. Since our main interest lies in the application of the Volterra functions, the same formal approach will be followed here. We suppose that the readers are familiar with the basic properties and techniques of the Laplace transformation which can be easily found in the literature (e.g. [15,19,22,31,33,38,66,74,76,83,84,90,92,93,96,97,107,117] and many others). Occasionally however, in order to make the presented material self-contained, the related general formulas of operational calculus will also be presented using the common notation of the direct Laplace transform, $F(s) = L\{f(t)\}$ and the inverse Laplace transform, $f(t) = L^{-1}\{F(s)\}$ of the function $f(t)$.

Evidently, the main attention will be directed to the $\nu(t,\alpha)$ and less to the $\mu(t,\beta,\alpha)$ functions considering that many results can be extended by a simple generalization or by use of the recurrence and other relations which exist between different functions.

4.1 Laplace Transforms of the Volterra Functions

The Laplace transforms of the Volterra functions of the real variable t are

$$L\{\nu(t)\} = \frac{1}{s\ln s} \tag{1.4.1}$$

$$L\{\nu(t,\alpha)\} = \frac{1}{s^{\alpha+1}\ln s} \tag{2.4.1}$$

$$L\{\mu(t,\beta)\} = \frac{1}{s(\ln s)^{\beta+1}} \tag{3.4.1}$$

$$L\{\mu(t,\beta,\alpha)\} = \frac{1}{s^{\alpha+1}(\ln s)^{\beta+1}} \tag{4.4.1}$$

$$\mathrm{Re}\,\alpha > -1 \quad , \quad \mathrm{Re}\,\beta > -1 \quad , \quad \mathrm{Re}\,s > 1$$

where s is a complex variable and the functions are assumed to be identically zero for negative values of t. Thus, the Laplace transforms of the Volterra functions are expressed in terms of elementary functions as the inverse powers of the logarithmic functions. These transforms, in part, have been already used, but in different contexts in (7.1), (12.2.3), (25.2.3) and (33.2.5).

Applying the operational property of the Laplace transformation

$$L\{f(at)\} = \frac{1}{a}F\left(\frac{s}{a}\right) \tag{5.4.1}$$

to (2.4.1) we have

$$L\{\nu(at,\alpha)\} = \frac{a^{\alpha}}{s^{\alpha+1}\ln(s/a)} \tag{6.4.1}$$

Introducing $a = \rho e^{i\theta}$ into (6.4.1), the Laplace transforms of the real and imaginary parts of the Volterra functions can be separated

$$L\{\mathrm{Re}\ \nu(\rho e^{i\theta}t,\alpha)\} = \frac{\rho^{\alpha}[\cos(\alpha\theta)\ln(s/\rho) - \theta\sin(\alpha\theta)]}{s^{\alpha+1}\{[\ln(s/\rho)]^2 + \theta^2\}} \tag{7.4.1}$$

$$L\{\mathrm{Im}\ \nu(\rho e^{i\theta}t,\alpha)\} = \frac{\rho^{\alpha}[\sin(\alpha\theta)\ln(s/\rho) + \theta\cos(\alpha\theta)]}{s^{\alpha+1}\{[\ln(s/\rho)]^2 + \theta^2\}} \tag{8.4.1}$$

and similarly

$$L\{\mathrm{Re}\,\mu(\rho e^{i\theta}t,\beta,\alpha)\} = \frac{\rho^{\alpha}\cos\{\alpha\theta + (\beta+1)\tan^{-1}[\frac{\theta}{\ln(s/\rho)}]\}}{s^{\alpha+1}\{[\ln(s/\rho)]^2 + \theta^2\}^{(\beta+1)/2}} \tag{9.4.1}$$

$$L\{\operatorname{Im}\mu(\rho e^{i\theta}t,\beta,\alpha)\} = \frac{\rho^{\alpha}\sin\{\alpha\theta+(\beta+1)\tan^{-1}[\frac{\theta}{\ln(s/\rho)}]\}}{s^{\alpha+1}\{[\ln(s/\rho)]^{2}+\theta^{2}\}^{(\beta+1)/2}} \tag{10.4.1}$$

Taking into account that

$$\int_{0}^{\infty}\frac{(t\rho)^{u+\alpha}\sin(\theta u)}{\Gamma(u+\alpha+1)}du = \{\cos(\alpha\theta)\operatorname{Im}\nu(\rho e^{i\theta}t,\alpha)-\sin(\alpha\theta)\operatorname{Re}\nu(\rho e^{i\theta}t,\alpha)\} \tag{11.4.1}$$

and

$$\int_{0}^{\infty}\frac{(t\rho)^{u+\alpha}\cos(\theta u)du}{\Gamma(u+\alpha+1)}du = \{\cos(\alpha\theta)\operatorname{Re}\nu(\rho e^{i\theta}t,\alpha)+\sin(\alpha\theta)\operatorname{Im}\nu(\rho e^{i\theta}t,\alpha)\} \tag{12.4.1}$$

it follows from (7.4.1) and (8.4.10) that

$$L\{\int_{0}^{\infty}\frac{(\rho t)^{u+\alpha}\sin(\theta u)}{\Gamma(u+\alpha+1)}du\} = \frac{\rho^{\alpha}\theta}{s^{\alpha+1}\{[\ln(s/\rho)]^{2}+\theta^{2}\}} \tag{13.4.1}$$

$$L\{\int_{0}^{\infty}\frac{(\rho t)^{u+\alpha}\cos(\theta u)}{\Gamma(u+\alpha+1)}du\} = \frac{\rho^{\alpha}\ln(s/\rho)}{s^{\alpha+1}\{[\ln(s/\rho)]^{2}+\theta^{2}\}} \tag{14.4.1}$$

and similarly

$$L\{\int_{0}^{\infty}\frac{(\rho t)^{u+\alpha}u^{\beta}\sin(\theta u)}{\Gamma(u+\alpha+1)}du = \frac{\Gamma(\beta+1)\rho^{\alpha}\sin\{(\beta+1)\tan^{-1}[\theta/\ln(s/\rho)]\}}{s^{\alpha+1}\{[\ln(s/\rho)]^{2}+\theta^{2}\}^{(\beta+1)/2}} \tag{15.4.1}$$

$$L\{\int_{0}^{\infty}\frac{(\rho t)^{u+\alpha}u^{\beta}\cos(\theta u)}{\Gamma(u+\alpha+1)}du = \frac{\Gamma(\beta+1)\rho^{\alpha}\cos\{(\beta+1)\tan^{-1}[\theta/\ln(s/\rho)]\}}{s^{\alpha+1}\{[\ln(s/\rho)]^{2}+\theta^{2}\}^{(\beta+1)/2}} \tag{16.4.1}$$

Introducing $\alpha = 0$, $\rho = 1$ and $\theta = \pi\lambda$ into (13.4.1) and (14.4.1) and taking into account equations (13.2.2) and (14.2.2), the Laplace transforms of functions $Tc(t,\lambda)$ and $Ts(t, \lambda)$ are

$$L\{Tc(t,\lambda)\} = \frac{\ln s}{s[(\pi\lambda)^2 + (\ln s)^2]} \tag{17.4.1}$$

$$L\{Ts(t,\lambda)\} = \frac{\pi\lambda}{s[(\pi\lambda)^2 + (\ln s)^2]} \tag{18.4.1}$$

The Ramanujan integral representation of $\nu(t)$ given in (1.2.3) can be written in the form

$$\nu(t) = e^t - L\left\{\frac{1}{u[\pi^2 + (\ln u)^2]}\right\} = e^t - F(t) \tag{19.4.1}$$

However, from (18.4.1)

$$L\{F(t)\} = \frac{1}{\pi} L\{Ts(t,1)\} \tag{20.4.1}$$

Thus, the Volterra function $\nu(t)$ can be expressed with the help of the Stieltjes transform (the iterated Laplace transform) as [13]

$$\nu(t) = e^t - \frac{1}{\pi}\int_0^\infty \frac{Ts(u,1)}{t+u}du \tag{21.4.1}$$

Other representation of $\nu(t)$ can be derived from

$$G(s) = L\{e^t - \frac{1}{2} - \nu(t)\} = \frac{1}{s-1} - \frac{1}{2s} - \frac{1}{s\ln s} \tag{22.4.1}$$

but from [28]

$$G(s) = 2\sum_{n=1}^{\infty} \frac{\ln s}{s[(2\pi n)^2 + (\ln s)^2]} \tag{23.4.1}$$

and using (17.4.1) $\nu(t)$ is given by

$$\nu(t) = e^t - \frac{1}{2} - 2\sum_{n=1}^{\infty} Tc(t,2n) \tag{24.4.1}$$

The rule of the Laplace transformation [88, 1.1.34]

$$L\{\int_t^\infty \frac{f(x)}{x}dx\} = \frac{1}{s}\int_0^s F(u)du \tag{25.4.1}$$

can be used to evaluate the Laplace transform of ν(a,t) where

$$\nu(a,t) = \int_0^\infty \frac{a^{u+t}}{\Gamma(u+t+1)}du = \int_t^\infty \frac{a^x}{x\Gamma(x)}dx \tag{26.4.1}$$

Starting with f(t) = $a^t/\Gamma(t)$

$$F(s) = \int_0^\infty e^{-st}\frac{a^t}{\Gamma(t)}dt = ae^{-s}\int_1^\infty \frac{(ae^{-s})^x}{\Gamma(x+1)}dx \tag{27.4.1}$$

and therefore

$$F(s) = ae^{-s}\nu(ae^{-s},-1) \tag{28.4.1}$$

Using (25.4.1) and (28.4.1) it follows that

$$L\{\nu(a,t)\} = \frac{a}{s}\int_0^s e^{-u}\nu(ae^{-u},-1)du \tag{29.4.1}$$

The change of variable of integration gives

$$L\{\nu(a,t)\} = \frac{1}{s}\int_{ae^{-s}}^a \nu(x,-1)dx \tag{30.4.1}$$

but ν(x,-1) = ν'(x) and therefore

$$L\{\nu(a,t)\} = \frac{1}{s}[\nu(a) - \nu(ae^{-s})] \tag{31.4.1}$$

$$a > 0$$

Since $L\{t^u\} = \Gamma(u+1)/s^{u+1}$

$$F(s) = L\{\int_0^\infty \frac{t^x \nu(a,x)}{\Gamma(x+1)} dx\} = \frac{1}{s}\int_0^\infty \frac{1}{s^u} \nu(a,u) du \qquad (32.4.1)$$

and

$$F(s) = \frac{1}{s}\int_0^\infty e^{-u \ln s} \nu(a,u) du \qquad (33.4.1)$$

Substitution of s by $\ln s$ in (31.4.1) gives

$$L\{\int_0^\infty \frac{t^x \nu(a,x)}{\Gamma(x+1)} dx\} = \frac{\nu(a) - \nu(\frac{a}{s})}{s \ln s} \qquad (34.4.1)$$

$$a > 0$$

This is a particular case of the general rule [88, 1.1.38]

$$L\{\int_0^\infty \frac{t^u}{\Gamma(u+1)} f(u) du\} = \frac{1}{s} F(\ln s) \qquad (35.4.1)$$

which will be considered later.

Using [88, 1.1.9]

$$L\{tf(t)\} = -\frac{d}{ds} F(s) \qquad (36.4.1)$$

and (28.4.1)

$$F(s) = L\{\frac{ta^t}{\Gamma(t)}\} = -\frac{d}{ds}\{ae^{-s} \nu(ae^{-s}, -1)\} \qquad (37.4.1)$$

it follows that

$$F(s) = ae^{-s} \nu(ae^{-s}, -1) + a^2 e^{-2s} \nu(ae^{-s}, -2) \qquad (38.4.1)$$

But from (25.4.1)

$$L\{\nu(a,t-1)\} = L\{\int_t^{\infty}\frac{a^x}{\Gamma(x)}dx\} = \frac{1}{s}\int_0^s F(u)du \tag{39.4.1}$$

and introducing $F(s)$ from (38.4.1) into (39.4.1) and performing the integration, the final result is

$$L\{\nu(a,t-1)\} = \frac{1}{s}[\nu(a,-1) - e^{-s}\nu(ae^{-s},-1)] \tag{40.4.1}$$

and the analogue of (34.4.1) is

$$L\{\int_0^{\infty}\frac{t^{x-1}\nu(a,x-1)}{\Gamma(x)}dx = \frac{s\nu(a,-1)-\nu(\frac{a}{s},-1)}{s\ln s} \tag{41.4.1}$$

$$a > 0$$

Once again, this is a particular case of the general rule [88, 1.1.36]

$$L\{\int_0^{\infty}\frac{t^{u-1}}{\Gamma(u)}f(u)\,du\} = F(\ln s) \tag{42.4.1}$$

The Laplace transform of *the Volterra integral function*

$$vi(t,\alpha) = \int_0^t \frac{1}{u}\nu(u,\alpha)\,du \tag{43.4.1}$$

$$\alpha > 0$$

can be evaluated using the Laplace transformation property

$$L\{\int_0^t \frac{1}{u}f(u)du\} = \frac{1}{s}\int_s^{\infty}F(x)dx \tag{44.4.1}$$

In the case under consideration

$$L\{vi(t,\alpha)\} = \frac{1}{s}\int_s^{\infty}\frac{dx}{x^{\alpha+1}\ln x} \tag{45.4.1}$$

and introducing a new variable $z = \alpha \ln x$ into (27.4.1), the Laplace transform is given as *the exponential integral*

$$L\{vi(t,\alpha)\} = \frac{1}{s}E_1(\alpha \ln s) = -\frac{1}{s}Ei(-\alpha \ln s) \tag{46.4.1}$$

$$\alpha > 0 \quad , \quad \operatorname{Re} s > 1$$

This result can be generalized if β is a positive integer, because then

$$L\{\mu i(t,n,\alpha)\} = \frac{1}{s(\ln s)^n}E_{n+1}(\alpha \ln s) \tag{47.4.1}$$

$$\alpha > 0 \quad , \quad \operatorname{Re} s > 1 \quad , \quad n = 0,1,2,3,\ldots$$

where the exponential integrals are defined as [1]

$$E_n(z) = \int_1^{\infty} u^{-n} e^{-zu} du$$

$$E_1(z) = -Ei(-z) \tag{48.4.1}$$

$$\operatorname{Re} z > 0 \quad , \quad n = 0,1,2,3,\ldots$$

Consecutive evaluation of $\mu i(t,n,\alpha)$ is possible from (29.4.1) using the recurrence relation of the exponential integrals

$$E_{n+1}(z) = \frac{1}{n}[e^{-z} - zE_n(z)] \tag{49.4.1}$$

$$n = 1,2,3,\ldots$$

From

$$L\{\mu i(t,1,\alpha)\} = \frac{1}{s \ln s}E_2(\alpha \ln s) \tag{50.4.1}$$

and

$$E_2(\alpha \ln s) = [\frac{1}{s^{\alpha}} - \alpha \ln s E_1(\alpha \ln s)] \tag{51.4.1}$$

it follows that

$$L\{\mu i(t,1,\alpha)\} = [\frac{1}{s^{\alpha} \ln s} - \frac{\alpha}{s}E_1(\alpha \ln s)] \tag{52.4.1}$$

Using (2.4.1) and (28.4.1), the inverse transform of (34.4.1) is

$$\mu i(t,1,\alpha)\} = \nu(t,\alpha) - \alpha \nu i(t,\alpha)$$
$$\alpha > 0 \tag{53.4.1}$$

In a similar way

$$\mu i(t,2,\alpha)\} = \frac{1}{2}\{\mu(t,1,\alpha) - \alpha\nu(t,\alpha) + \alpha^2 \nu i(t,\alpha)\}$$
$$\alpha > 0 \tag{54.4.1}$$

4.2 Application of the Laplace Transform Properties

Using *the product theorem* of the Laplace transformation, i.e. if $F(s) = L\{f(t)\}$ and $G(s) = L\{g(t)\}$ then

$$L\{f(t) * g(t)\} = L\{\int_0^t f(t-u)g(u)du\} = F(s) \cdot G(s) = H(s) \tag{1.4.2}$$

it is evident from (6.4.1) - (16.4.1) that if $g(t) = L^{-1}\{G(s)\}$ is known, the Laplace transforms of the type

$$H(s) = \frac{G(s)}{s^{\alpha+1}(\ln s + a)^{\beta+1}} = \frac{G(s)}{s^{\alpha+1}[\ln(se^a)]^{\beta+1}} \tag{2.4.2}$$

and

$$H(s) = \frac{G(s)}{s^{\alpha+1}[(\ln s + a)^2 + b^2]^{\beta+1}} = \frac{G(s)}{s^{\alpha+1}\{[\ln(se^a)]^2 + b^2\}^{\beta+1}} \tag{3.4.2}$$

have inverses which are expressible in terms of the convolution integrals containing the Volterra functions.

If the Laplace transform of $\nu(t,\alpha)$ from (2.4.1) is presented in the form

$$\frac{1}{s^{\alpha+1}\ln s} = \frac{1}{s^{\alpha_1}} \cdot \frac{1}{s^{\alpha_2+1}\ln s}$$
$$\alpha = \alpha_1 + \alpha_2 \tag{4.4.2}$$

then the convolution theorem (1.4.2) gives

$$\nu(t,\alpha) = \frac{1}{\Gamma(\alpha_1)} \int_0^t (t-u)^{\alpha_1 - 1} \nu(u,\alpha_2)\, du \qquad \alpha = \alpha_1 + \alpha_2 \tag{5.4.2}$$

Introducing $u = t \sin^2\theta$, the function $\nu(t,\alpha)$ is represented by

$$\nu(t,\alpha) = \frac{2t^{\alpha_1}}{\Gamma(\alpha_1)} \int_0^{\pi/2} \sin\theta (\cos\theta)^{\alpha_1 - 1} \nu[t(\sin\theta)^2, \alpha_2]\, d\theta \qquad \alpha = \alpha_1 + \alpha_2 \tag{6.4.2}$$

and similarly from

$$\frac{1}{s^{\alpha+1}(\ln s)^{\beta+1}} = \frac{1}{s^{\alpha_1}} \cdot \frac{1}{s^{\alpha_2+1}(\ln s)^{\beta+1}} \tag{7.4.2}$$

$$\frac{1}{s^{\alpha+1}(\ln s)^{\beta+1}} = \frac{1}{s^{\alpha_1+1}(\ln s)^{\beta_1+1}} \cdot \frac{1}{s^{\alpha_2+1}(\ln s)^{\beta_2+1}} \qquad \alpha = \alpha_1 + \alpha_2 \quad , \quad \beta = \beta_1 + \beta_2 \tag{8.4.2}$$

we have

$$\mu(t,\beta,\alpha) = \frac{2t^{\alpha_1}}{\Gamma(\alpha_1)} \int_0^{\pi/2} \sin\theta (\cos\theta)^{\alpha_1 - 1} \mu[t(\sin\theta)^2, \beta, \alpha_2]\, d\theta \qquad \alpha = \alpha_1 + \alpha_2 \tag{9.4.2}$$

$$\mu(t,\beta,\alpha) = \int_0^t \mu(t-u, \beta_1, \alpha_1) \mu(u, \beta_2 - 1, \alpha_2 - 1)\, du \qquad \alpha = \alpha_1 + \alpha_2 \quad , \quad \beta = \beta_1 + \beta_2 \tag{10.4.2}$$

and

$$\mu(t,1,\alpha+1) = \int_0^t \nu(t-u, \alpha_1) \nu(u, \alpha_2 - 1)\, du \tag{11.4.2}$$

$$\mu(t,\beta+1,\alpha+1)=\int_0^t \nu(t-u,\alpha_1)\mu(u,\beta,\alpha_2)du$$
$$\alpha=\alpha_1+\alpha_2 \tag{12.4.2}$$

The Laplace transform of function $f(t)$ multiplied by powers of t is expressed by the successive differentiation of the transform $F(s)$

$$L\{t^n f(t)\}=(-1)^n \frac{d^n F(s)}{ds^n} \tag{13.4.2}$$

From (2.4.1) with n = 1

$$L\{t\,\nu(t,\alpha)\}=-\frac{d}{ds}[\frac{1}{s^{\alpha+1}\ln s}]=\frac{\alpha+1}{s^{\alpha+2}\ln s}+\frac{1}{s^{\alpha+2}(\ln s)^2} \tag{14.4.2}$$

The inversion of (14.4.2) and using (2.4.1) and (3.4.1) gives immediately the recurrence relation (1.2.2)

$$\mu(t,1,\alpha)=t\,\nu(t,\alpha)-(\alpha+1)\nu(t,\alpha+1) \tag{15.4.2}$$

Similarly from

$$t\mu(t,\beta,\alpha)=-L^{-1}\{\frac{d}{ds}[\frac{1}{s^{\alpha+1}(\ln s)^{\beta+1}}]\} \tag{16.4.2}$$

we have

$$t\mu(t,\beta,\alpha)=(\alpha+1)\mu(t,\beta,\alpha+1)+(\beta+1)\mu(t,\beta+1,\alpha+1) \tag{17.4.2}$$

Considering that for positive values of parameter α, the functions $\nu(t,\alpha)$ and $\mu(t,\beta\alpha)$ approach zero as $t\to 0$, these recurrence relations permit easily to show that for $-1<\alpha<0$ we have

$$\lim_{t\to 0}[t\nu(t,\alpha)]=0 \tag{18.4.2}$$

$$\lim_{t\to 0}[t\,\mu(t,\beta,\alpha)]=0 \tag{19.4.2}$$

These results were already established previously, but by a different way (see for example (42.2.5)). Evidently, by a different arrangement of (14.4.2) and (16.4.2), the recurrence relations can be written in the alternative, compact form

$$t\nu(t,\alpha) = \nu(t,\alpha) * [(\alpha+1) + \nu(t)] \tag{20.4.2}$$

$$t\mu(t,\beta,\alpha) = \mu(t,\beta,\alpha) * [(\alpha+1) + (\beta+1)\nu(t)] \tag{21.4.2}$$

The second derivative of the Laplace transform of $\nu(t,\alpha)$ is

$$\frac{d^2}{ds^2}\left[\frac{1}{s^{\alpha+1}\ln s}\right] = \frac{(\alpha+1)(\alpha+2)}{s^{\alpha+3}\ln s} + \frac{2\alpha+3}{s^{\alpha+3}(\ln s)^2} + \frac{2}{s^{\alpha+3}(\ln s)^3} \tag{22.4.2}$$

and its inverse gives the following relation between the Volterra functions

$$t^2\nu(t,\alpha) = \\ (\alpha+1)(\alpha+2)\nu(t,\alpha+2) + (2\alpha+3)\mu(t,1,\alpha+2) + 2\mu(t,2,\alpha+2) \tag{23.4.2}$$

In a similar way it is possible to obtain

$$t^2\mu(t,\beta,\alpha) = \\ (\alpha+1)(\alpha+2)\mu(t,\beta,\alpha+2) + (2\alpha+3)(\beta+1)\mu(t,\beta+1,\alpha+2) + \\ (\beta+1)(\beta+2)\mu(t,\beta+2,\alpha+2) \tag{24.4.2}$$

These relations can be presented in the compact form as

$$t^2\nu(t,\alpha) = \nu(t,\alpha+1) * [(\alpha+1)(\alpha+2) + (2\alpha+3)\nu(t) + 2\mu(t,1)] \tag{25.4.2}$$

$$t^2\mu(t,\beta,\alpha) = \mu(t,\beta,\alpha+1) * [(\alpha+1)(\alpha+2) + (2\alpha+3)(\beta+1)\nu(t) \\ +(\beta+1)(\beta+2)\mu(t,1)] \tag{26.4.2}$$

Let us evaluate the Laplace transform of

$$F(s) = \int_0^\infty e^{-st}\left\{\int_\alpha^{\alpha+1} a^{-u}\frac{\gamma(u,t)}{\Gamma(u)}du\right\}dt \\ a > 0 \tag{27.4.2}$$

where $\gamma(u,t)$ is *the incomplete gamma function* (the *Prym function* in French terminology)

$$\gamma(u,t) = \int_0^t e^{-x}x^{u-1}dx \quad ; \quad \mathrm{Re}(u) > 0 \tag{28.4.2}$$

Inserting (28.4.2) into (27.4.2) and inverting the order of integration we have

$$F(s)=\frac{a}{[a(s+1)]^{\alpha+1}\ln[a(s+1)]}+\frac{a-1}{s[a(s+1)]^{\alpha+1}\ln[a(s+1)]} \tag{29.4.2}$$

The inverse transform of (29.4.2) can be determined using [88, 2.1.3]

$$L^{-1}\{F[a(s+1)]\}=\frac{1}{a}e^{-\tau}f(\frac{\tau}{a}) \tag{30.4.2}$$

and therefore

$$\int_{\alpha}^{\alpha+1} a^{-u}\frac{\gamma(u,\tau)}{\Gamma(u)}du = e^{-\tau}\nu(\frac{\tau}{a},\alpha)+\frac{(a-1)}{a}\int_{0}^{\tau/a} e^{-x}\nu(\frac{x}{a},\alpha)dx \tag{31.4.2}$$

Denoting $t = a\tau$ and $x = au$, (31.4.2) takes the form

$$\int_{\alpha}^{\alpha+1} a^{-u}\frac{\gamma(u,at)}{\Gamma(u)}du = e^{-at}\nu(t,\alpha)+(a-1)\int_{0}^{t} e^{-u}\nu(u,\alpha)du \tag{32.4.2}$$

$$a>0$$

For a = 1, integral (31.4.2) becomes

$$\nu(t,\alpha)=e^{t}\int_{\alpha}^{\alpha+1}\frac{\gamma(u,t)}{\Gamma(u)}du \tag{33.4.2}$$

and can be considered as a new definition of the Volterra function $\nu(t,\alpha)$ [14,30].

It follows immediately from (33.4.2) that the asymptotic behaviour $\nu(t,\alpha) \sim e^{t}$ as $t\to\infty$, a result which was given already in (1.2.4).

If $\alpha = 0$, then

$$\nu(t)=e^{t}\int_{0}^{1}\frac{\gamma(u,t)}{\Gamma(u)}du \tag{34.4.2}$$

But $\gamma(u,t) = \Gamma(u) - \Gamma(u,t)$, and using the Ramanujan integral representation of $\nu(t)$, the following equality of integrals is established

$$\int_{0}^{1}\frac{\Gamma(u,t)}{\Gamma(u)}du = \int_{0}^{\infty}\frac{e^{-t(u+1)}}{u[\pi^{2}+(\ln u)^{2}]}du \tag{35.4.2}$$

Introducing $x = \ln u$ into the right-hand side integral, (35.4.2) becomes

$$\int_0^1 \frac{\Gamma(u,t)}{\Gamma(u)} du = \int_{-\infty}^{\infty} \frac{e^{-t(e^x+1)}}{\pi^2 + x^2} dx \tag{36.4.2}$$

There are many *substitution formulas* in which variable s is replaced in the Laplace transform by some function of s. For example [88, 2.1.18]

$$L^{-1}\{\frac{1}{\sqrt{s}} F(\sqrt{s})\} = \frac{1}{\sqrt{\pi t}} \int_0^{\infty} e^{-u^2/4t} f(u) du \tag{37.4.2}$$

Considering the case of ν(*t*)

$$L^{-1}\{\frac{1}{\sqrt{s}} \cdot \frac{1}{\sqrt{s} \ln \sqrt{s}}\} = L^{-1}\{\frac{2}{s \ln s}\} = 2\nu(t) \tag{38.4.2}$$

and therefore from (37.4.2)

$$\frac{1}{2\sqrt{\pi t}} \int_0^{\infty} e^{-u^2/4t} \nu(u) du = \nu(t) \tag{39.4.2}$$

The same procedure gives immediately

$$\frac{1}{2^{\beta+1}\sqrt{\pi t}} \int_0^{\infty} e^{-u^2/4t} \mu(u,\beta,2\alpha) du = \mu(t,\beta,\alpha) \tag{40.4.2}$$

$$\mathrm{Re}\,\alpha > -1 \quad , \quad \mathrm{Re}\,\beta > -1$$

For $\beta = 0$ and α being a positive integer, (40.4.2) reduces to

$$\frac{1}{2\sqrt{\pi t}} \int_0^{\infty} e^{-u^2/4t} \nu(u,2n) du = \nu(t,n) \tag{41.4.2}$$

$$n = 0,1,2,3,\ldots$$

These formulas are out of numerous solutions of integral equations that are associated with the Volterra functions. Other integral formulas will be considered later in Chapter 5.

Using (25.2.2)

$$\frac{d^n \nu(t,2n)}{dt^n} = \nu(t,n) \tag{42.4.2}$$

it follows that $f(t) = \nu(t,2n)$ is a solution of the integro-differential equation [66]

$$2\sqrt{\pi t}\, f^{(n)}(t) = \int_0^\infty e^{-u^2/4t} f(u)\,du \qquad (43.4.2)$$
$$n = 0,1,2,3\ldots$$

Similarly $\mu(t,\beta,2n)$ solves

$$2^{\beta+1}\sqrt{\pi t}\, f^{(n)}(t) = \int_0^\infty e^{-u^2/4t} f(u)\,du \qquad (44.4.2)$$
$$n = 0,1,2,3\ldots$$

Colombo [27] also considered the following integro-differential equation

$$t f'(t) - n f(t) = \int_0^t f(u) f^{(n+1)}(t-u)\,du = f(t) * f^{(n+1)}(t) \qquad (45.4.2)$$
$$f(0) = f'(0) = f''(0) = \ldots = f^{(n)}(0) = 0$$

and the transformed form (45.4.2) is

$$s\frac{dF(s)}{ds} + s^{n+1}[F(s)]^2 + (n+1)F(s) = 0 \qquad (46.4.2)$$

Introducing $G(s) = 1/F(s)$, (45.4.2) becomes

$$s\frac{dG(s)}{ds} - (n+1)G(s) = s^{n+1} \qquad (47.4.2)$$

and this linear differential equation has the following solution, $G(s) = s^{n+1}\ln(s/C)$, where C is the integration constant and

$$F(s) = \frac{1}{s^{n+1}\ln(s/C)} \qquad (48.4.2)$$

Using (6.4.1), the inverse transform of (48.4.2) is $f(t) = C^{-n}\nu(Ct,n)$ and therefore from (45.4.2)

$$t\nu'(t,n) - n\nu(t,n) = \nu(t,n) * \nu^{(n+1)}(t,n) \qquad (49.4.2)$$

This result can be generalized to non-integer values of ρ and σ

$$t\nu'(t,n) - n\nu(t,n) = \nu^{(\rho)}(t,n) * \nu^{(\sigma)}(t,n)$$
$$\rho + \sigma = n+1 \quad , \quad n = 1,2,3,\ldots \tag{50.4.2}$$

when the convolution is extended to the fractional integrals.

4.3. Laplace Transforms Expressible in Terms of the Volterra Functions

There are numerous cases in which the Laplace transforms are expressed in terms of the Volterra functions. This happens with quite different functions $f(t)$ of simple or composite argument. It is noteworthy that the integral representations of the Volterra functions can be considered as the Laplace transforms of elementary functions. For example the integral representations of Ramanujan (9.1) and Touchard (12.1) - (14.1) can be written as

$$L\{\frac{1}{t[\pi^2 + (\ln t)^2]}\} = e^s - \nu(s) \tag{1.4.3}$$

$$L\{\frac{\ln t}{\sqrt{t}\,[\pi^2 + (\ln t)^2]}\} = \pi[\nu(s, -\frac{1}{2}) - e^s] \tag{2.4.3}$$

$$L\{\tan^{-1}(\frac{\ln t}{\pi})\} = \frac{\pi}{s}[e^s - \frac{1}{2} - \nu(s)] \tag{3.4.3}$$

$$L\{\tan^{-1}(\frac{\pi}{\ln t})\} = \frac{\pi}{s}[\nu(s) - 2\sinh s]$$
$$\mathrm{Re}\, s > 0 \tag{4.4.3}$$

However, it must be pointed out that e^s and $\nu(s)$, when considered separately, are not the Laplace transforms. Only the difference of these functions in (1.4.3) is the Laplace transform and the same can be said about (2.4.3) - (4.4.3).

If the Laplace transform of function $f(t,\lambda)$ is of the form

$$L\{f(t,\lambda)\} = F_1(s)[F_2(s)]^\lambda \tag{5.4.3}$$

then

$$L\{\int_\alpha^\infty \frac{f(t,u)}{\Gamma(u+1)} du\} = F_1(s)\nu[F_2(s), \alpha]$$
$$\mathrm{Re}\,\alpha > -1 \tag{6.4.3}$$

Similarly if

$$L\{f(t,\lambda,\alpha)\} = F_1(s)[F_2(s)]^{\lambda+\alpha} \tag{7.4.3}$$

then

$$L\{\int_0^\infty \frac{u^\beta f(t,u,\alpha)}{\Gamma(u+\alpha+1)} du\} = \Gamma(\beta+1)F_1(s)\mu[F_2(s),\beta,\alpha] \tag{8.4.3}$$
$$\operatorname{Re}\alpha > -1 \quad , \quad \operatorname{Re}\beta > -1$$

Thus, the Volterra functions of composite arguments are the Laplace transforms of improper integrals (6.4.3) and (8.4.3). For example from [44, 4.13.34]

$$L\{\gamma(\lambda,at)\} = \frac{\Gamma(\lambda)}{s}\left(\frac{a}{s+a}\right)^\lambda \tag{9.4.3}$$
$$a > 0 \quad , \quad \operatorname{Re}\lambda > -1$$

and using (6.4.3)

$$L\{\int_\alpha^\infty \frac{\gamma(u,at)}{\Gamma(u)\Gamma(u+1)} du\} = \frac{1}{s}\nu[\frac{a}{s+a},\alpha] \tag{10.4.3}$$
$$\operatorname{Re}\alpha > -1 \quad , \quad \operatorname{Re}(s-a) > 0$$

Similarly, from [88, 1.14.1]

$$L\{J_\lambda(at)\} = \frac{1}{(s^2+a^2)^{1/2}}\left[\frac{a}{s+(s^2+a^2)^{1/2}}\right]^\lambda \tag{11.4.3}$$
$$a > 0 \quad , \quad \operatorname{Re}\lambda > -1$$

the Laplace transform of the integral with respect to the order of the Bessel function is

$$L\{\int_\alpha^\infty \frac{J_u(at)}{\Gamma(u+1)} du\} = \frac{1}{(s^2+a^2)^{1/2}}\nu[\frac{a}{s+(s^2+a^2)^{1/2}},\alpha] \tag{12.4.3}$$
$$a > 0 \quad , \quad \operatorname{Re}\alpha > -1 \quad , \quad \operatorname{Re}s > 0$$

In a third example the use of (6.4.3) is devoted to the Volterra functions themselves. Since from (4.4.1)

$$L\{\mu(t,\lambda,\lambda)\} = \frac{1}{s\ln s} \cdot \frac{1}{(s\ln s)^{\lambda}} \qquad (13.4.3)$$

$$\operatorname{Re}\lambda > -1$$

it follows immediately that

$$L\{\int_0^{\infty} \frac{\mu(t,u,u)}{\Gamma(u+1)} du\} = \frac{1}{s\ln s} \nu(\frac{1}{s\ln s}) \qquad (14.4.3)$$

Reciprocal gamma functions are an important group of functions which are associated with the Volterra functions via their Laplace transforms.

In the simplest case

$$L\{\frac{1}{\Gamma(t+1)}\} = \int_0^{\infty} \frac{e^{-st}}{\Gamma(t+1)} dt = \int_0^{\infty} \frac{(e^{-s})^t}{\Gamma(t+1)} dt \qquad (15.4.3)$$

and therefore

$$L\{\frac{1}{\Gamma(t+1)}\} = \nu(e^{-s}) \qquad (16.4.3)$$

In a similar way it is possible to obtain

$$L\{\frac{1}{\Gamma(t+\alpha+1)}\} = e^{\alpha s} \nu(e^{-s},\alpha) \qquad (17.4.3)$$

$$L\{\frac{t^{\beta}}{\Gamma(t+\alpha+1)}\} = \Gamma(\beta+1) e^{\alpha s} \mu(e^{-s},\beta,\alpha) \qquad (18.4.3)$$

$$\operatorname{Re}\alpha > -1 \quad , \quad \operatorname{Re}\beta > -1$$

Applying [88, 1.8]

$$L\{a^t f(t)\} = F(s - \ln a) \qquad (19.4.3)$$

$$a > 0$$

to (17.4.3) and (18.4.3) we have

$$L\{\frac{a^{t+\alpha}}{\Gamma(t+\alpha+1)}\}=e^{\alpha s}\nu(ae^{-s},\alpha) \qquad (20.4.3)$$

$$L\{\frac{a^{t+\alpha}t^{\beta}}{\Gamma(t+\alpha+1)}\}=\Gamma(\beta+1)e^{\alpha s}\mu(ae^{-s},\beta,\alpha) \qquad (21.4.3)$$

or in the general form

$$L\{\frac{a^{t+\alpha}}{\Gamma(bt+\alpha+1)}\}=\frac{a^{\alpha(b-1)/b}e^{\alpha s/b}}{b}\nu(a^{1/b}e^{-s/b},\alpha) \qquad (22.4.3)$$

$$L\{\frac{a^{t+\alpha}t^{\beta}}{\Gamma(bt+\alpha+1)}\}=\frac{\Gamma(\beta+1)a^{\alpha(b-1)/b}e^{\alpha s/b}}{b^{\beta+1}}\mu(a^{1/b}e^{-s/b},\beta,\alpha) \qquad (23.4.3)$$

$$a>0 \quad , \quad b>0 \quad , \quad \mathrm{Re}\,\alpha>-1 \quad , \quad \mathrm{Re}\,\beta>-1$$

Now, a number of new relations will be established by the use of the Laplace transforms of the reciprocal gamma functions. In the case of $f(t)=1/\Gamma(t+1)$, the Laplace transforms are

$$L\{f(t)\}=L\{\frac{1}{\Gamma(t+1)}\}=\nu(e^{-s}) \qquad (24.4.3)$$

$$L\{t^{\beta}f(t)\}=L\{t^{\beta}\frac{1}{\Gamma(t+1)}\}=\Gamma(\beta+1)\mu(e^{-s},\beta) \qquad (25.4.3)$$

On the other hand, if the original function $f(t)$ is multiplied by t^{β} with $0<\beta<1$, then the Laplace transform of the product $t^{\beta}f(t)$ is given by [92,93]

$$L\{t^{\beta}f(t)\}=-\frac{1}{\Gamma(1-\beta)}\int_0^{\infty}u^{-\beta}[\frac{dF(s+u)}{d(s+u)}]du \qquad (26.4.3)$$

In our case from (24.4.3)

$$F(s+u)=\nu[e^{-(s+u)}]=\int_0^{\infty}\frac{[e^{-(s+u)}]^x}{\Gamma(x+1)}dx \qquad (27.4.3)$$

and the inner derivative in (26.4.3) is

$$\frac{dF(s+u)}{d(s+u)} = -\int_0^\infty \frac{x[e^{-(s+u)}]^x}{\Gamma(x+1)}\,dx = -\mu(e^{-(s+u)},1) \tag{28.4.3}$$

Introducing (25.4.3) and (28.4.3) into (26.4.3) we have

$$\Gamma(\beta+1)\mu(e^{-s},\beta) = \frac{1}{\Gamma(1-\beta)}\int_0^\infty u^{-\beta}\mu(e^{-(s+u)},1)\,du \tag{29.4.3}$$

and denoting $t = e^{-s}$, (29.4.3) becomes

$$\mu(t,\beta) = \frac{1}{\Gamma(1-\beta)\Gamma(\beta+1)}\int_0^\infty u^{-\beta}\mu(te^{-u},1)\,du \qquad 0<\beta<1 \tag{30.4.3}$$

As pointed out by Parodi [89], this relation can be treated as a solution of the singular integral equation

$$\mu(t,x) = \frac{1}{\Gamma(1-x)\Gamma(x+1)}\int_0^\infty \xi^{-x} f(t,\xi)\,d\xi \tag{31.4.3}$$

In a similar way it is possible to obtain

$$\mu(t,\beta,\alpha) = \frac{1}{\Gamma(1-\beta)\Gamma(\beta+1)}\int_0^\infty u^{-\beta}e^{\alpha u}\mu(te^{-u},1,\alpha)\,du \qquad 0<\beta<1 \tag{32.4.3}$$

Considering the Laplace transforms of $f(t) = 1/\Gamma(t+1)$ and $g(t) = t^{\beta}f(t)$

$$L\{f(t)\} = L\{t^{-\beta}g(t)\} = \nu(e^{-s}) \tag{33.4.3}$$

$$L\{g(t)\} = \Gamma(\beta+1)\mu(e^{-s},\beta) \tag{34.4.3}$$

and if the original function $g(t)$ is multiplied by $t^{-\beta}$ with $0 < \beta < 1$, then the Laplace transform of the product $t^{-\beta}g(t)$ is given by [92,93]

$$L\{t^{-\beta}g(t)\} = \frac{1}{\Gamma(\beta)}\int_0^\infty u^{-\beta}G(s+u)\,du \tag{35.4.3}$$

Introducing (33.4.3) and (34.4.3) into (35.4.3) we have

$$\nu(e^{-s}) = \frac{\Gamma(\beta+1)}{\Gamma(\beta)} \int_0^\infty u^{\beta-1} \mu(te^{-(s+u)}, \beta) du \tag{36.4.3}$$

and denoting $t = e^{-s}$, (36.4.3) becomes

$$\nu(t) = \beta \int_0^\infty u^{\beta-1} \mu(te^{-u}, \beta)\, du \qquad 0 < \beta < 1 \tag{37.4.3}$$

and the generalization of (37.4.3) is

$$\nu(t,\alpha) = \beta \int_0^\infty u^{\beta-1} e^{\alpha u} \mu(te^{-u}, \beta, \alpha)\, du \qquad 0 < \beta < 1 \tag{38.4.3}$$

If the variable t is the order of special function, by using (22.4.3) or (23.4.3), the Laplace transforms become series of the Volterra functions. Two typical examples related to the Bessel functions are given here, while few others will be considered later.

The modified Bessel function of the first kind of order λ is expressed by the series [1,80,118]

$$I_\lambda(x) = \sum_{n=0}^{\infty} \frac{(x/2)^{\lambda+2n}}{n!\Gamma(\lambda+n+1)} \tag{39.4.3}$$

or denoting $t = \lambda$

$$I_t(x) = \sum_{n=0}^{\infty} \frac{(x/2)^n}{n!} \left[\frac{(x/2)^{t+n}}{\Gamma(t+n+1)}\right] \tag{40.4.3}$$

Using (20.4.3) the Laplace transform of (40.4.3) is therefore

$$L\{I_t(x)\} = \sum_{n=0}^{\infty} \frac{(x/2)^n}{n!} e^{sn} \nu(\frac{x}{2} e^{-s}, n) \tag{41.4.3}$$

If x in (40.4.3) is replaced by $2x^{1/2}$, then

$$(x/2)^{t/2} I_t(2\sqrt{x}) = \sum_{n=0}^{\infty} \frac{1}{n!} [\frac{x^{t+n}}{\Gamma(t+n+1)}] \tag{42.4.3}$$

and

$$L\{(x/2)^{t/2} I_t(2\sqrt{x})\} = \sum_{n=0}^{\infty} \frac{e^{sn} \nu(xe^{-s}, n)}{n!} \tag{43.4.3}$$

If the Laplace variable s is replaced by ln s in

$$L\{f(t)\} = \int_0^{\infty} e^{-st} f(t) dt = F(s) \tag{44.4.3}$$

then provided that the integrals exist [88, 1.1.38 and 1.1.36]

$$L\{\int_0^{\infty} \frac{t^u}{\Gamma(u+1)} f(u)\, du\} = \frac{F(\ln s)}{s} \qquad \mathrm{Re}\, s > 1 \tag{45.4.3}$$

and

$$L\{\int_0^{\infty} \frac{t^{u-1}}{\Gamma(u)} f(u)\, du\} = F(\ln s) \qquad \mathrm{Re}\, s > 1 \tag{46.4.3}$$

These operational rules were already mentioned in (35.4.1) and (42.4.1). Here they are used together with (17.4.3) and (18.4.3) to give

$$L\{\int_0^{\infty} \frac{t^{u-1}}{\Gamma(u)\Gamma(u+\alpha+1)} du\} = s^{\alpha} \nu(\frac{1}{s}, \alpha) \tag{47.4.3}$$

$$L\{\int_0^{\infty} \frac{t^{u-1} u^{\beta}}{\Gamma(u)\Gamma(u+\alpha+1)} du\} = \Gamma(\beta+1) s^{\alpha} \mu(\frac{1}{s}, \beta, \alpha) \tag{48.4.3}$$

and

$$L\{\int_0^\infty \frac{t^u}{\Gamma(u+1)\Gamma(u+\alpha+1)}du\} = s^{\alpha-1}\nu(\frac{1}{s},\alpha) \tag{49.4.3}$$

$$L\{\int_0^\infty \frac{t^u u^\beta}{\Gamma(u+1)\Gamma(u+\alpha+1)}du\} = \Gamma(\beta+1)s^{\alpha-1}\mu(\frac{1}{s},\beta,\alpha) \tag{50.4.3}$$

Combining these expressions for α = 0

$$L\{\int_0^\infty \frac{t^{u-1}(t-u)}{\Gamma(u+1)^2}du\} = (\frac{1}{s}-1)\nu(\frac{1}{s}) \tag{51.4.3}$$

$$L\{\int_0^\infty \frac{t^{u-1}u^\beta(t-u)}{\Gamma(u+1)^2}du\} = \Gamma(\beta+1)(\frac{1}{s}-1)\mu(\frac{1}{s},\beta) \tag{52.4.3}$$

Other substitution formulas which include ln *s* are [3]

$$L\{\int_0^\infty \nu(t,u+\alpha)f(u)du\} = \frac{F(\ln s)}{s^{\alpha+1}\ln s} \tag{53.4.3}$$

$\mathrm{Re}\,\alpha > -1 \quad , \quad \mathrm{Re}\,s > 1$

and

$$L\{\int_0^\infty \mu(t,\beta,u+\alpha)f(u)du\} = \frac{F(\ln s)}{s^{\alpha+1}(\ln s)^{\beta+1}} \tag{54.4.3}$$

$\mathrm{Re}\,\alpha > -1 \quad , \quad \mathrm{Re}\,\beta > -1 \quad , \quad \mathrm{Re}\,s > 1$

More general expressions of similar type are

$$L\{\int_0^\infty \mu(t,\beta u,\alpha u)f(u)du\} = \frac{F\{\ln[s^\alpha(\ln s)^\beta]\}}{s\ln s} \tag{55.4.3}$$

$\alpha \ge 0 \quad , \quad \beta > 0 \quad , \quad \mathrm{Re}\,s > 1$

$$L\{\int_0^\infty \mu(t,\beta u-1,\alpha u)f(u)du\} = \frac{1}{s}F\{\ln[s^\alpha(\ln s)^\beta]\} \tag{56.4.3}$$

$\alpha \ge 0 \quad , \quad \beta > 0 \quad , \quad \mathrm{Re}\,s > 1$

$$L\{\int_0^\infty \mu(t,\beta u-1,\alpha u-1)f(u)du\} = F\{\ln[s^\alpha(\ln s)^\beta]\} \tag{57.4.3}$$
$$\alpha > 0 \quad , \quad \beta > 0 \quad , \quad \mathrm{Re}\, s > 1$$

These substitution formulas are mainly applied to evaluate improper integrals of the Volterra functions, but are also useful in the determination of the Laplace transforms. e.g. the introduction of $f(t) = 1/\Gamma(t+1)$ and $F(s) = \nu(e^{-s})$ into (55.4.3) leads to

$$L\{\int_0^\infty \frac{\mu(t,\beta u,\alpha u)}{\Gamma(u+1)}du\} = \frac{1}{s\ln s}\nu[\frac{1}{s^\alpha(\ln s)^\beta}] \tag{58.4.3}$$

The result was already established for $\alpha = \beta = 1$ in (14.4.3), but in a different way. Similarly, from (17.4.3) and (55.4.3)

$$L\{\int_0^\infty \frac{\mu(t,\beta u,\alpha u)}{\Gamma(u+\lambda+1)}du\} = s^{\alpha\lambda-1}(\ln s)^{\beta\lambda-1}\nu[\frac{1}{s^\alpha(\ln s)^\beta},\lambda] \tag{59.4.3}$$
$$\alpha \geq 0 \quad , \quad \beta > 0 \quad , \quad \lambda > 0$$

Finally, the Laplace transforms of the Volterra functions in which variable t is replaced by the square root of t will be considered.

Let us start with

$$L\{\frac{1}{\sqrt{t}}\nu(a\sqrt{t})\} = L\{\frac{1}{\sqrt{t}}\int_0^\infty \frac{(a\sqrt{t})^u}{\Gamma(u+1)}du\} \tag{60.4.3}$$
$$a > 0$$

Inversion of the order of integration gives

$$L\{\frac{1}{\sqrt{t}}\nu(a\sqrt{t})\} = \frac{1}{\sqrt{s}}\int_0^\infty \frac{\Gamma(\frac{u+1}{2})}{\Gamma(u+1)}(\frac{a^2}{s})^{u/2}du \tag{61.4.3}$$

but by using *the duplication formula of gamma function* [1]

$$\Gamma(u+1) = \frac{2^u}{\sqrt{\pi}}\Gamma(\frac{u+1}{2})\Gamma(\frac{u}{2}+1) \tag{62.4.3}$$

we have

$$L\{\frac{1}{\sqrt{t}}\nu(a\sqrt{t})\} = \frac{1}{\sqrt{s}}\int_0^\infty \frac{(\frac{a^2}{4s})^{u/2}}{\Gamma(\frac{u}{2}+1)}du = 2\sqrt{\frac{\pi}{s}}\,\nu(\frac{a^2}{4s}) \qquad (63.4.3)$$

and similarly

$$L\{\frac{1}{\sqrt{t}}\nu(a\sqrt{t},2\alpha)\} = 2\sqrt{\frac{\pi}{s}}\,\nu(\frac{a^2}{4s},\alpha) \qquad (64.4.3)$$

$a > 0 \quad , \quad \mathrm{Re}\,\alpha > -1$

$$L\{\frac{1}{\sqrt{t}}\mu(a\sqrt{t},\beta,2\alpha)\} = 2^{\beta+1}\sqrt{\frac{\pi}{s}}\,\mu(\frac{a^2}{4s},\beta,\alpha) \qquad (65.4.3)$$

$a > 0 \quad , \quad \mathrm{Re}\,\alpha > -1 \quad , \quad \mathrm{Re}\,\beta > -1$

and

$$L\{\nu(a\sqrt{t},\alpha)\} = \frac{\sqrt{\pi}\,a}{s^{3/2}}\nu(\frac{a^2}{4s},\frac{\alpha-1}{2}) \qquad (66.4.3)$$

$a > 0 \quad , \quad \mathrm{Re}\,\alpha > -1$

$$L\{\mu(a\sqrt{t},\beta,\alpha)\} = \frac{2^\beta\sqrt{\pi}\,a}{s^{3/2}}\mu(\frac{a^2}{4s},\beta,\frac{\alpha-1}{2}) \qquad (67.4.3)$$

$a > 0 \quad , \quad \mathrm{Re}\,\alpha > -1 \quad , \quad \mathrm{Re}\,\beta > -1$

4.4. Integral Formulas Involving the Volterra Functions

There are various types of integral relations that can be established with the aid of operational calculus. Many of them were already mentioned in different contexts, some new ones are considered here.

If the Laplace transform

$$G(\lambda,s) = L\{t^\lambda g(t)\} \quad , \quad \mathrm{Re}\,\lambda > -1 \qquad (1.4.4)$$

is known, it is possible to present the Laplace transform of the product of the Volterra function and $g(t)$ in the form

$$\int_0^{\infty} e^{-st} g(t)\, v(at,\alpha)dt = \int_0^{\infty} \frac{a^{u+\alpha} G(u+\alpha,s)}{\Gamma(u+\alpha+1)} du \tag{2.4.4}$$

$$a > 0 \quad , \quad \operatorname{Re}\alpha > -1 \quad , \quad \operatorname{Re}s > 0$$

In some cases, the right-hand side integral can be evaluated in the closed form, but usually the equality of integrals in (2.4.4) leads to some interesting relationships between the Volterra functions and other special functions.

For example, from [88, 1.6.37]

$$L\{t^{\lambda} \ln t\} = \frac{\Gamma(\lambda+1)}{s^{\lambda+1}} [\psi(\lambda+1) - \ln s] \tag{3.4.4}$$

$$\operatorname{Re}\lambda > -1$$

and using (2.4.4)

$$\int_0^{\infty} e^{-st} \ln t\, v(at,\alpha)dt = \frac{1}{a} \int_0^{\infty} \left(\frac{a}{s}\right)^{u+\alpha+1} [\psi(u+\alpha+1) - \ln s]\, du \tag{4.4.4}$$

$$\operatorname{Re}\alpha > -1 \quad , \quad \operatorname{Re}(s-a) > 0$$

The above equality of integrals can be expected, considering that the Volterra functions and the digamma functions are related via the gamma function. However, it is rather surprising to find that integrals with respect to the order of *the parabolic cylinder function* $D_{\lambda}(z)$ or *the modified Bessel functions of the second kind* $K_{\lambda}(z)$ [1,80] can be expressed in terms of integrals of the Volterra functions.

From [88, 1.5.46]

$$L\{t^{\lambda} e^{-b^2t^2/2}\} = \frac{\Gamma(\lambda+1)}{b^{\lambda+1}} e^{(s/2b)^2} D_{-(\lambda+1)}\left(\frac{s}{b}\right) \tag{5.4.4}$$

$$\operatorname{Re}b > 0 \quad , \quad \operatorname{Re}\lambda > -1$$

and (2.4.4)

$$\int_0^{\infty} e^{-st-b^2t^2/2} v(at,\alpha)dt = \frac{e^{(s/2b)^2}}{a} \int_0^{\infty} \left(\frac{a}{b}\right)^{u+\alpha+1} D_{-(u+\alpha+1)}\left(\frac{s}{b}\right) du \tag{6.4.4}$$

$$a > 0 \quad , \quad \operatorname{Re}b > 0 \quad , \quad \operatorname{Re}\alpha > -1 \quad , \quad \operatorname{Re}s > 0$$

which for $\alpha = 0$ and $a = b = 1$ reduces to

$$\int_0^\infty e^{-st-t^2/2}\nu(t)dt = e^{s^2/4}\int_1^\infty D_{-x}(s)dx \tag{7.4.4}$$

Similarly from [88, 1.5.34]

$$L\{t^\lambda e^{-b/t}\} = 2(\frac{b}{s})^{(\lambda+1)/2}K_{\lambda+1}(2\sqrt{bs}) \tag{8.4.4}$$
$$\mathrm{Re}\, b > 0$$

and (2.4.4) it follows that

$$\int_0^\infty e^{-st-b/t}\nu(at,\alpha)dt = \frac{2}{a}\int_0^\infty \frac{(a\sqrt{b/s})^{u+\alpha+1}}{\Gamma(u+\alpha+1)}K_{u+\alpha+1}(2\sqrt{bs})du \tag{9.4.4}$$
$$a>0 \quad , \quad \mathrm{Re}\, b>0 \quad , \quad \mathrm{Re}\,\alpha > -1 \quad , \quad \mathrm{Re}\, s>0$$

which for α = 0 and a = 1 takes the form

$$\int_0^\infty e^{-st-b/t}\nu(t)dt = 2\int_1^\infty \frac{(b/s)^{x/2}}{\Gamma(x)}K_x(2\sqrt{bs})dx \tag{10.4.4}$$

The second group of the integral equalities can be established if the Laplace transforms of the form

$$F(\lambda,s) = L\{[f(t)]^\lambda\} \tag{11.4.4}$$
$$\mathrm{Re}\,\lambda > -1$$

are known, because then

$$\int_0^\infty e^{-st}\nu[af(t),\alpha]dt = \int_0^\infty \frac{a^{u+\alpha}F(u+\alpha,s)}{\Gamma(u+\alpha+1)}du \tag{12.4.4}$$
$$a>0 \quad , \quad \mathrm{Re}\,\alpha > -1 \quad , \quad \mathrm{Re}\, s>0$$

and

$$\int_0^\infty e^{-st}\mu[af(t),\beta,\alpha]dt = \int_0^\infty \frac{a^{u+\alpha}u^\beta F(u+\alpha,s)}{\Gamma(u+\alpha+1)}du \tag{13.4.4}$$
$$a>0 \quad , \quad \mathrm{Re}\,\alpha > -1 \quad , \quad \mathrm{Re}\,\beta > -1 \quad , \quad \mathrm{Re}\, s>1$$

Thus, in the left-hand side integrals, variable t in the Volterra functions is replaced by $f(t)$. The integral equalities (12.4.4) and (13.4.4) can be generalized to give

$$\int_0^\infty e^{-st} g(t) \nu[af(t), \alpha] dt = \int_0^\infty \frac{a^{u+\alpha} H(u+\alpha, s)}{\Gamma(u+\alpha+1)} du \tag{14.4.4}$$

$$\int_0^\infty e^{-st} g(t) \mu[af(t), \beta, \alpha] dt = \int_0^\infty \frac{a^{u+\alpha} u^\beta H(u+\alpha, s)}{\Gamma(u+\alpha+1)} du \tag{15.4.4}$$

where

$$H(\lambda, s) = L\{g(t)[f(t)]^\lambda\} \qquad \mathrm{Re}\,\lambda > -1 \tag{16.4.4}$$

These formulas will be illustrated by considering once again integrals with respect to the order of $D_\lambda(z)$ and $K_\lambda(z)$ functions. From [88, 1.3.9]

$$L\{\frac{1}{(b+t)^{3/2}} (\frac{t}{b+t})^\lambda\} = \frac{2^{\lambda+1}\Gamma(\lambda+1) e^{bs/2}}{\sqrt{b}} D_{-2(\lambda+1)}(\sqrt{2bs}) \qquad \mathrm{Re}\,\lambda > -1 \;, \quad |\arg b| < \pi \tag{17.4.4}$$

and using (14.4.4)

$$\int_0^\infty \frac{e^{-st}}{(b+t)^{3/2}} \nu(\frac{at}{b+t}, \alpha) dt = \frac{e^{bs/2}}{a} \int_0^\infty (2a)^{u+\alpha+1} D_{-2(u+\alpha+1)}(\sqrt{2bs}) du \qquad a > 0 \;, \quad b > 0 \;, \quad \mathrm{Re}\,\alpha > -1 \;, \quad \mathrm{Re}\, s > 0 \tag{18.4.4}$$

In particular, for $\alpha = 0$, a = b = 1/2 and denoting $s = c^2$, (18.4.4) becomes

$$\int_0^\infty \frac{e^{-c^2 t}}{(1+2t)^{3/2}} \nu(\frac{t}{1+2t}) dt = \frac{e^{c^2/4}}{\sqrt{2}} \int_1^\infty D_{-2x}(c) dx \tag{19.4.4}$$

Similarly from [48, 1.3.16]

$$L\{\frac{1}{\sqrt{2bt+t^2}}(2bt+t^2)^{\lambda}\}=\frac{\Gamma(\lambda+\frac{1}{2})}{\sqrt{\pi}}(\frac{2b}{s})^{\lambda}e^{bs}K_{\lambda}(bs) \tag{20.4.4}$$

$$\operatorname{Re}\lambda>-\frac{1}{2} \quad , \quad |\arg b|<\pi$$

we have

$$\int_0^{\infty}\frac{e^{-st}\nu(2abt+at^2,\alpha)}{\sqrt{2bt+t^2}}dt=\frac{e^{bs}}{\sqrt{\pi}}\int_0^{\infty}\frac{\Gamma(u+\alpha+\frac{1}{2})}{\Gamma(u+\alpha+1)}(\frac{2ab}{s})^{u+\alpha}K_{u+\alpha}(bs)du \tag{21.4.4}$$

$$a>0 \quad , \quad b>0 \quad , \quad \operatorname{Re}\alpha>-\frac{1}{2} \quad , \quad \operatorname{Re}s>0$$

which for a = 1 and b = 1/2 is

$$\int_0^{\infty}\frac{e^{-st}\nu(t+t^2,\alpha)}{\sqrt{t+t^2}}dt=\frac{e^{s/2}}{\sqrt{\pi}}\int_{\alpha}^{\infty}\frac{\Gamma(x+\frac{1}{2})}{\Gamma(x+1)}(\frac{1}{s})^{x}K_{x}(\frac{s}{2})dx \tag{22.4.4}$$

In some cases, the right-hand side integrals in (12.4.4) - (15.4.4) can be evaluated in the closed form, e.g. using [88, 1.5.8]

$$L\{(1-e^{-bt})^{\lambda}\}=\frac{1}{b}B(\lambda+1,\frac{s}{b}) \tag{23.4.4}$$

$$b>0 \quad , \quad \operatorname{Re}\lambda>-1$$

equation (12.4.4) becomes

$$\int_0^{\infty}e^{-st}\nu[a(1-e^{-bt}),\alpha]dt=\frac{1}{b}\int_0^{\infty}\frac{a^{u+\alpha}B(u+\alpha+1,\frac{s}{b})}{\Gamma(u+\alpha+1)}du \tag{24.4.4}$$

But *the beta function* is given by [1]

$$B(u+\alpha+1,\frac{s}{b})=\frac{\Gamma(\frac{s}{b})\Gamma(u+\alpha+1)}{\Gamma(u+\alpha+\frac{s}{b}+1)} \tag{25.4.4}$$

and therefore

$$\int_0^\infty e^{-st}\nu[a(1-e^{-bt}),\alpha]dt = \frac{\Gamma(\frac{s}{b})}{b}\int_0^\infty \frac{a^{u+\alpha}}{\Gamma(u+\alpha+\frac{s}{b}+1)}du \tag{26.4.4}$$

which can be written as the Laplace transform

$$L\{\nu[a(1-e^{-bt}),\alpha]\} = \frac{\Gamma(\frac{s}{b})}{a^{s/b}b}\nu[a,\alpha+\frac{s}{b}] \tag{27.4.4}$$

$$a>0 \quad , \quad b>0 \quad , \quad \mathrm{Re}\,\alpha > -1 \quad , \quad \mathrm{Re}\,s>0$$

or by changing the variable of integration as

$$\int_0^1 (1-x)^{\lambda-1}\nu(ax,\alpha)dx = \frac{\Gamma(\lambda)}{a^\lambda}\nu(a,\alpha+\lambda) \tag{28.4.4}$$

$$a>0 \quad , \quad \mathrm{Re}\,\alpha > -1 \quad , \quad \mathrm{Re}\,\lambda>0$$

Two additional integral formulas are useful in the evaluation of integrals of the Volterra functions

$$\int_0^\infty e^{\alpha t}f(t)\nu(ae^{-t},\alpha)dt = \int_0^\infty \frac{a^{u+\alpha}F(u)}{\Gamma(u+\alpha+1)}du \tag{29.4.4}$$

$$a>0 \quad , \quad \mathrm{Re}\,\alpha > -1$$

$$\int_0^\infty e^{\alpha t}f(t)\mu(ae^{-t},\beta,\alpha)dt = \int_0^\infty \frac{a^{u+\alpha}u^\beta F(u)}{\Gamma(u+\alpha+1)}du \tag{30.4.4}$$

$$a>0 \quad , \quad \mathrm{Re}\,\alpha > -1 \quad , \quad \mathrm{Re}\,\beta>-1$$

For example, from the introduction of $f(t) = t^{\lambda-1}$ and $F(s) = \Gamma(\lambda)/s^\lambda$ into (30.4.4), the well-known result [43] can be established

$$\int_0^\infty t^{\lambda-1}e^{\alpha t}\mu(ae^{-t},\beta,\alpha)dt = \frac{\Gamma(\lambda)\Gamma(\beta-\lambda+1)}{\Gamma(\beta+1)}\mu(a,\beta-\lambda,\alpha) \tag{31.4.4}$$

$$a>0 \quad , \quad \mathrm{Re}\,\alpha > -1 \quad , \quad \mathrm{Re}\,\beta>-1 \quad , \quad \mathrm{Re}\,\lambda>0$$

which by changing the variable of integration can be written as

$$\int_0^1 \frac{1}{x^{\alpha+1}}[\ln(\frac{1}{x})]^{\lambda-1}\mu(ax,\beta,\alpha)dx = \frac{\Gamma(\lambda)\Gamma(\beta-\lambda+1)}{\Gamma(\beta+1)}\mu(a,\beta-\lambda,\alpha) \tag{32.4.4}$$

Many other integral equalities, finite and improper integrals based on (2.4.4), (12.4.4) - (15.4.4) and (29.4.4) - (30.4.4) are presented in Chapter 5.

4.5. Interrelations between the Volterra Functions and other Special Functions

Operational calculus is a source for many relations among various special functions, and the same can be expected when the Volterra functions are considered. Some relations were already met while dealing with different aspects of the Laplace transforms of the Volterra functions. Additional examples illustrating an extensive list of special functions is given in this section.

Starting with the digamma function and using the Laplace transform [88, 1.6.37]

$$L\{t^{\lambda}(\ln t)^2\} = \frac{\Gamma(\lambda+1)}{s^{\lambda+1}}\{[\psi(\lambda+1)-\ln s]^2+\psi'(\lambda+1)\}$$
$$\mathrm{Re}\,\lambda > -1 \quad , \quad \mathrm{Re}\,s > 0 \tag{1.4.5}$$

(2.4.4) then becomes

$$L\{(\ln t)^2 \nu(at,\alpha)\} = \frac{1}{a}\int_0^\infty (\frac{a}{s})^{u+\alpha+1}\{[\psi(u+\alpha+1)-\ln s]^2+\psi'(u+\alpha+1)\}du$$
$$\mathrm{Re}\,\alpha > -1 \quad , \quad \mathrm{Re}(s-a) > 0 \tag{2.4.5}$$

For a = 1, after some manipulations

$$L\{(\ln t)^2 \nu(t,\alpha)\} = \frac{\ln s}{s^{\alpha+1}} - \frac{2\ln s}{s^{\alpha+1}}\int_0^\infty \psi(u+\alpha+1)du +$$
$$\frac{1}{s^{\alpha+1}}\int_0^\infty s^{-u}[\psi(u+\alpha+1)^2+\psi'(u+\alpha+1)]du \tag{3.4.5}$$

But

$$\int_0^\infty s^{-u}\psi(u+\alpha+1)du = \frac{\psi(\alpha+1)}{\ln s} + \frac{1}{\ln s}\int_0^\infty s^{-u}\psi'(u+\alpha+1)du \tag{4.4.5}$$

and therefore

$$L\{(\ln t)^2 v(t,\alpha)\} = \frac{\ln s}{s^{\alpha+1}} - \frac{2\psi(\alpha+1)}{s^{\alpha+1}} + \frac{1}{s^{\alpha+1}} \int_0^\infty s^{-u}[\psi(u+\alpha+1)^2 - \psi'(u+\alpha+1)]du \tag{5.4.5}$$

Considering that

$$L^{-1}\{\frac{1}{s^{\alpha+1}}\} = \frac{t^\alpha}{\Gamma(\alpha+1)} \tag{6.4.5}$$

$$L^{-1}\{\frac{\ln s}{s^{\alpha+1}}\} = \frac{t^\alpha}{\Gamma(\alpha+1)}[\psi(\alpha+1) - \ln t] \tag{7.4.5}$$

it is possible to present (5.4.5) in the form

$$L\{(\ln t)^2 v(t,\alpha) + \frac{t^\alpha}{\Gamma(\alpha+1)}[\psi(\alpha+1) + \ln t]\} = \frac{1}{s^{\alpha+1}} \int_0^\infty s^{-u}[\psi(u+\alpha+1)^2 - \psi'(u+\alpha+1)]du \tag{8.4.5}$$

$$\operatorname{Re}\alpha > -1 \quad , \quad \operatorname{Re}s > 1$$

and the corresponding expression based on (4.4.4) is

$$L\{(\ln t)v(t,\alpha) + \frac{t^\alpha}{\Gamma(\alpha+1)}\} = \frac{1}{s^{\alpha+1}} \int_0^\infty s^{-u}\psi(u+\alpha+1)du \tag{9.4.5}$$

Laplace transforms of the Volterra function integrals

$$F_n(s) = \int_0^\infty e^{-st}\{\int_{\ln n}^{\ln(n+1)} v(t,x)dx\}dt \tag{10.4.5}$$

$$n = 1,2,3,\ldots$$

can be evaluated by inverting the order of integration and using (2.4.1)

$$F_n(s)=\int_{\ln n}^{\ln(n+1)}\frac{dx}{s^{x+1}\ln s}=\frac{1}{s(\ln s)^2}[\frac{1}{n^{\ln s}}-\frac{1}{(n+1)^{\ln s}}] \tag{11.4.5}$$

but from (11.4.5) it is possible to construct *the Riemann zeta function* $\zeta(z)$ [1,43,80]

$$\sum_{n=1}^{\infty}n[\frac{1}{n^{\ln s}}-\frac{1}{(n+1)^{\ln s}}]=\sum_{n=1}^{\infty}\frac{1}{n^{\ln s}}=\zeta(\ln s) \tag{12.4.5}$$

and therefore from (10.4.5), (11.4.5) and (12.4.5)

$$L\{\sum_{n=1}^{\infty}\int_{\ln n}^{\ln(n+1)}\nu(t,x)dx\}=\sum_{n=1}^{\infty}nF_n(s)=\frac{1}{s(\ln s)^2}\zeta(\ln s) \tag{13.4.5}$$

$$\operatorname{Re}s>e$$

An alternative way to establish (13.4.5) or similar expressions is to use (53.4.3) and introduce a function $f(t)$ which is built up of the *shifted unit step functions*

$$f(t)=\sum_{n=1}^{\infty}n\{u[t-\ln n]-u[t-\ln(n+1)]\} \tag{14.4.5}$$

Since

$$L\{u[t-a]\}=\frac{e^{-as}}{s} \tag{15.4.5}$$

the Laplace transform of $f(t)$ is

$$F(s)=\frac{1}{s}\sum_{n=1}^{\infty}n\{e^{-s\ln n}-e^{-s\ln(n+1)}\}=\frac{1}{s}\sum_{n=1}^{\infty}\frac{1}{n^s}=\frac{1}{s}\zeta(s) \tag{16.4.5}$$

$$\operatorname{Re}s>1$$

But from (53.4.3)

$$L\{\int_0^{\infty}\nu(t,x)f(x)dx\}=\frac{F(\ln s)}{s\ln s} \tag{17.4.5}$$

and

$$\int_0^{\infty} v(t,x)\{u[x-\ln n]-u[x-\ln(n+1)]\}dx = \int_{\ln n}^{\ln(n+1)} v(t,x)dx \tag{18.4.5}$$

the previous result (13.4.5) is immediately reached.

Let us introduce the following function

$$f(t) = \sum_{n=0}^{\infty} \omega^n u[t-\ln(n+\alpha)] \qquad \alpha > 0 \ , \ |\omega| \le 1 \tag{19.4.5}$$

Using (15.4.5) the Laplace transform of this function is

$$F(s) = \frac{1}{s}\sum_{n=0}^{\infty} \frac{\omega^n}{(n+\alpha)^s} = \frac{1}{s}\Phi(\omega,s,\alpha) \tag{20.4.5}$$

where $\Phi(\omega,s,\alpha)$ is *the Lerch function* [43,88]. Applying the operational rule (17.4.5) to (19.4.5)

$$L\{\sum_{n=0}^{\infty} \omega^n \int_{\ln(n+\alpha)}^{\infty} v(t,x)dx\} = \frac{1}{s(\ln s)^2}\Phi(\omega,\ln s,\alpha) \qquad \alpha > 0 \ , \ |\omega| \le 1 \ , \ \operatorname{Re} s > e \tag{21.4.5}$$

This is a more general result than that given in (13.4.5) because the Riemann zeta and related functions are particular cases of the Lerch function. For example, considering that $\Phi(1,z,1) = \zeta(z)$, it follows from (21.4.5) that

$$L\{\sum_{n=0}^{\infty} \int_{\ln(n+1)}^{\infty} v(t,x)dx\} = \frac{1}{s(\ln s)^2}\zeta(\ln s) \tag{22.4.5}$$

which is identical with (13.4.5) because

$$\sum_{n=0}^{\infty} \int_{\ln(n+1)}^{\infty} v(t,x)dx = \sum_{n=1}^{\infty} n \int_{\ln n}^{\ln(n+1)} v(t,x)dx \tag{23.4.5}$$

Other related cases of the Lerch function are

$$\omega\Phi(\omega,z,1)=F(\omega,z)=\sum_{n=1}^{\infty}\frac{\omega^n}{n^z} \tag{24.4.5}$$

$$\Phi(-1,z,1)=(1-2^{1-z})\zeta(z)=\sum_{n=1}^{\infty}\frac{(-1)^{n-1}}{n^z} \tag{25.4.5}$$

$$2^{-z}\Phi(1,z,\frac{1}{2})=(1-2^{-z})\zeta(z)=\sum_{n=0}^{\infty}\frac{1}{(2n+1)^z} \tag{26.4.5}$$

$$2^{-z}\Phi(-1,z,\frac{1}{2})=\sum_{n=0}^{\infty}\frac{(-1)^n}{(2n+1)^z} \tag{27.4.5}$$

$$\Phi(1,z,\alpha)=\zeta(z,\alpha)=\sum_{n=1}^{\infty}\frac{1}{(n+\alpha)^z} \tag{28.4.5}$$

where F(ω,z) is *the Jonquière function* [43] and ζ(z,α) is the *generalized zeta function* [1,43,80]. Introducing these particular cases of the Lerch function into (21.4.5) we have

$$L\{\sum_{n=0}^{\infty}\omega^{n+1}\int_{\ln(n+1)}^{\infty}\nu(t,x)dx\}=\frac{1}{s(\ln s)^2}F(\omega,\ln s) \tag{29.4.5}$$

$$L\{\sum_{n=0}^{\infty}(-1)^n\int_{\ln(n+1)}^{\infty}\nu(t,x)dx\}=\frac{(1-2^{1-\ln s})}{s(\ln s)^2}\zeta(\ln s) \tag{30.4.5}$$

$$L\{\sum_{n=0}^{\infty}\int_{\ln(n+1/2)}^{\infty}\nu(t,x)dx\}=\frac{(2^{\ln s}-1)}{s(\ln s)^2}\zeta(\ln s) \tag{31.4.5}$$

$$L\{\sum_{n=0}^{\infty}(-1)^n\int_{\ln(n+1/2)}^{\infty}\nu(t,x)dx\}=\frac{2^{\ln s}}{s(\ln s)^2}\Phi(-1,\ln s,\frac{1}{2}) \tag{32.4.5}$$

$$L\{\sum_{n=0}^{\infty}\int_{\ln(n+\alpha)}^{\infty}\nu(t,x)dx\}=\frac{1}{s(\ln s)^2}\zeta(\ln s,\alpha) \tag{33.4.5}$$

Other expressions of this type but associated with *the functions of number theory* [43] can be derived by replacing $f(t)$ in (19.4.5) with

$$f(t) = \sum_{n=1}^{\infty} g(n) u[t - \ln n] \tag{34.4.5}$$

and then

$$F(s) = \frac{1}{s} \sum_{n=1}^{\infty} \frac{g(n)}{n^s} \tag{35.4.5}$$

where g(n) is the arithmetical function defined for all positive integers n. Only two such functions are considered here

$$\sum_{n=1}^{\infty} \frac{\lambda(n)}{n^s} = \frac{\zeta(2s)}{\zeta(s)} \qquad \operatorname{Re} s > 1 \tag{36.4.5}$$

$$\sum_{n=1}^{\infty} \frac{\mu(n)}{n^s} = \frac{1}{\zeta(s)} \qquad \operatorname{Re} s > 1 \tag{37.4.5}$$

where λ(n) is *the Liouville function* and μ(n) is *the Möbius function* [43].

Thus, introducing (34.4.5) and (35.4.5) into (17.4.5)

$$L\{\sum_{n=1}^{\infty} \lambda(n) \int_{\ln n}^{\infty} v(t,x) dx\} = \frac{\zeta(2 \ln s)}{s (\ln s)^2 \zeta(\ln s)} \tag{38.4.5}$$

$$L\{\sum_{n=1}^{\infty} \mu(n) \int_{\ln n}^{\infty} v(t,x) dx\} = \frac{1}{s (\ln s)^2 \zeta(\ln s)} \qquad \operatorname{Re} s > e \tag{39.4.5}$$

The generalized zeta function appears in the following Laplace transform [44, 4.5.10]

$$L\{\frac{t^{\lambda}}{1 - e^{-t}}\} = \Gamma(\lambda + 1) \zeta(\lambda, s) \qquad \operatorname{Re} \lambda > -1 \quad , \quad \operatorname{Re} s > 0 \tag{40.4.5}$$

and applying (2.4.4) we have

$$L\{\frac{1}{1-e^{-t}}\nu(t,\alpha)\} = \int_0^\infty \zeta(u+\alpha+1,s)du \qquad (41.4.5)$$

$$\text{Re}\,\alpha > -1 \quad , \quad \text{Re}\,s > 1$$

This integral equality can be easily verified by performing the inversion of (35.4.5). If the integration and summation are interchanged in the right-hand side integral then

$$\int_0^\infty \zeta(u+\alpha+1,s)du = \int_0^\infty \{\sum_{n=0}^\infty \frac{1}{(n+s)^{u+\alpha+1}}\}du =$$
$$\sum_{n=0}^\infty \frac{1}{(n+s)^{u+\alpha+1}} \int_0^\infty \frac{du}{(n+s)^u} = \sum_{n=0}^\infty \frac{1}{(n+s)^{\alpha+1}\ln(n+s)} \qquad (42.4.5)$$

and the term by term inversion of (36.4.5) gives the expected result.

$$L^{-1}\{\sum_{n=0}^\infty \frac{1}{(n+s)^{\alpha+1}\ln(n+s)}\} = \sum_{n=0}^\infty e^{-nt}\nu(t,\alpha) = \frac{\nu(t,\alpha)}{1-e^{-t}} \qquad (43.4.5)$$

As already pointed out, the Laplace transforms of the reciprocal gamma functions are expressed in terms of the Volterra functions (see (15.4.3) - (23.4.3)). This fact connects them with *the Mittag-Leffler function* and related functions [2,43,64,65].

The Mittag-Leffler function is defined by the series

$$E_\alpha(x) = \sum_{n=0}^\infty \frac{x^n}{\Gamma(\alpha n+1)} \qquad (44.4.5)$$

and other functions of this type are

$$E_{\alpha,\beta}(x) = \sum_{n=0}^\infty \frac{x^n}{\Gamma(\alpha n+\beta)} \qquad (45.4.5)$$

$${}_1E_\alpha(x) = \sum_{n=0}^\infty \frac{x^n}{n!\,\Gamma(\alpha n+1)} \qquad (46.4.5)$$

$${}_1E_{\alpha,\beta}(x) = \sum_{n=0}^\infty \frac{x^n}{n!\,\Gamma(\alpha n+\beta)} \qquad (47.4.5)$$

$$ {}_mE_{\alpha,\beta}(x) = \sum_{n=0}^{\infty} \frac{x^n}{(n!)^m \Gamma(\alpha n + \beta)} \tag{48.4.5} $$

$$ m = 1, 2, 3, \ldots $$

For particular values of parameters α and β, these functions can be reduced to the elementary or special functions (for details see Appendix A).

Laplace transforms with regard to the variables and parameters of the Mittag-Leffler type functions are given by

$$ L\{E_{t+\alpha}(a) - 1\} = e^{\alpha s} \sum_{n=1}^{\infty} \frac{a^n}{n} \nu(e^{-s/n}, \alpha n) \tag{49.4.5} $$

$$ L\{E_{t+\alpha,\beta}(a) - \frac{1}{\Gamma(\beta)}\} = e^{\alpha s} \sum_{n=1}^{\infty} \frac{a^n}{n} e^{(\beta-1)s/n} \nu(e^{-s/n}, \omega) \tag{50.4.5} $$

$$ \omega = \alpha n + \beta - 1 $$

$$ L\{E_{t+\alpha,t+\beta}(a)\} = \sum_{n=0}^{\infty} \frac{a^n}{(n+1)} e^{\omega s/(n+1)} \nu(e^{-s/(n+1)}, \omega) \tag{51.4.5} $$

and

$$ L\{E_{t+\alpha}(ab^t t^\rho) - 1\} = e^{\alpha s} \sum_{n=1}^{\infty} \frac{\Gamma(\rho n + 1)}{n^{\rho n+1}} (\frac{a}{b^\alpha})^n \mu(be^{-s/n}, \rho n, \alpha n) \tag{52.4.5} $$

$$ L\{E_{t+\alpha,\beta}(ab^t t^\rho) - \frac{1}{\Gamma(\beta)}\} = $$
$$ e^{\alpha s} b^{1-\beta} \sum_{n=1}^{\infty} \frac{\Gamma(\rho n + 1)(ab^\alpha)^n}{n^{\rho n+1}} e^{(\beta-1)s/n} \mu(be^{-s/n}, \rho n, \omega) \tag{53.4.5} $$

$$ L\{E_{t+\alpha,t+\beta}(ab^t t^\rho)\} = $$
$$ \sum_{n=0}^{\infty} \frac{\Gamma(\rho n + 1)}{(n+1)^{\rho n+1}} [\frac{a}{b^{\omega/(n+1)}}]^n e^{\omega s/(n+1)} \mu(b^{n/(n+1)} e^{-s/(n+1)}, \rho n, \omega) \tag{54.4.5} $$

The Laplace transforms of ${}_mE_{t+\alpha,t+\beta}(a)$ and ${}_mE_{t+\alpha,t+\beta}(ab^t t^\rho)$ functions are omitted because they can easily be written by introducing in above expressions the factor $1/(n!)^m$, m = 1,2,3, ... under the sign of the sum. The function ${}_1E_{\alpha,\beta}(t)$ is also called *the Wright function* and denoted as $\phi(\alpha,\beta;t)$ [43]. This function may be regarded as a kind of generalized Bessel function because

$$J_\lambda(t) = (\frac{t}{2})^\lambda \phi(1, \lambda+1; -\frac{t^2}{4}) \tag{55.4.5}$$

$$I_\lambda(t) = (\frac{t}{2})^\lambda \phi(1, \lambda+1; \frac{t^2}{4}) \tag{56.4.5}$$

The Bessel functions were encountered earlier (see (12.4.3), (43.4.3), (9.4.4) and (21.4.4)), but few more examples will be considered here. If the Laplace transform of $\nu(t,\alpha)$ is written in the form

$$L\{\nu(t,\alpha)\} = \frac{1}{s^{\alpha+1}\ln s} = -\frac{1}{s^{\alpha+2}\{(\frac{1}{s})\ln(\frac{1}{s})\}} \tag{57.4.5}$$

then using the operational rule [88, 2.1.22]

$$L^{-1}\{\frac{1}{s^{2\lambda+1}}F(\frac{1}{s})\} = \int_0^\infty (\frac{t}{u})^\lambda J_{2\lambda}(2\sqrt{tu}) f(u) du \tag{58.4.5}$$

we have

$$\nu(t,\alpha) = -\int_0^\infty (\frac{t}{u})^{(\alpha+1)/2} J_{\alpha+1}(2\sqrt{tu}) \nu(u) du \tag{59.4.5}$$

which was already given by Humbert and Poli [67] for $\alpha = 0$.

It is noteworthy that the Volterra functions are mainly involved with operations regarding the order of Bessel and related functions. Using the original function-transform pairs [88, 1.14.24, 1.15.24]

$$L\{t^{\lambda/2} J_\lambda(a\sqrt{t})\} = \frac{1}{s}(\frac{a}{2s})^\lambda e^{-a^2/4s} \tag{60.4.5}$$

$$L\{t^{\lambda/2} I_\lambda(a\sqrt{t})\} = \frac{1}{s}(\frac{a}{2s})^\lambda e^{a^2/4s} \qquad \mathrm{Re}\,\lambda > -1 \quad , \quad \mathrm{Re}\,s > 0 \tag{61.4.5}$$

and performing integration under the sign of integrals

$$L\{\int_{\alpha}^{\infty} t^{\lambda/2} J_{\lambda}(a\sqrt{t})d\lambda\} = \frac{e^{-a^2/4s}}{s}\int_{\alpha}^{\infty}(\frac{a}{2s})^{\lambda}d\lambda = \frac{e^{-a^2/4s}}{s}\cdot\frac{1}{(\frac{2s}{a})^{\alpha}\ln(\frac{2s}{a})} \quad (62.4.5)$$

However

$$L^{-1}\{\frac{e^{-a^2/4s}}{s}\} = J_0(a\sqrt{t}) \quad (63.4.5)$$

$$L^{-1}\{\frac{e^{a^2/4s}}{s}\} = I_0(a\sqrt{t}) \quad (64.4.5)$$

$$L^{-1}\{\frac{1}{(\frac{2s}{a})^{\alpha}\ln(\frac{2s}{a})}\} = \frac{a}{2}\nu(\frac{at}{2},\alpha-1) \quad (65.4.5)$$

and therefore from the convolution theorem

$$\int_{\alpha}^{\infty} t^{\lambda/2} J_{\lambda}(a\sqrt{t})d\lambda = \frac{a}{2}\int_{0}^{t} J_0(a\sqrt{u})\nu[\frac{a}{2}(t-u),\alpha-1]du \quad (66.4.5)$$

$$\int_{\alpha}^{\infty} t^{\lambda/2} I_{\lambda}(a\sqrt{t})d\lambda = \frac{a}{2}\int_{0}^{t} I_0(a\sqrt{u})\nu[\frac{a}{2}(t-u),\alpha-1]du \quad (67.4.5)$$

Similar expressions hold also for the Struve functions

$$\int_{\alpha}^{\infty} t^{\lambda/2} L_{\lambda}(a\sqrt{t})d\lambda = \frac{a}{2}\int_{0}^{t} L_0(a\sqrt{u})\nu[\frac{a}{2}(t-u),\alpha-1]du \quad (68.4.5)$$

$$\int_{\alpha}^{\infty} t^{\lambda/2} L_{\lambda}(a\sqrt{t})d\lambda = \frac{a}{2}\int_{0}^{t} L_0(a\sqrt{u})\nu[\frac{a}{2}(t-u),\alpha-1]du \quad (69.4.5)$$

Since (60.4.5) and (61.4.5) have the form of (5.4.3) then from (6.4.3)

$$L\{\int_{\alpha}^{\infty}\frac{t^{\lambda/2} J_{\lambda}(a\sqrt{t})}{\Gamma(\lambda+1)}d\lambda\} = \frac{1}{s}e^{-a^2/4s}\nu(\frac{a}{2s},\alpha) \quad (70.4.5)$$

$$L\{\int_{\alpha}^{\infty}\frac{t^{\lambda/2}I_{\lambda}(a\sqrt{t})}{\Gamma(\lambda+1)}d\lambda\}=\frac{1}{s}e^{a^2/4s}\nu(\frac{a}{2s},\alpha) \quad (71.4.5)$$

and from [88, 1.16.27-28]

$$L\{\int_{\alpha}^{\infty}\frac{t^{\lambda/2}H_{\lambda}(a\sqrt{t})}{\Gamma(\lambda+1)}d\lambda\}=-\frac{i}{s}e^{-a^2/4s}erf(\frac{ia}{2\sqrt{s}})\nu(\frac{a}{2s},\alpha) \quad (72.4.5)$$

$$L\{\int_{\alpha}^{\infty}\frac{t^{\lambda/2}L_{\lambda}(a\sqrt{t})}{\Gamma(\lambda+1)}d\lambda\}=\frac{1}{s}e^{a^2/4s}erf(\frac{a}{2\sqrt{s}})\nu(\frac{a}{2s},\alpha) \quad (73.4.5)$$

Other cases of this type are [88, 2.41-2, 2.47-8]

$$L\{\int_{\alpha}^{\infty}\frac{a^{\lambda}J_{\lambda}(at)}{\Gamma(\lambda+1)}d\lambda\}=\frac{1}{\sqrt{s^2+a^2}}\nu(\frac{a^2}{s+\sqrt{s^2+a^2}},\alpha) \quad (74.4.5)$$

$$L\{\int_{\alpha}^{\infty}\frac{a^{\lambda}I_{\lambda}(at)}{\Gamma(\lambda+1)}d\lambda\}=\frac{1}{\sqrt{s^2-a^2}}\nu(\frac{a^2}{s+\sqrt{s^2-a^2}},\alpha) \quad (75.4.5)$$

$$L\{\int_{\alpha}^{\infty}\frac{a^{\lambda}}{\Gamma(\lambda+1)}(\frac{\lambda}{t})J_{\lambda}(at)d\lambda\}=\nu(\frac{a^2}{s+\sqrt{s^2+a^2}},\alpha) \quad (76.4.5)$$

$$L\{\int_{\alpha}^{\infty}\frac{a^{\lambda}}{\Gamma(\lambda+1)}(\frac{\lambda}{t})I_{\lambda}(at)d\lambda\}=\nu(\frac{a^2}{s+\sqrt{s^2-a^2}},\alpha) \quad (77.4.5)$$

and from [88, 2.27-8, 2.51-2]

$$L\{\int_{\alpha}^{\infty}\frac{a^{\lambda}}{\Gamma(\lambda+1)}[J_{\lambda-1}(at)-J_{\lambda+1}(at)]d\lambda\}=\frac{(2s/a)}{\sqrt{s^2+a^2}}\nu(\frac{a^2}{s+\sqrt{s^2+a^2}},\alpha) \quad (78.4.5)$$

$$L\{\int_{\alpha}^{\infty}\frac{a^{\lambda}}{\Gamma(\lambda+1)}[I_{\lambda-1}(at)+I_{\lambda+1}(at)]d\lambda\}=\frac{(2s/a)}{\sqrt{s^2-a^2}}\nu(\frac{a^2}{s+\sqrt{s^2-a^2}},\alpha) \quad (79.4.5)$$

$$L\{\int_{\alpha}^{\infty}\frac{1}{\Gamma(\lambda)\Gamma(\lambda+1)}(\frac{t}{2a})^{\lambda-1/2}J_{\lambda-1/2}(at)d\lambda\}=\frac{1}{\sqrt{\pi}}\nu(\frac{1}{s^2+a^2},\alpha) \quad (80.4.5)$$

$$L\{\int_{\alpha}^{\infty}\frac{1}{\Gamma(\lambda)\Gamma(\lambda+1)}(\frac{t}{2a})^{\lambda-1/2}I_{\lambda-1/2}(at)d\lambda\}=\frac{1}{\sqrt{\pi}}\nu(\frac{1}{s^2-a^2},\alpha) \tag{81.4.5}$$

Considering that

$$L^{-1}\{\frac{1}{\sqrt{s^2+a^2}}\}=J_0(at) \tag{82.4.5}$$

and taking into account (74.4.5) and (76.4.5), the convolution theorem gives

$$\int_{\alpha}^{\infty}\frac{a^{\lambda}}{\Gamma(\lambda+1)}J_{\lambda}(at)d\lambda=\int_0^t J_0[a(t-u)]\{\int_{\alpha}^{\infty}\frac{a^{\lambda}}{\Gamma(\lambda+1)}(\frac{\lambda}{u})J_{\lambda}(au)d\lambda\}du \tag{83.4.5}$$

or by changing the order of integration in the right-hand side integral we have

$$\int_{\alpha}^{\infty}\frac{a^{\lambda}}{\Gamma(\lambda+1)}\{\lambda\int_0^t J_0[a(t-u)]J_{\lambda}(au)\frac{du}{u}-J_{\lambda}(at)\}d\lambda=0 \tag{84.4.5}$$

The expression in the parenthesis represents the well-known convolution integral

$$\frac{1}{\lambda}J_{\lambda}(at)=\int_0^t J_0[a(t-u)]J_{\lambda}(au)\frac{du}{u} \tag{85.4.5}$$

Similarly from

$$L^{-1}\{\frac{1}{\sqrt{s^2-a^2}}\}=I_0(at) \tag{86.4.5}$$

and using (75.4.5) and (77.4.5) it is possible to obtain

$$\frac{1}{\lambda}I_{\lambda}(at)=\int_0^t I_0[a(t-u)]I_{\lambda}(au)\frac{du}{u} \tag{87.4.5}$$

Evidently, these convolution integrals (85.4.5) and (87.4.5) can be directly derived by inverting the corresponding Laplace transforms of the Bessel functions with the help of the product theorem. The same operations can be performed with (78.4.5) and (79.4.5), for example

$$L^{-1}\{\frac{s}{\sqrt{s^2+a^2}}\} = \delta(t) - aJ_1(at) \tag{88.4.5}$$

where $\delta(t)$ is *the Dirac delta function* and the convolution theorem as applied to (76.4.5) and (78.4.5) gives

$$\begin{aligned}&\int_{\alpha}^{\infty}\frac{a^{\lambda}}{\Gamma(\lambda+1)}[J_{\lambda-1}(at)-J_{\lambda+1}(at)]d\lambda = \\ &\frac{2}{a}\int_{\alpha}^{\infty}\frac{a^{\lambda}}{\Gamma(\lambda+1)}\{\lambda\int_{0}^{t}[[\delta(u)-J_1(au)]J_{\lambda}[a(t-u)]\frac{du}{t-u}\}d\lambda\end{aligned} \tag{89.4.5}$$

But in this case

$$\int_{0}^{t}[\delta(u)J_{\lambda}[a(t-u)]\frac{du}{t-u} = \frac{1}{t}J_{\lambda}(at) \tag{90.4.5}$$

and using the recurrence relation [1]

$$J_{\lambda-1}(at)+J_{\lambda+1}(at) = \frac{2\lambda}{at}J_{\lambda}(at) \tag{91.4.5}$$

the corresponding convolution integral from (89.4.5) is

$$\frac{1}{\lambda}J_{\lambda+1}(at) = \int_{0}^{t}J_1[a(t-u)]J_{\lambda}(au)\frac{du}{u} \tag{92.4.5}$$

In a similar way it is possible to obtain from (79.4.5)

$$\frac{1}{\lambda}I_{\lambda+1}(at) = \int_{0}^{t}I_1[a(t-u)]I_{\lambda}(au)\frac{du}{u} \tag{93.4.5}$$

Properties of the Volterra functions can serve to establish asymptotic behavior of integrals with respect to the order of the Bessel functions. The Laplace transforms of the Bessel fuctions $J_{\lambda}(t)$ and $I_{\lambda}(t)$ are

$$\begin{aligned}&L\{J_{\lambda}(t)\} = \frac{1}{\sqrt{s^2+1}[s+\sqrt{s^2+1}]^{\lambda}} \\ &\mathrm{Re}\,\lambda > -1 \quad , \quad \mathrm{Re}\,s > 0\end{aligned} \tag{94.4.5}$$

$$L\{I_{\lambda}(t)\} = \frac{1}{\sqrt{s^2-1}[s+\sqrt{s^2-1}]^{\lambda}} \tag{95.4.5}$$

$$\operatorname{Re}\lambda > -1 \quad , \quad \operatorname{Re}s > 1$$

Integrating on both sides and changing the order of integration

$$F(s) = L\{\int_{\alpha}^{\infty} I_{\lambda}(t)d\lambda\} = \frac{1}{\sqrt{s^2-1}} \int_{\alpha}^{\infty} \frac{d\lambda}{[s+\sqrt{s^2-1}]^{\lambda}} = \frac{1}{\sqrt{s^2-1}[s+\sqrt{s^2-1}]^{\alpha} \ln[s+\sqrt{s^2-1}]} \tag{96.4.5}$$

$$G(s) = L\{\int_{\alpha}^{\infty} I_{\lambda}(t)d\lambda\} = \frac{1}{\sqrt{s^2-1}} \int_{\alpha}^{\infty} \frac{d\lambda}{[s+\sqrt{s^2-1}]^{\lambda}} = \frac{1}{\sqrt{s^2-1}[s+\sqrt{s^2-1}]^{\alpha} \ln[s+\sqrt{s^2-1}]} \tag{97.4.5}$$

Performing the inversion, the integrals can be evaluated by the contour integration and are expressed by integrals of elementary functions (for details see [4]). Here we will limit ourselves to expressions representing the approximations of these integrals for $t \to 0$ and $t \to \infty$. The limiting behavior of $F(s)$ and $G(s)$ is

$$F(s \to 0) \square \frac{1}{s(s+1)^{\alpha}} \tag{98.4.5}$$

$$F(s \to \infty) \square \frac{2}{(2s)^{\alpha+1} \ln(2s)} \tag{99.4.5}$$

$$G(s \to 1) \square \frac{1}{2(s-1)} - \frac{\alpha}{\sqrt{2(s-1)}} \tag{100.4.5}$$

$$G(s \to \infty) \square \frac{2}{(2s)^{\alpha+1} \ln(2s)} \tag{101.4.5}$$

and therefore the inverse transforms of (98.4.5) and (100.4.5) approximate the integrals for $t \to \infty$

$$\int_{\alpha}^{\infty} J_{\lambda}(t)d\lambda \approx 1-\frac{\Gamma(\alpha,t)}{\Gamma(\alpha)} \tag{102.4.5}$$

$$\int_{\alpha}^{\infty} I_{\lambda}(t)d\lambda \approx \frac{e^{t}}{2}\{1-\alpha\sqrt{\frac{2}{\pi t}}\} \tag{103.4.5}$$

and for $t \to 0$, we have

$$\int_{\alpha}^{\infty} C_{\lambda}(t)d\lambda \approx \nu(\frac{t}{2},\alpha) \tag{104.4.5}$$

where $C_{\lambda}(t)$ denotes $J_{\lambda}(t)$ or $I_{\lambda}(t)$. However, the asymptotic behavior of the Volterra function for $t \to 0$ is known from (7.2.4) and therefore

$$\int_{\alpha}^{\infty} C_{\lambda}(t)d\lambda \approx \frac{(t/2)^{\alpha}}{\Gamma(\alpha+1)\ln(2/t)}\{1-\frac{\psi(\alpha+1)}{\ln(2/t)}+\ldots\} \tag{105.4.5}$$

Now, we will consider integrals with respect to the order of the Struve functions. Since [88, 1.16.27]

$$L\{t^{\lambda/2}H_{\lambda}(a\sqrt{t})\} = -\frac{i}{s}(\frac{a}{2s})^{\lambda} e^{-a^{2}/4s} erf(\frac{ia}{2\sqrt{s}}) \tag{106.4.5}$$

we have

$$L\{\int_{0}^{\alpha} t^{\lambda/2}H_{\lambda}(a\sqrt{t})d\lambda\} = \frac{ie^{-a^{2}/4s}}{s\ln(2s/a)} erf(\frac{ia}{2\sqrt{s}})[(\frac{a}{2s})^{\alpha}-1] \tag{107.4.5}$$

From the convolution theorem, as applied to (107.4.5), it is possible to express the Struve integral in a number of equivalent forms.

From [58, 2.15.74, 2.15.80]

$$L^{-1}\{\frac{1}{\sqrt{s}} e^{-a^{2}/4s} erf(\frac{ia}{2\sqrt{s}})\} = \frac{i}{\sqrt{\pi t}}\sin(a\sqrt{t}) \tag{108.4.5}$$

$$L^{-1}\{\frac{1}{s^{\alpha}} e^{-a^{2}/4s} erf(\frac{ia}{2\sqrt{s}})\} = i(\frac{a}{2})^{1-\alpha} t^{(\alpha-1)/2} H_{\alpha-1}(a\sqrt{t}) \tag{109.4.5}$$

and using (65.4.5) it follows that

$$\int_0^{\alpha} t^{\lambda/2} H_\lambda(a\sqrt{t})d\lambda = \sqrt{\frac{a}{2\pi}}\int_0^t \frac{\sin[a\sqrt{t-u}]}{\sqrt{t-u}}\{\nu(\frac{at}{2},-\frac{1}{2})-\nu(\frac{at}{2},\alpha-\frac{1}{2})\}du \tag{110.4.5}$$

For α → ∞, (107.4.5) becomes

$$L\{\int_0^{\infty} t^{\lambda/2} H_\lambda(a\sqrt{t})d\lambda\} = -\frac{ie^{-a^2/4s}}{s\ln(2s/a)} erf(\frac{ia}{2\sqrt{s}}) \tag{111.4.5}$$

and therefore

$$\int_0^{\infty} t^{\lambda/2} H_\lambda(a\sqrt{t})d\lambda = \sqrt{\frac{a}{2\pi}}\int_0^t \frac{\sin[a\sqrt{t-u}]}{\sqrt{t-u}}\nu(\frac{at}{2},-\frac{1}{2})du \tag{112.4.5}$$

Introducing $t = 1$ and $u = 1 - x^2$

$$\int_0^{\infty} H_\lambda(a)d\lambda = \sqrt{\frac{2a}{\pi}}\int_0^1 \sin(ax)\nu[\frac{a}{2}(1-x^2),-\frac{1}{2}]dx \tag{113.4.5}$$

or with $x = \cos\omega$

$$\int_0^{\infty} H_\lambda(a)d\lambda = \sqrt{\frac{2a}{\pi}}\int_0^{\pi/2} \sin\omega\sin(a\cos\omega)\nu[\frac{a}{2}(\sin\omega)^2,-\frac{1}{2}]d\omega \tag{114.4.5}$$

If the inverse transform (109.4.5) is used, then the convolution theorem gives

$$\int_0^{\alpha} t^{\lambda/2} H_\lambda(a\sqrt{t})d\lambda = \frac{a}{2}\nu(\frac{au}{2})*[u^{-1/2}H_{-1}(a\sqrt{u})-u^{(\alpha-1)/2}H_{\alpha-1}(a\sqrt{u})] \tag{115.4.5}$$

where the Struve function $H_{-1}(a\sqrt{u})$ can be replaced by

$$H_{-1}(a\sqrt{u}) = \frac{2}{\pi} - H_1(a\sqrt{u}) \tag{116.4.5}$$

For α → ∞ and $t = 1$, (115.4.5) becomes

$$\int_0^\infty H_\lambda(a)d\lambda = a\int_0^1 H_{-1}(ax)\,\nu[\frac{a}{2}(1-x^2)]dx \tag{117.4.5}$$

Considering that [1, 12.1.3, 6.1.12]

$$H_\lambda(z) = \sum_{n=0}^{\infty} \frac{(-1)^n (z/2)^{2n+\lambda+1}}{\Gamma(n+3/2)\Gamma(n+\lambda+3/2)} \tag{118.4.5}$$

where

$$\Gamma(n+3/2) = \frac{1\cdot 3\cdot 5\cdot 7\ldots(2n+1)}{2^{n+1}}\sqrt{\pi} \tag{119.4.5}$$

the integral in (112.4.5) can be developed in series of the Volterra functions by introducing (118.4.5) into (112.4.5) and inverting the order of integration and summation

$$\int_0^\infty t^{\lambda/2} H_\lambda(a\sqrt{t})d\lambda = \sqrt{\frac{2a}{\pi}}\sum_{n=0}^{\infty}\frac{(-1)^n a^n \nu(at/2, n+1/2)}{1\cdot 3\cdot 5\cdot 7\ldots(2n+1)} \tag{120.4.5}$$

Putting $t = 1$ into (120.4.5) leads to the series for the integral (113.4.5).

Corresponding integrals of the modified Struve function can be derived taking into account that [1, 12.2.1]

$$L_\lambda(z) = -ie^{-i\pi\lambda/2} H_\lambda(iz) \tag{121.4.5}$$

and therefore

$$\int_0^\alpha t^{\lambda/2} L_\lambda(a\sqrt{t})d\lambda = \sqrt{\frac{a}{2\pi}}\int_0^t \frac{\sinh[a\sqrt{t-u}]}{\sqrt{t-u}}\{\nu(\frac{at}{2},-\frac{1}{2}) - \nu(\frac{at}{2},\alpha-\frac{1}{2})\}du \tag{122.4.5}$$

$$\int_0^\infty L_\lambda(a)d\lambda = \sqrt{\frac{2a}{\pi}}\int_0^1 \sinh(ax)\nu[\frac{a}{2}(1-x^2),-\frac{1}{2}]dx \tag{123.4.5}$$

and

$$\int_0^{\varepsilon} t^{\lambda/2} L_{\lambda}(a\sqrt{t})d\lambda = \frac{a}{2}\int_0^t \{u^{-1/2}L_{-1}(a\sqrt{u}) - u^{(\alpha-1)/2}H_{\alpha-1}(a\sqrt{u})\}\, v[\frac{a}{2}(t-u)]du \tag{124.4.5}$$

The Volterra function series, analogous to that of (120.4.5) is

$$\int_0^{\infty} t^{\lambda/2} L_{\lambda}(a\sqrt{t})d\lambda = \sqrt{\frac{2a}{\pi}}\sum_{n=0}^{\infty}\frac{a^n v(at/2, n+1/2)}{1\cdot 3\cdot 5\cdot \ldots (2n+1)} \tag{125.4.5}$$

Examples of integrals of t*he Lommel function* $S_{\lambda,\mu}(z)$ and *the Anger-Weber function* $\mathbf{J}_{\lambda}(z)$ [1,43] come from [44, 4.3.25-28]

$$\int_0^{\infty} e^{-st} v[a(\sqrt{t^2+1}\pm t), \alpha]dt = \frac{1}{s}\int_0^{\infty}\frac{a^{u+\alpha}}{\Gamma(u+\alpha+1)}\{S_{1,u+\alpha}(s) \pm (u+\alpha)S_{0,u+\alpha}(s)\}du \tag{126.4.5}$$

$$\int_0^{\infty}\frac{e^{-st}}{\sqrt{t^2+1}} v[a(\sqrt{t^2+1}-t), \alpha]dt = \int_0^{\infty}\frac{a^{u+\alpha}}{\Gamma(u+\alpha+1)}\{S_{0,u+\alpha}(s) - (u+\alpha)S_{-1,u+\alpha}(s)\}du \tag{127.4.5}$$

and

$$\int_0^{\infty}\frac{e^{-st}}{\sqrt{t^2+1}} v[a(\sqrt{t^2+1}+t), \alpha]dt = \pi\int_0^{\infty}\frac{a^{u+\alpha}}{\Gamma(u+\alpha+1)\sin[\pi(u+\alpha)]}\{\mathbf{J}_{-(u+\alpha)}(s) - J_{-(u+\alpha)}(s)\}du \tag{128.4.5}$$

Two integrals with the respect to the order of the parabolic functions were already given in (6.4.4) and (18.4.4), from many available integral equalities, two additional examples which are based on [88, 1.5.40, 2.4.22] are presented

$$\int_0^\infty \frac{e^{-st}}{\sqrt{b+t}} v[\frac{at}{\sqrt{b}+\sqrt{b+t}}, \alpha]dt = 2e^{bs/2} \int_\alpha^\infty (\frac{a}{2\sqrt{s}})^x D_{-(x+1)}(\sqrt{2bs})dx \qquad (129.4.5)$$

$$L\{\int_0^\infty \frac{(2t)^{(u-1)/2} e^{-at/2}}{\Gamma(u+1)} D_{-u}(\sqrt{2at})du\} = \sqrt{\frac{\pi}{2(s+a)}} v[\frac{\sqrt{s+a}-\sqrt{a}}{s}, \alpha] \qquad (130.4.5)$$

Integrals of *the Whittaker functions* $W_{\lambda,\mu}(z)$ and $M_{\lambda,\mu}(z)$ are the next group of integrals that will be considered. Using the Laplace transform [88, 1.3.22]

$$L\{t^\lambda (t+b)^\rho\} = \frac{\Gamma(\lambda+1)}{s} (\frac{b}{s})^{(\lambda+\rho)/2} e^{bs/2} W_{(\rho-\lambda)/2,(\lambda+\rho+1)/2}(bs)$$
$$\mathrm{Re}\,\lambda > -1 \quad , \quad \mathrm{Re}\,s > 0 \qquad (131.4.5)$$

we have from (2.4.4)

$$\int_0^\infty e^{-st}(t+b)^\rho v(at,\alpha)dt =$$
$$\frac{e^{bs/2}}{s} (\frac{b}{s})^{\rho/2} \int_0^\infty (a\sqrt{\frac{b}{s}})^{u+\alpha} W_{(\rho-u-\alpha)/2,(u+\alpha+\rho+1)/2}(bs)du \qquad (132.4.5)$$

For a = b = 1 and ρ = 0, the integral reduces to

$$\int_0^\infty e^{-st} v(t,\alpha)dt = \frac{e^{s/2}}{s} \int_0^\infty (\frac{1}{\sqrt{s}})^{u+\alpha} W_{-(u+\alpha)/2,(u+\alpha+1)/2}(s)du \qquad (133.4.5)$$

Introducing $x = u+\alpha$, and taking into account that the left-hand side integral is the Laplace transform of the Volterra function $v(t,\alpha)$, the Whittaker integral with respect to the order can be written in the closed form

$$\int_\alpha^\infty (\frac{1}{s})^{x/2} W_{-x/2,(x+1)/2}(s)dx = \frac{e^{-s/2}}{s^\alpha \ln s}$$
$$\mathrm{Re}\,\alpha > -1 \quad , \quad \mathrm{Re}\,s > 1 \qquad (134.4.5)$$

From the Laplace transform [88, 1.5.57]

$$L\{(1-e^{-t})^{\lambda}e^{be^{-t}}\}=\frac{\Gamma(s)\Gamma(\lambda+1)e^{b/2}}{\Gamma(\lambda+s+1)b^{(\lambda+s+1)/2}}M_{(\lambda-s+1)/2,(\lambda+s+1)/2}(b) \tag{135.4.5}$$

$$\mathrm{Re}\,\lambda>-1 \quad , \quad \mathrm{Re}\,s>0$$

it follows that

$$\int_0^{\infty} e^{-st+be^{-t}}\nu[a(1-e^{-t}),\alpha]dt = \frac{\Gamma(s)e^{b/2}}{b^{(s+1)/2}}\int_0^{\infty}\frac{1}{\Gamma(u+\alpha+s+1)}\left(\frac{a}{\sqrt{b}}\right)^{u+\alpha}M_{(u+\alpha-s+1)/2,(u+\alpha+s+1)/2}(b)du \tag{136.4.5}$$

Introducing $\tau = e^{-t}$ and $x = u+\alpha+s$ into (136.4.5), the following integral equality is established

$$\int_0^{1}\tau^{s-1}e^{b\tau}\nu[a(1-\tau),\alpha]d\tau = \frac{\Gamma(s)e^{b/2}}{a^{s}\sqrt{b}}\int_{\alpha+s}^{\infty}\frac{1}{\Gamma(x+1)}\left(\frac{a}{\sqrt{b}}\right)^{x}M_{(x+1)/2-s,(x+1)/2}(b)\,dx \tag{137.4.5}$$

Using [102, 1.2.35, 1.2.37]

$$L\{\frac{e^{-t/2}}{t^{1/4}}M_{\lambda-1/4,1/4}(t)\}=\frac{\sqrt{\pi}}{2s}\left(\frac{s}{s+1}\right)^{\lambda} \tag{138.4.5}$$

$$L\{\frac{e^{-t/2}}{t^{3/4}}M_{\lambda-1/4,-1/4}(t)\}=\sqrt{\frac{\pi}{s}}\left(\frac{s}{s+1}\right)^{\lambda} \tag{139.4.5}$$

we have from (6.4.3)

$$L\{\int_0^{\infty}\frac{e^{-t/2}}{\Gamma(u+1)t^{1/4}}M_{u-1/4,1/4}(t)du\}=\frac{\sqrt{\pi}}{2s}\nu\left(\frac{s}{s+1},\alpha\right) \tag{140.4.5}$$

$$L\{\int_0^{\infty}\frac{e^{-t/2}}{\Gamma(u+1)t^{3/4}}M_{u-1/4,-1/4}(t)du\}=\sqrt{\frac{\pi}{s}}\nu\left(\frac{s}{s+1},\alpha\right) \tag{141.4.5}$$

and as a consequence of (138.4.5) and (139.4.5) the following convolution integral can be evaluated

$$\int_0^t \frac{e^{u/2}}{(t-u)^{3/4}} M_{\lambda-1/4,-1/4}(t-u)\frac{du}{\sqrt{u}} = \frac{2\sqrt{\pi}}{t^{1/4}} M_{\lambda-1/4,1/4}(t) \tag{142.4.5}$$

Introducing $u = t(\cos\omega)^2$, (142.4.5) takes the form

$$\int_0^{\pi/2} \frac{e^{t(\cos\omega)^2/2}}{\sqrt{\sin\omega}} M_{\lambda-1/4,-1/4}[t(\sin\omega)^2]d\omega = \sqrt{\pi t} M_{\lambda-1/4,1/4}(t) \tag{143.4.5}$$

Now, a number of examples connecting the Volterra functions with various *orthogonal polynomials* and *generalized hypergeometric functions* will be considered.

From [88, 1.11, 19, 22, 46, 55] and (2.4.4.) we have

$$\int_0^\infty e^{-st} t^{n/2} He_n(\sqrt{2t})\nu(at,\alpha)dt = \frac{n!}{s(2s)^{n/2}} \int_\alpha^\infty (\frac{a}{s})^x C_n^{x+1}(\frac{1}{\sqrt{s}})dx \tag{144.4.5}$$

$$\int_0^\infty e^{-st}(1+t)^\rho P_n^{b,c}(1+2t)\nu(at,\alpha)dt =$$
$$\frac{e^{s/2}}{n!s^{(b+c+2)/2}} \int_\alpha^\infty \frac{\Gamma(n+x+1)a^x}{\Gamma(x+1)} W_{\rho-x,n+(\rho+x+1)/2}(b)dx \tag{145.4.5}$$

$$\int_0^\infty e^{-st} t^n L_n^\beta(t)\nu(at,\alpha)dt =$$
$$\frac{1}{s^{n+1}} \int_\alpha^\infty \frac{\Gamma(n+x)}{\Gamma(x+1)} (\frac{a}{s})^x P_n^{\beta,x-\beta}(1-\frac{2}{s})dx \tag{146.4.5}$$

$$\int_0^\infty e^{-st} L_n^\beta(t)\nu(at,\alpha)dt =$$
$$\frac{\Gamma(n+\beta+1)}{n!\Gamma(\beta+1)s} \int_\alpha^\infty (\frac{a}{s})^x {}_2F_1(-n,x+1;\beta+1;\frac{1}{s})dx \tag{147.4.5}$$

where $C_n^\alpha(z)$, $He_n(z)$, $P_n^{\alpha,\beta}(z)$ and $L_n^\alpha(z)$ are *the Gugenbauer, Hermite, Jacobi and Laguerre polynomials* respectively [1,80] and n is a positive integer. Applying (6.4.3) and [88, 2.9.15, 26, 27; 102, 2.9.2. 1.3] it is possible to obtain

$$L\{\int_{\alpha}^{\infty}\frac{t^{u-1}(1+t)^{n}}{\Gamma(u+1)\Gamma(u+n)}P_{n}^{a,u-1}(\frac{t-1}{t+1})du\}=\frac{1}{s^{n}}L_{n}^{a}(s)\nu(\frac{1}{s},\alpha) \quad (148.4.5)$$

$$L\{\int_{\alpha}^{\infty}\frac{t^{u-1}}{\Gamma(u)\Gamma(u+1)}{}_{1}F_{2}(-n,\beta+1,u,at)du\}=\frac{n!\Gamma(\beta+1)}{\Gamma(n+\beta+1)}L_{n}^{\beta}(\frac{a}{s})\nu(\frac{1}{s},\alpha) \quad (149.4.5)$$

$$L\{\int_{\alpha}^{\infty}\frac{t^{u-1}e^{-bt}}{\Gamma(u)\Gamma(u+1)}\,{}_{0}F_{2}[-n,n+2\beta;\beta+\frac{1}{2},u;\frac{(b-a)t}{2}]du\}= \quad (150.4.5)$$
$$nB(n,2\beta)C_{n}^{\beta}(\frac{s+a}{s+b})\nu(\frac{1}{s+b},\alpha)$$

$$\int_{\alpha}^{\infty}\frac{t^{u-1}e^{-bt}}{\Gamma(u)\Gamma(u+1)}\,{}_{2}F_{2}[-n,n+2\beta;1,u;\frac{(b-a)t}{2}]du\}=P_{n}(\frac{s+a}{s+b})\nu(\frac{1}{s+b},\alpha) \quad (151.4.5)$$

where $P_n(z)$ is *the Legendre polynomial* [1,80].

Laplace transforms which include only various hypergeometric functions and coming from [88, 2.7.15,16,52,53,2.15.82; 102, 2.2.36] and (6.4.3) are

$$L\{\int_{\alpha}^{\infty}\frac{t^{\rho-1}}{\Gamma(u+1)}{}_{1}F_{1}(u;\rho;at)du\}=\frac{\Gamma(\rho)}{s^{\rho}}\nu(\frac{1}{s-a},\alpha) \quad (152.4.5)$$

$$L\{\int_{\alpha}^{\infty}\frac{t^{u-1/2}}{\Gamma(u+\frac{1}{2})\Gamma(u+1)}{}_{2}F_{1}(1,\frac{1}{2};u+\frac{1}{2};\rho;-\frac{t}{a})du\}= \quad (153.4.5)$$
$$\sqrt{\pi a}\,e^{as}erfc(\sqrt{as})\nu(\frac{1}{s},\alpha)$$

and

$$L\{\int_{\alpha}^{\infty}\frac{t^{u-1}}{\Gamma(u)\Gamma(u+1)}\,{}_{0}F_{2}(;\frac{1}{2};u;-\frac{a^{2}t}{2})du\}=\cos(\frac{a}{\sqrt{s}})\nu(\frac{1}{s},\alpha) \quad (154.4.5)$$

$$L\{\int_{\alpha}^{\infty}\frac{t^{u-1/2}}{\Gamma(u+\frac{1}{2})\Gamma(u+1)}{}_{0}F_{2}(;\frac{3}{2};u;-\frac{a^{2}t}{2})du\}=\frac{1}{a}\sin(\frac{a}{\sqrt{s}})\nu(\frac{1}{s},\alpha) \quad (155.4.5)$$

$$L\{\int_{\alpha}^{\infty}\frac{t^{u-1}}{\Gamma(u)\Gamma(u+1)}\,{}_2F_2(-n,n+1;\frac{1}{2};u;t)du\}=\cos[2n\sin^{-1}(\frac{1}{\sqrt{s}})]\nu(\frac{1}{s},\alpha) \quad (156.4.5)$$

$$L\{\int_{\alpha}^{\infty}\frac{t^{u-1}}{\Gamma(u)\Gamma(u+1)}\,{}_2F_2(-n,n+1;\frac{3}{2};u;t)du\}=$$
$$\frac{1}{2n+1}\sin[(2n+1)\sin^{-1}(\frac{1}{\sqrt{s}})]\nu(\frac{1}{s},\alpha) \quad (157.4.5)$$

Barrucand and Colombo [12,14,30] considered the following expansion

$$\frac{x}{\ln(1-x)}=-1+\sum_{n=1}^{\infty}\mathrm{i}_n x^n$$
$$|x|<1 \quad (158.4.5)$$

where the coefficients i_n are related to *the Bernoulli numbers* [1,80] and have been used by Gregory towards 1670 in some summation formulas. The nomenclature is in this case rather confusing, the coefficients are called *the Gregory numbers* [12,14,30], but sometimes also as *the Cauchy numbers* [66]. If the Laplace transform of the Volterra function $\nu(t)$ is written in the form resembling (158.4.5)

$$L\{\nu(t)\}=\frac{1}{s\ln s}=-\frac{1}{s-1}\cdot\{\frac{1-1/s}{\ln[1-(1-1/s)]}\}$$
$$\mathrm{Re}\,s>1 \quad (159.4.5)$$

then

$$\frac{1}{s\ln s}=\frac{1}{s-1}-\frac{1}{s}\sum_{n=0}^{\infty}\mathrm{i}_{n+1}\left(1-\frac{1}{s}\right)^n \quad (160.4.5)$$

but the Laplace transform of the Laguerre polynomials is

$$L\{L_n(t)\}=\frac{1}{s}\left(1-\frac{1}{s}\right)^n$$
$$n=0,1,2,3,\ldots \quad (161.4.5)$$

and therefore the inverse transform of (160.4.5) is

$$\nu(t) = e^t - \sum_{n=0}^{\infty} \mathrm{i}_{n+1} L_n(t) \tag{162.4.5}$$

If the Ramanujan integral representation of ν(t) given in (1.2.3) is compared with (162.4.5) , the above series is equal to the improper integral

$$\int_0^{\infty} \frac{e^{-tu}}{u[\pi^2 + (\ln u)^2]} du = \sum_{n=0}^{\infty} \mathrm{i}_{n+1} L_n(t) \tag{163.4.5}$$

If the orthogonality relation satisfied by the Laguerre polynomials [1,80]

$$\int_0^{\infty} e^{-t} L_m(t) L_n(t) dt = \begin{cases} 0 & \text{for} \quad m \neq n \\ 1 & \text{for} \quad m = n \end{cases} \tag{164.4.5}$$

is applied to (163.4.5) then

$$\begin{gathered} \mathrm{i}_{n+1} = \int_0^{\infty} e^{-t} L_n(t) \{ \int_0^{\infty} \frac{e^{-tu}}{u[\pi^2 + (\ln u)^2]} du \} dt = \\ \int_0^{\infty} \frac{1}{u[\pi^2 + (\ln u)^2]} \{ \int_0^{\infty} e^{-(u+1)t} L_n(t) dt \} du \end{gathered} \tag{165.4.5}$$

Now, using (161.4.5) with $s = u+1$, the Gregory numbers are expressed by

$$\mathrm{i}_{n+1} = \int_0^{\infty} \frac{u^{n-1}}{(u+1)^{n+1} [\pi^2 + (\ln u)^2]} du = \int_0^{\infty} \frac{1}{(x+1)^{n+1} [\pi^2 + (\ln x)^2]} dx \tag{166.4.5}$$

Other representation is derived if the orthogonality relation (164.4.5) is applied directly to (162.4.5), because then

$$\mathrm{i}_{n+1} = \int_0^{\infty} [1 - e^{-t} \nu(t)] L_n(t) dt \tag{167.4.5}$$

The improper integrals in (166.4.5) and (167.4.5) can be evaluated by different methods, but it is easier to obtain them from the recurrence relation for the Gregory numbers. From the Taylor expansion

$$\frac{\ln(1-x)}{x} = -1 - \sum_{n=1}^{\infty} \frac{x^n}{n+1} \tag{168.4.5}$$

and (158.4.5) it follows that

$$(-1 + \sum_{n=1}^{\infty} \mathrm{i}_n x^n)(-1 - \sum_{n=1}^{\infty} \frac{x^n}{n+1}) = 1 \tag{169.4.5}$$

By equating the coefficients in (169.4.5)

$$\mathrm{i}_n = \frac{1}{n+1} - \sum_{m=1}^{n-1} \frac{\mathrm{i}_m}{n-m+1} \tag{170.4.5}$$

$$n = 1, 2, 3, \ldots$$

The first eight Gregory numbers are $\mathrm{i}_1 = 1/2$, $\mathrm{i}_2 = 1/12$, $\mathrm{i}_3 = 1/24$, $\mathrm{i}_4 = 19/720$, $\mathrm{i}_5 = 3/160$, $\mathrm{i}_6 = 863/60480$, $\mathrm{i}_7 = 275/24192$ and $\mathrm{i}_8 = 33953/3628800$.

4.6. Volterra Functions and Other Integral Transformations

The Mellin transformation, the exponential Fourier transformation, the unilateral Laplace-Abel (one-sided Laplace) transformation and the bilateral (two-sided) Laplace transformation are respectively [29,44,80,107]

$$M\{f(t); s\} = \int_0^{\infty} t^{s-1} f(t) dt \tag{1.4.6}$$

$$F_E\{f(t); s\} = \int_{-\infty}^{\infty} e^{-ist} f(t) dt \tag{2.4.6}$$

$$L_I\{f(t); s\} = \int_0^{\infty} e^{-st} f(t) dt = L\{f(t); s\} \tag{3.4.6}$$

$$L_{II}\{f(t); s\} = \int_{-\infty}^{\infty} e^{-st} f(t) dt \tag{4.4.6}$$

and are related to each other in the following way [44]

$$M\{f(t);s\} = F_E\{f(e^t);is\} = L\{f(e^t);-s\} + L\{f(e^{-t});s\} = L_{II}\{f(e^{-t});s\} \tag{5.4.6}$$

Thus, *the Mellin transform* is the substituted *two-sided Laplace transform* which may be expressed by *the exponential Fourier transform* in the complex domain or by the sum of the substituted ordinary Laplace transforms. The Mellin and Laplace transforms are also associated with *the Stieltjes* and *Hankel transformations*, and with *the Riemann-Liouville* and *the Weyl fractional integrals* [44]. *The factorial transform* introduced by Fraser [48,49] is also closely related to the Laplace transform. Interrelations between various integral transforms primarily serve to enlarge a number of available transform pairs and integral equalities and in this context the Volterra functions will be considered.

Considering the Mellin transformation of the function $\nu(e^{-t})$, for which the result of reversing the order of integration in (1.4.6) is

$$M\{\nu(e^{-t})\} = \Gamma(s)\int_0^\infty \frac{u^{-s}}{\Gamma(u+1)}du = \Gamma(s)\Gamma(1-s)\mu(1,-s) = \frac{\pi}{\sin(\pi s)}\mu(1,-s) \tag{6.4.6}$$

$$0 < \operatorname{Re} s < 1$$

and

$$M\{\nu(e^{-t},\alpha)\} = \Gamma(s)\int_\alpha^\infty \frac{u^{-s}}{\Gamma(u+1)}du \tag{7.4.6}$$

These expressions can be generalized

$$M\{\nu(ae^{-bt})\} = \frac{\pi}{b^s \sin(\pi s)}\mu(a,-s) \tag{8.4.6}$$

$$M\{\nu(ae^{-bt},\alpha)\} = \frac{\Gamma(s)}{b^s}\int_\alpha^\infty \frac{a^u u^{-s}}{\Gamma(u+1)}du \tag{9.4.6}$$

$$a > 0 \quad , \quad b > 0 \quad , \quad 0 < \operatorname{Re} s < 1$$

Applying the operational rules of the Mellin transformation [44, 6.1.7-8]

$$M\{t^\lambda f(at^\rho)\} = \frac{1}{\rho a^{(s+\lambda)/\rho}} M(\frac{s+\lambda}{\rho}) \tag{10.4.6}$$

$$M\{t^\lambda f(at^{-\rho})\} = \frac{a^{(s+\lambda)/\rho}}{\rho} M(-\frac{s+\lambda}{\rho}) \tag{11.4.6}$$

$$a > 0 \quad , \quad \rho > 0$$

to (6.4.6) we have

$$M\{t^{\lambda}\nu(e^{-t^{\rho}})\}=\frac{1}{\rho}\Gamma(\frac{s+\lambda}{\rho})\Gamma(1-\frac{s+\lambda}{\rho})\mu[1,-(s+\lambda)/\rho]$$
$$0<\mathrm{Re}(\frac{s+\lambda}{\rho})<1 \tag{12.4.6}$$

$$M\{t^{\lambda}\nu(e^{-1/t^{\rho}})\}=\frac{1}{\rho}\Gamma(-\frac{s+\lambda}{\rho})\Gamma(1+\frac{s+\lambda}{\rho})\mu[1,(s+\lambda)/\rho]$$
$$-1<\mathrm{Re}(\frac{s+\lambda}{\rho})<0 \tag{13.4.6}$$

For $\lambda = -1$ and $\rho = 1$, the Mellin transform (13.4.6) becomes

$$M\{\frac{1}{t}\nu(e^{-1/t})\}=\Gamma(s)\Gamma(1-s)\mu(1,s-1)=\frac{\pi}{\sin(\pi s)}\mu(1,s-1)$$
$$0<\mathrm{Re}\, s<1 \tag{14.4.6}$$

If $M\{f_1(t)\} = M_1(s)$ and $M\{f_2(t)\} = M_2(s)$ then [25, 6.1.13-14]

$$M\{t^{\lambda}\int_0^{\infty}u^{\sigma}f_1(tu)f_2(u)du\}=M_1(s+\lambda)M_2(1-s-\lambda+\sigma) \tag{15.4.6}$$

$$M\{t^{\lambda}\int_0^{\infty}u^{\sigma}f_1(\frac{t}{u})f_2(u)du\}=M_1(s+\lambda)M_2(s+\lambda+\sigma+1) \tag{16.4.6}$$

Introducing the Mellin transform pairs (6.4.6) and (14.4.6) in these expressions

$$M\{t^{\lambda}\int_0^{\infty}u^{\sigma-1}\nu(e^{-tu})\nu(e^{-1/u})du\}=$$
$$\Gamma(s+\lambda)\Gamma(1-s-\lambda)\Gamma(s+\lambda-\sigma)\Gamma(1-s-\lambda-\sigma)\mu(1,-s-\lambda)\mu(1,\sigma-s-\lambda) \tag{17.4.6}$$

$$M\{t^{\lambda}\int_0^{\infty}u^{\sigma-1}\nu(e^{-t/u})\nu(e^{-1/u})du\}=$$
$$\Gamma(-s-\lambda)\Gamma(1+s+\lambda)\Gamma(s+\lambda+\sigma)\Gamma(1-s-\lambda-\sigma)\mu(1,-s-\lambda)\mu(1,s+\lambda+\sigma) \tag{18.4.6}$$

which can be slightly simplified for $\sigma = 0$

$$M\{t^{\lambda}\int_0^{\infty}\frac{1}{u}\nu(e^{-tu})\nu(e^{-1/u})du\} = [\Gamma(s+\lambda)\Gamma(1-s-\lambda)\mu(1,-s-\lambda)]^2 \tag{19.4.6}$$

and

$$M\{t^{\lambda}\int_0^{\infty}\frac{1}{u}\nu(e^{-t/u})\nu(e^{-1/u})du\} = \\ \Gamma(s+\lambda)\Gamma(-s-\lambda)\Gamma(1+s+\lambda)\Gamma(1-s-\lambda)\mu(1,-s-\lambda)\mu(1,s+\lambda) \tag{20.4.6}$$

The Volterra functions also represent the Mellin transforms of the reciprocal gamma functions

$$M\{\frac{t^{\beta}a^{t+\alpha}}{\Gamma(t+\alpha+1)}\} = \int_0^{\infty}\frac{t^{s+\beta-1}a^{t+\alpha}}{\Gamma(t+\alpha+1)}dt = \Gamma(s+\beta)\mu(a,s+\beta-1,\alpha) \\ \mathrm{Re}\,\alpha > -1 \quad , \quad \mathrm{Re}\,\beta > -1 \quad , \quad \mathrm{Re}(s+\beta) > 0 \tag{21.4.6}$$

If the Mellin transform exists and the order of integration may be inverted, then

$$M\{f(t)\nu(t,\alpha)\} = \int_0^{\infty}t^{s-1}f(t)\{\int_0^{\infty}\frac{t^{u+\alpha}}{\Gamma(u+\alpha+1)}du\}dt = \\ \int_0^{\infty}\frac{1}{\Gamma(u+\alpha+1)}\{\int_0^{\infty}t^{u+s+\alpha-1}f(t)dt\}du = \int_0^{\infty}\frac{M(u+s+\alpha)}{\Gamma(u+\alpha+1)}du \tag{22.4.6}$$

and similarly

$$M\{f(t)\mu(t,\beta,\alpha)\} = \frac{1}{\Gamma(\beta+1)}\int_0^{\infty}\frac{u^{\beta}M(u+s+\alpha)}{\Gamma(u+\alpha+1)}du \tag{23.4.6}$$

For example, from [44, 6.3.15]

$$M\{e^{-t^{\rho}}\} = \frac{1}{\rho}\Gamma(\frac{s}{\rho}) \\ \rho > 0 \quad , \quad \mathrm{Re}\,s > 0 \tag{24.4.6}$$

and using (22.4.6) we have

$$\int_0^\infty t^{s-1} e^{-t^\rho} \nu(t,\alpha) dt = \frac{1}{\rho} \int_0^\infty \frac{\Gamma(\frac{u+s+\alpha}{\rho})}{\Gamma(u+\alpha+1)} du \tag{25.4.6}$$

$$\operatorname{Re}\alpha > -1 \quad , \quad \rho > 1 \quad , \quad 0 < \operatorname{Re}(\frac{s+\alpha}{\rho}) < 1$$

For $\alpha = 0$, $\rho = 2$ and $s = 1$, (25.4.6) leads to

$$\int_0^\infty e^{-t^2} \nu(t) dt = \frac{1}{2} \int_0^\infty \frac{\Gamma(\frac{u+1}{2})}{\Gamma(u+1)} du \tag{26.4.6}$$

Using the duplication formula of gamma function from (62.4.3) and (26.4.6)

$$\int_0^\infty e^{-t^2} \nu(t) dt = \sqrt{\pi} \int_0^\infty \frac{(\frac{1}{2})^u}{\Gamma(\frac{u}{2}+1)} du = \sqrt{\pi} \int_0^\infty \frac{(\frac{1}{4})^x}{\Gamma(x+1)} dx = \sqrt{\pi}\, \nu(\frac{1}{4}) \tag{27.4.6}$$

the result already presented in (39.4.2).

Introducing $\alpha = 0$, $\rho = 3$ and $s = 1$ into (25.4.6) gives

$$\int_0^\infty e^{-t^3} \nu(t) dt = \frac{1}{3} \int_0^\infty \frac{\Gamma(\frac{u+1}{3})}{\Gamma(u+1)} du = \frac{2\pi}{\sqrt{3}} \int_0^\infty \frac{(\frac{1}{27})^x}{\Gamma(x+\frac{2}{3})\Gamma(x+1)} dx \tag{28.4.6}$$

where *the triplication formula of gamma function* [1]

$$\frac{\Gamma(z)}{\Gamma(3z)} = \frac{2\pi}{3^{3z-1/2}\Gamma(z+\frac{1}{3})\Gamma(z+\frac{2}{3})} \tag{29.4.6}$$

was used.

Similarly from [43 1.13.3]

$$M\{erfc(t)\} = \frac{1}{\sqrt{\pi}\, s} \Gamma(\frac{s+1}{2}) \tag{30.4.6}$$

$$\operatorname{Re} s > 0$$

and (22.4.6) the following integral equality is established

$$\int_0^\infty t^{s-1} erfc(t) \nu(t,\alpha) dt = \frac{1}{\sqrt{\pi}} \int_0^\infty \frac{\Gamma(\frac{u+s+\alpha+1}{2})}{(u+s+\alpha)\Gamma(u+\alpha+1)} du \tag{31.4.6}$$

The right-hand side integral can be evaluated in the closed form for α = 0 and s = 1. From

$$\int_0^\infty \frac{\Gamma(\frac{u}{2}+1)}{(u+1)\Gamma(u+1)} du = \frac{1}{2} \int_0^\infty \frac{\Gamma(\frac{u}{2})}{(u+1)\Gamma(u)} du \tag{32.4.6}$$

and [1]

$$\frac{\Gamma(\frac{u}{2})}{\Gamma(u)} = \frac{2\sqrt{\pi}}{2^u \Gamma(\frac{u+1}{2})} \tag{33.4.6}$$

we have

$$\begin{gathered} \int_0^\infty erfc(t)\, \nu(t) dt = \frac{1}{2} \int_0^\infty \frac{(\frac{1}{2})^u}{\Gamma(\frac{u}{2}+\frac{3}{2})} du = \\ 2\int_0^\infty \frac{(\frac{1}{4})^{x+1/2}}{\Gamma(x+\frac{3}{2})} dx = 2\nu(\frac{1}{4},\frac{1}{2}) = 0.636644\ldots \end{gathered} \tag{34.4.6}$$

If $\nu(t,\alpha)$ in (22.4.6) is replaced by $\nu(e^{-t},\alpha)$, and the integral

$$\begin{gathered} G(u+\alpha,s) = \int_0^\infty t^{s-1} e^{-t(u+\alpha)} f(t) dt = \\ M\{e^{-t(u+\alpha)} f(t); s\} = L\{t^{s-1} f(t); u+\alpha\} \end{gathered} \tag{35.4.6}$$

is treated either as the Mellin transform or the Laplace transform, then (22.4.6) becomes

$$M\{f(t)\nu(e^{-t},\alpha)\} = \int_0^\infty \frac{G(u+\alpha,s)}{\Gamma(u+\alpha+1)} du \tag{36.4.6}$$

For example, introducing the Laplace transform [88, 1.5.46]

$$L\{t^{\lambda-1}e^{-t^2}\}=\frac{\Gamma(\lambda)}{2^{\lambda/2}}e^{s^2/8}D_{-\lambda}(\frac{s}{\sqrt{2}}) \qquad (37.4.6)$$
$$\operatorname{Re}\lambda>0 \quad , \quad \operatorname{Re}s>0$$

into (36.4.6) leads to

$$\int_0^\infty t^{s-1}e^{-t^2}\nu(e^{-t},\alpha)dt=\frac{\Gamma(s)}{2^{s/2}}\int_0^\infty\frac{e^{(u+\alpha)^2/8}}{\Gamma(u+\alpha+1)}D_{-s}(\frac{u+\alpha}{\sqrt{2}})du \qquad (38.4.6)$$
$$\operatorname{Re}\alpha>-1 \quad , \quad \operatorname{Re}s>0$$

In some cases, the parabolic cylinder functions in (38.4.6) can be reduced to *the repeated integrals of the error function* [1,45] and to the modified Bessel functions of the second kind, for example

$$\int_0^\infty t^n e^{-t^2}\nu(e^{-t})dt=\sqrt{\pi}\,n!\int_0^\infty\frac{e^{u^2}}{\Gamma(2u+1)}i^n erfc(u)du \qquad (39.4.6)$$
$$n=0,1,2,3,\ldots$$

$$\int_0^\infty t^{-1/2}e^{-t^2}\nu(e^{-t})dt=\frac{1}{2}\int_0^\infty\frac{\sqrt{u}e^{u^2/8}}{\Gamma(u+1)}K_{1/4}(\frac{u^2}{8})du \qquad (40.4.6)$$

$$\int_0^\infty t^{1/2}e^{-t^2}\nu(e^{-t})dt=\frac{1}{8}\int_0^\infty\frac{u^{3/2}e^{u^2/8}}{\Gamma(u+1)}\{K_{3/4}(\frac{u^2}{8})-K_{1/4}(\frac{u^2}{8})\}du \qquad (41.4.6)$$

If the integral in (35.4.6) is considered as the Mellin transform, then from [25, 6.3.9]

$$M\{\frac{e^{-\lambda t}}{e^t-a}\}=\Gamma(s)\Phi(a,s,\lambda+1) \qquad (42.4.6)$$
$$\operatorname{Re}\lambda>-1 \quad , \quad |\arg(1-a)|<\pi \quad , \quad \operatorname{Re}s>0$$

and (36.4.6)

$$\int_0^\infty\frac{t^{s-1}}{e^t-a}\nu(e^{-t},\alpha)dt=\Gamma(s)\int_0^\infty\frac{\Phi(a,s,u+\alpha+1)}{\Gamma(u+\alpha+1)}du \qquad (43.4.6)$$

For a = 1, the Lerch function in (43.4.6) is replaced by the generalized zeta function

$$\int_0^\infty \frac{t^{s-1}}{e^t-1}\nu(e^{-t},\alpha)dt = \Gamma(s)\int_0^\infty \frac{\zeta(s,u+\alpha+1)}{\Gamma(u+\alpha+1)}du \tag{44.4.6}$$

$$\mathrm{Re}\,\alpha > -1 \quad , \quad \mathrm{Re}\,s > 1$$

Similar integral formulas based on [43, 1.3.21, 1.3.24, 1.6.13]

$$M\{\frac{e^{-\lambda t}}{e^t+1}\} = \frac{\Gamma(s)}{2^s}\{\zeta(s,\frac{\lambda+1}{2}) - \zeta(s,\frac{\lambda}{2}+1)\} \tag{45.4.6}$$

$$M\{\frac{e^{-\lambda t}}{e^{2t}-1}\} = \frac{\Gamma(s)}{2^s}\zeta(s,\frac{\lambda}{2}+1)\} \tag{46.4.6}$$

$$\mathrm{Re}\,\lambda > -1 \quad , \quad \mathrm{Re}\,s > 1$$

$$M\{\frac{e^{-\lambda t}}{(e^t-1)^2}\} = \Gamma(s)\{\zeta(s-1,\lambda+1) - (\lambda+1)\zeta(s,\lambda+1)\} \tag{47.4.6}$$

$$\mathrm{Re}\,\lambda > -2 \quad , \quad \mathrm{Re}\,s > 2$$

are

$$\int_0^\infty \frac{t^{s-1}}{e^t+1}\nu(e^{-t},\alpha)dt = \frac{\Gamma(s)}{2^s}\int_0^\infty \frac{[\zeta(s,\frac{u+\alpha+1}{2}) - \zeta(s,\frac{u+\alpha}{2}+1)]}{\Gamma(u+\alpha+1)}du \tag{48.4.6}$$

$$\int_0^\infty \frac{t^{s-1}}{e^{2t}-1}\nu(e^{-t},\alpha)dt = \frac{\Gamma(s)}{2^s}\int_0^\infty \frac{\zeta(s,\frac{u+\alpha}{2}+1)}{\Gamma(u+\alpha+1)}du \tag{49.4.6}$$

$$\mathrm{Re}\,\alpha > -1 \quad , \quad \mathrm{Re}\,s > 1$$

$$\int_0^\infty \frac{t^{s-1}}{(e^t-1)^2}\nu(e^{-t},\alpha)dt =$$

$$\Gamma(s)\int_0^\infty \frac{[\zeta(s-1,u+\alpha+1) - (u+\alpha+1)\zeta(s,u+\alpha+1)]}{\Gamma(u+\alpha+1)}du \tag{50.4.6}$$

$$\mathrm{Re}\,\alpha > -1 \quad , \quad \mathrm{Re}\,s > 2$$

The combination of (48.4.6) with (49.4.6) and (44.4.6) with (50.4.6) gives

$$\int_0^\infty \frac{t^{s-1}e^t}{e^{2t}-1}\nu(e^{-t},\alpha)dt = \frac{\Gamma(s)}{2^s}\int_0^\infty \frac{\zeta(s,\frac{u+\alpha+1}{2})}{\Gamma(u+\alpha+1)}du \tag{51.4.6}$$

$$\int_0^\infty \frac{t^{s-1}e^t}{(e^t-1)^2}\nu(e^{-t},\alpha)dt = \\ \Gamma(s)\int_0^\infty \frac{[\zeta(s-1,u+\alpha+1)-(u+\alpha)\zeta(s,u+\alpha+1)]}{\Gamma(u+\alpha+1)}du \tag{52.4.6}$$

If the Mellin transforms are in the form

$$M\{h(t,\lambda)\} = M\{g(t)[f(t)]^\lambda\} = M(\lambda,s) \\ \operatorname{Re}\lambda > -1 \tag{53.4.6}$$

and the involved integrals exist, then

$$\int_0^\infty t^{s-1}g(t)\nu[af(t),\alpha]dt = \int_0^\infty \frac{a^{u+\alpha}M(u+\alpha,s)}{\Gamma(u+\alpha+1)}du \\ a>0 \tag{54.4.6}$$

which is similar to the expression dealing with the Laplace transformation presented in (12.4.4).

Since [43, 1.2.20]

$$M\{h(t,\lambda)\} = b^{\lambda+1}B(s,\lambda+1) \\ h(t,\lambda) = \begin{cases} b-t & \text{for} \quad t<b \\ 0 & \text{for} \quad t>b \end{cases} \\ \operatorname{Re}\lambda > -1 \quad , \quad \operatorname{Re}s>0 \tag{55.4.6}$$

it follows from (54.4.6) that

$$\int_0^b t^{s-1}\nu[a(b-t)]dt = \frac{\Gamma(s)}{a^s}\int_0^\infty \frac{(ab)^{u+s}}{\Gamma(u+s+1)}du = \frac{\Gamma(s)}{a^s}\nu(ab,s) \tag{56.4.6}$$

Replacing t in (56.4.6) by $x=$ b-t we have

$$\int_0^b t^{s-1}\nu[a(b-t)]dt = \int_0^b (b-x)^{s-1}\nu(ax)dx = \frac{\Gamma(s)}{a^s}\nu(ab,s) \tag{57.4.6}$$

or in the equivalent form

$$\int_0^1 t^{s-1}\nu[a(1-t)]dt = \int_0^1 (1-x)^{s-1}\nu(ax)dx = \frac{\Gamma(s)}{a^s}\nu(a,s) \tag{58.4.6}$$

which is a particular case of (28.4.4).

If s is a positive integer, the integral (57.4.6) becomes

$$\int_0^b (b-x)^n \nu(ax)dx = \frac{n!}{a^{n+1}}\nu(ab,n+1) \qquad n = 0,1,2,3,\ldots \tag{59.4.6}$$

and from (59.4.6) the integrals of the Volterra functions can be evaluated successively. For example the first three integrals are

$$\int_0^b \nu(ax)dx = \frac{1}{a}\nu(ab,1) \tag{60.4.6}$$

$$\int_0^b x\,\nu(ax)dx = \frac{b}{a}\nu(ab,1) - \frac{1}{a^2}\nu(ab,2) \tag{61.4.6}$$

$$\int_0^b x^2\nu(ax)dx = \frac{2}{a^3}\nu(ab,3) - 2\frac{b}{a^2}\nu(ab,2) - \frac{b^2}{a}\nu(ab,1) \tag{62.4.6}$$

Introducing a = 1 and s = n+3/2 into (58.4.6), then we have

$$\int_0^1 t^{n+3/2}\nu(1-t)dt = \Gamma(n+\frac{3}{2})\nu(1,n+\frac{3}{2}) \qquad n = 0,1,2,3,\ldots \tag{63.4.6}$$

and the sum of m terms from (63.4.6) gives

$$\int_0^1 t^{1/2}\left[\frac{t^m-1}{t-1}\right]\nu(1-t)dt = \frac{\sqrt{\pi}}{2}\sum_{k=1}^{m}\frac{1\cdot 3\cdot 5\cdot \ldots \cdot(2k-1)}{2^{k-1}}\nu(1,k+\frac{1}{2}) \tag{64.4.6}$$

If the integral transform of function $h(t,\lambda)$ with respect to a kernel $K(t,s)$ exists

$$T\{h(t,\lambda)\} = \int_0^\infty K(t,s)h(t,\lambda)dt = R(\lambda,s)$$

$$h(t,\lambda) = g(t)[f(t)^\lambda] \qquad (65.4.6)$$

$$\operatorname{Re}\lambda > -1$$

and inversion of the order of integration is permissible, then

$$\int_0^\infty K(t,s)g(t)\nu[af(t),\alpha]dt = \int_0^\infty \frac{a^{u+\alpha}R(u+\alpha,s)}{\Gamma(u+\alpha+1)}du \qquad (66.4.6)$$

$$a > 0 \quad , \quad \operatorname{Re}\alpha > -1$$

Similar formulas were already considered in the case of the Laplace and Mellin transformations in (12.4.4) and (54.4.6) respectively. *The Fourier cosine and sine transforms* will be presented to show usefulness of (66.4.6).

From [44, 1.3.8, 2.3.7]

$$f(t) = \begin{cases} (b^2-t^2)^{\lambda-1/2} & \text{for} \quad 0 < t < b \\ 0 \quad \text{for} \quad t > b \end{cases} \qquad (67.4.6)$$

$$\int_0^b \cos(st)(b^2-t^2)^{\lambda-1/2}dt = 2^{\lambda-1}\sqrt{\pi}\,\Gamma(\lambda+\frac{1}{2})(\frac{b}{s})^\lambda J_\lambda(bs) \qquad (68.4.6)$$

$$\int_0^b \sin(st)(b^2-t^2)^{\lambda-1/2}dt = 2^{\lambda-1}\sqrt{\pi}\,\Gamma(\lambda+\frac{1}{2})(\frac{b}{s})^\lambda H_\lambda(bs)$$

$$\operatorname{Re}\lambda > -\frac{1}{2} \qquad (69.4.6)$$

and using (66.4.6), then

$$\int_0^b \cos(st)\,\nu[c(b^2-t^2),-\frac{1}{2}]dt = \frac{1}{2}\sqrt{\frac{\pi}{c}}\int_0^\infty (\frac{2bc}{s})^u J_u(bs)du \qquad (70.4.6)$$

$$\int_0^b \sin(st)\nu[c(b^2-t^2),-\frac{1}{2}]dt = \frac{1}{2}\sqrt{\frac{\pi}{c}}\int_0^\infty (\frac{2bc}{s})^u H_u(bs)du \qquad (71.4.6)$$

Introducing a = s, b = 1 and c = a/2 into (70.4.6) and (71.4.6), the improper integrals with respect to the order of the Bessel and Struve functions are

$$\int_0^1 \cos(at)\nu[\frac{a}{2}(1-t^2),-\frac{1}{2}]dt = \sqrt{\frac{\pi}{2a}}\int_0^\infty J_u(a)du \tag{72.4.6}$$

$$\int_0^1 \sin(at)\,\nu[\frac{a}{2}(1-t^2),-\frac{1}{2}]dt = \sqrt{\frac{\pi}{2a}}\int_0^\infty H_u(a)du \tag{73.4.6}$$

The last integral was already derived, but through the application of the Laplace transformation and was presented in (113.4.5). The Volterra function integral in (72.4.6) can be expressed in terms of the integrals of elementary functions because [4]

$$\int_0^\infty J_u(a)du = \frac{1}{2}+\frac{1}{\pi}\int_0^\pi \sin(a\sin x)\frac{dx}{x} - \int_0^\infty \frac{e^{-a\sinh x}}{\pi^2+x^2}dx \tag{74.4.6}$$

Similarly from [44, 1.3.14, 2.3.13]

$$f(t) = \begin{cases} (2bt-t^2)^{\lambda-1/2} & \text{for} \quad 0<t<2b \\ 0 \quad \text{for} \quad t>2b \end{cases} \tag{75.4.6}$$

$$\int_0^{2b} \cos(st)(2bt-t^2)^{\lambda-1/2}dt = 2^\lambda\sqrt{\pi}\,\Gamma(\lambda+\frac{1}{2})(\frac{b}{s})^\lambda \cos(bs)J_\lambda(bs) \tag{76.4.6}$$

$$\int_0^{2b} \sin(st)(2bt-t^2)^{\lambda-1/2}dt = 2^\lambda\sqrt{\pi}\,\Gamma(\lambda+\frac{1}{2})(\frac{b}{s})^\lambda \sin(bs)J_\lambda(bs) \tag{77.4.6}$$

and applying (66.4.6) we have

$$\int_0^{2b} \cos(st)\,\nu[c(2bt-t^2),-\frac{1}{2}]dt = \sqrt{\frac{\pi}{c}}\cos(bs)\int_0^\infty (\frac{2bc}{s})^u J_u(bs)du \tag{78.4.6}$$

$$\int_0^{2b} \sin(st)\,\nu[c(2bt-t^2),-\frac{1}{2}]dt = \sqrt{\frac{\pi}{c}}\sin(bs)\int_0^\infty (\frac{2bc}{s})^u J_u(bs)du \tag{79.4.6}$$

For a = s, b = 1 and c = a/2, these integral equalities take the form

$$\int_0^1 \cos(2at)\, \nu[2at(1-t), -\frac{1}{2}]dt = \sqrt{\frac{\pi}{2a}} \cos a \int_0^\infty J_u(a) du \quad (80.4.6)$$

$$\int_0^1 \sin(2at)\, \nu[2at(1-t), -\frac{1}{2}]dt = \sqrt{\frac{\pi}{2a}} \sin a \int_0^\infty J_u(a) du \quad (81.4.6)$$

By comparison of (72.4.6), (80.4.6) and (81.4.6), it follows that

$$\int_0^1 \cos(at)\, \nu[\frac{a}{2}(1-t^2), -\frac{1}{2}]dt = \frac{1}{\cos a} \int_0^1 \cos(2at)\, \nu[2at(1-t), -\frac{1}{2}]dt = \frac{1}{\sin a} \int_0^1 \sin(2at)\, \nu[2at(1-t), -\frac{1}{2}]dt \quad (82.4.6)$$

An additional example of an integral with respect to the order of the modified Bessel function of the second kind is derived from the following Fourier cosine transform [44, 1.3.7]

$$\int_0^\infty \frac{\cos(st)}{(a^2+t^2)^{\lambda+1/2}} dt = \frac{\sqrt{\pi}}{\Gamma(\lambda+\frac{1}{2})} (\frac{s}{2a})^\lambda K_\lambda(as)$$

$$\operatorname{Re} a > 0 \quad , \quad \operatorname{Re}\lambda > -\frac{1}{2} \quad (83.4.6)$$

using (66.4.6) and the duplication formula of gamma function

$$\int_0^\infty \frac{\cos(st)}{\sqrt{a^2+t^2}} \nu(\frac{b}{a^2+t^2}) dt = \int_0^\infty \frac{(\frac{2bs}{a})^u}{\Gamma(2u+1)} K_u(as) du \quad (84.4.6)$$

The Riemann-Liouville and the Weyl fractional integrals are defined by [44]

$$I_+^\rho f(a) = \frac{1}{\Gamma(\rho)} \int_0^a f(t)(a-t)^{\rho-1} dt = R(\rho; a) \quad (85.4.6)$$

$$I_-^\rho f(a) = \frac{1}{\Gamma(\rho)} \int_a^\infty f(t)(t-a)^{\rho-1} dt = W(\rho; a) \quad (86.4.6)$$

Multiplying both sides of (85.4.6) and (86.4.6) by dρ = d(u+α+1) and integrating from zero to infinity,

$$\int_0^a f(t)\,\nu(a-t,\alpha)dt = \int_0^\infty R(u+\alpha+1;a)du \tag{87.4.6}$$

$$\int_a^\infty f(t)\,\nu(t-a,\alpha)dt = \int_0^\infty W(u+\alpha+1;a)du \tag{88.4.6}$$

$$\mathrm{Re}\,\alpha > -1$$

provided that the order of integration may be inverted.

When the right-hand side integrals in (87.4.6) and (88.4.6) can be evaluated in the closed form, then from [44, 13.1.7]

$$\frac{1}{\Gamma(\rho)}\int_0^a t^{\lambda-1}(a-t)^{\rho-1}dt = \frac{\Gamma(\lambda)}{\Gamma(\lambda+\rho)}a^{\lambda+\rho-1} \tag{89.4.6}$$

$$\mathrm{Re}\,\lambda > 0 \quad , \quad \mathrm{Re}\,\rho > 0$$

and (87.4.6) we have

$$\int_0^a t^{\lambda-1}\,\nu(a-t,\alpha)dt = \Gamma(\lambda)\int_0^\infty \frac{a^{u+\alpha+\lambda}}{\Gamma(u+\alpha+\lambda+1)}du = \Gamma(\lambda)\nu(a,\alpha+\lambda) \tag{90.4.6}$$

$$\mathrm{Re}\,\alpha > -1 \quad , \quad \mathrm{Re}\,\lambda > 0$$

which reduces to (57.4.6) for α = 0.

The next example of the Riemann-Liouville fractional integral is [44, 13.1.16]

$$\frac{1}{\Gamma(\rho)}\int_0^a \frac{e^{-b/t}}{t^2}\left(\frac{a-t}{t}\right)^{\rho-1}dt = \frac{1}{a}\left(\frac{a}{b}\right)^{\rho}e^{-b/a} \tag{91.4.6}$$

$$b > 0 \quad , \quad \mathrm{Re}\,\rho > 0$$

and in this case (87.4.6) gives

$$\int_0^a \frac{e^{-b/t}}{t^2} \nu(\frac{a-t}{t},\alpha)dt = \frac{e^{-b/a}}{b\ln(b/a)}(\frac{a}{b})^{\alpha} \tag{92.4.6}$$

$$\mathrm{Re}\,\alpha > -1 \quad , \quad b > a > 0$$

The change of variable of integration from t to $x = (a-t)/t$, restores the Laplace transform of thc Volterra function $\nu(x,\alpha)$

$$\int_0^\infty e^{-bx/a} \nu(x,\alpha)dx = \frac{1}{(b/a)^{\alpha+1}\ln(b/a)} \tag{93.4.6}$$

$$\mathrm{Re}\,\alpha > -1 \quad , \quad b > a > 0$$

The Weyl fractional integral [44, 13.2.11]

$$\frac{1}{\Gamma(\rho)} \int_a^\infty e^{-bt}(a-t)^{\rho-1}dt = \frac{e^{-ab}}{b^{\rho}} \tag{94.4.6}$$

$$b > 0 \quad , \quad \mathrm{Re}\,\rho > 0$$

together with (88.4.6) leads to

$$\int_a^\infty e^{-bt} \nu(a-t,\alpha)dt = \frac{e^{-ab}}{b^{\alpha+1}\ln b} \tag{95.4.6}$$

$$\mathrm{Re}\,\alpha > -1 \quad , \quad a > 0, b > 1$$

From [44, 13.1.24]

$$\frac{1}{\Gamma(\rho)} \int_0^a t^{\lambda-1}\ln t(a-t)^{\rho-1}dt = \frac{\Gamma(\lambda)}{\Gamma(\lambda+\rho)} a^{\lambda+\rho-1}[\ln a + \psi(\lambda) - \psi(\lambda+\rho)] \tag{96.4.6}$$

$$\mathrm{Re}\,\lambda > 0 \quad , \quad \mathrm{Re}\,\rho > 0$$

and (87.4.6) we have

$$\int_0^a t^{\lambda-1}\ln t\,\nu(a-t,\alpha)dt = \tag{97.4.6}$$

$$\Gamma(\lambda)\int_0^\infty \frac{a^{u+\alpha+\lambda}}{\Gamma(u+\alpha+\lambda+1)}[\ln a + \psi(\lambda) - \psi(u+\alpha+\lambda+1)]du$$

But [3]

$$\int_0^\infty \frac{a^{u+\alpha+\lambda}\,\psi(u+\alpha+\lambda+1)}{\Gamma(u+\alpha+\lambda+1)}du = \frac{a^{\alpha+\lambda}}{\Gamma(\alpha+\lambda+1)} + \nu(a,\alpha+\lambda)\ln a \tag{98.4.6}$$

and therefore

$$\int_0^a t^{\lambda-1}\ln t\,\nu(a-t,\alpha)dt = \Gamma(\lambda)\{\psi(\lambda)\,\nu(a,\alpha+\lambda) - \frac{a^{\alpha+\lambda}}{\Gamma(\alpha+\lambda+1)}\} \tag{99.4.6}$$

$$\operatorname{Re}\alpha > -1 \quad , \quad \operatorname{Re}\lambda > 0$$

Now, consider the application of (87.4.6) to the Bessel and Struve functions. From [44, 13.1.25, 13.1.31]

$$\frac{1}{\Gamma(\rho)}\int_0^a \sin(bt)[t(a-t)^{\rho-1}]dt = \sqrt{\pi}(\frac{a}{b})^{\rho-1/2}\sin(\frac{ab}{2})J_{\rho-1/2}(\frac{ab}{2}) \tag{100.4.6}$$

$$\frac{1}{\Gamma(\rho)}\int_0^a \cos(bt)[t(a-t)^{\rho-1}]dt = \sqrt{\pi}(\frac{a}{b})^{\rho-1/2}\cos(\frac{ab}{2})J_{\rho-1/2}(\frac{ab}{2}) \tag{101.4.6}$$

$$\operatorname{Re}\rho > 0$$

we have

$$\int_0^a \sin(bt)\,\nu[t(a-t),\alpha]dt = \sqrt{\pi}(\frac{a}{b})^{\alpha+1/2}\sin(\frac{ab}{2})\int_0^\infty (\frac{a}{b})^u J_{u+\alpha+1/2}(\frac{ab}{2})du \tag{102.4.6}$$

$$\int_0^a \cos(bt)\,\nu[t(a-t),\alpha]dt = \sqrt{\pi}(\frac{a}{b})^{\alpha+1/2}\cos(\frac{ab}{2})\int_0^\infty (\frac{a}{b})^u J_{u+\alpha+1/2}(\frac{ab}{2})du \tag{103.4.6}$$

$$\operatorname{Re}\alpha > -1$$

For $\alpha = -1/2$ these integral equalities can be reduced to (81.4.6) and (80.4.6) respectively.

Introducing the Riemann-Liouville fractional integral [44, 13.1.27]

$$\frac{1}{\Gamma(\rho)}\int_0^a \sin(b\sqrt{t})(a-t)^{\rho-1}dt = \sqrt{\pi}(\frac{2\sqrt{a}}{b})^{\rho-1/2}J_{\rho+1/2}(b\sqrt{a}) \tag{104.4.6}$$

it is possible to obtain

$$\int_0^a \sin(b\sqrt{t})\, \nu(a-t,\alpha)dt = \sqrt{\pi a}(\frac{2\sqrt{a}}{b})^{\alpha+1/2} \int_0^\infty (\frac{2\sqrt{a}}{b})^u J_{u+\alpha+3/2}(b\sqrt{a})du \qquad (105.4.6)$$

$$\mathrm{Re}\,\alpha > -1$$

and by changing the variable of integration with $\alpha = -1/2$ and $b = 2\sqrt{a}$ we have

$$\int_0^1 t\sin(at)\, \nu[\frac{a}{2}(1-t^2), -\frac{1}{2}]dt = \sqrt{\frac{\pi}{a}} \int_0^\infty J_{u+1}(a)du \qquad (106.4.6)$$

The Volterra function integral in (106.4.6) can be expressed in terms of the integrals of elementary functions because [4]

$$\int_0^\infty J_{u+1}(a)du = \int_1^\infty J_x(a)dx = \frac{1}{2} + \frac{1}{\pi}\int_0^\pi \sin(a\sin y - y)\frac{dy}{y} + \int_0^\infty \frac{e^{-y-a\sinh y}}{\pi^2 + y^2}dy \qquad (107.4.6)$$

A generalization of (105.4.6) comes from the following Riemann-Liouville integral [44, 13.1.63]

$$\frac{1}{\Gamma(\rho)}\int_0^a t^{\lambda/2} J_\lambda(b\sqrt{t})(a-t)^{\rho-1}dt = (\frac{2}{b})^\rho a^{(\lambda+\rho)/2} J_{\lambda+\rho}(b\sqrt{a}) \qquad (108.4.6)$$

$$\mathrm{Re}\,\lambda > -1 \quad , \quad \mathrm{Re}\,\rho > 0$$

and using (87.4.6) we have

$$\int_0^a t^{\lambda/2} J_\lambda(b\sqrt{t})\, \nu(a-t,\alpha)dt = (\frac{2\sqrt{a}}{b})^{\alpha+1} a^{\lambda/2} \int_0^\infty (\frac{2\sqrt{a}}{b})^u J_{u+\alpha+\lambda+1}(b\sqrt{a})du \qquad (109.4.6)$$

$$\mathrm{Re}\,\alpha > -1 \quad , \quad \mathrm{Re}\,\lambda > -1$$

or in the form

$$\int_0^1 t^{\lambda+1} J_\lambda(abt)\, \nu[a^2(1-t^2),\alpha]dt = \frac{1}{2a^2}(\frac{2a}{b})^{\alpha+1} \int_0^\infty (\frac{2a}{b})^u J_{u+\alpha+\lambda+1}(ab)du \qquad (110.4.6)$$

For $b = 2a$ (110.4.6) becomes

$$\int_0^1 t^{\lambda+1} J_\lambda(at)\, \nu[\frac{a}{2}(1-t^2),\alpha]dt = \frac{1}{a}\int_0^\infty J_{u+\alpha+\lambda+1}(a)du \qquad (111.4.6)$$

As expected for $\alpha = 1/2$ and $\lambda = \pm 1/2$ with

$$J_{-1/2}(at) = \sqrt{\frac{2}{\pi at}}\cos(at) \tag{112.4.6}$$

$$J_{1/2}(at) = \sqrt{\frac{2}{\pi at}}\sin(at) \tag{113.4.6}$$

the integral equality (111.4.6) reduces to (106.4.6) and (72.4.6) respectively.

For $\alpha = 0$ and $\lambda = 0$ it gives

$$\int_0^1 t J_0(at)\, \nu[\frac{a}{2}(1-t^2)]dt = \frac{1}{a}\int_0^\infty J_{u+1}(a)du \tag{114.4.6}$$

It follows from (106.4.6) and (114.4.6) that

$$\int_0^1 t\sin(at)\, \nu[\frac{a}{2}(1-t^2), -\frac{1}{2}]dt = \sqrt{\frac{\pi a}{2}}\int_0^1 t J_0(at)\, \nu[\frac{a}{2}(1-t^2)]dt \tag{115.4.6}$$

Similar results are available for the modified Bessel functions from [44, 13.1.83]

$$\frac{1}{\Gamma(\rho)}\int_0^a t^{\lambda/2} I_\lambda(b\sqrt{t})(a-t)^{\rho-1}dt = (\frac{2}{b})^\rho a^{(\lambda+\rho)/2} I_{\lambda+\rho}(b\sqrt{a})$$
$$\operatorname{Re}\lambda > -1 \quad , \quad \operatorname{Re}\rho > 0 \tag{116.4.6}$$

with use of (87.4.6)

$$\int_0^a t^{\lambda/2} I_\lambda(b\sqrt{t})\, \nu(a-t,\alpha)dt =$$
$$(\frac{2\sqrt{a}}{b})^{\alpha+1} a^{\lambda/2}\int_0^\infty (\frac{2\sqrt{a}}{b})^u I_{u+\alpha+\lambda+1}(b\sqrt{a})du \tag{117.4.6}$$
$$\operatorname{Re}\alpha > -1 \quad , \quad \operatorname{Re}\lambda > -1$$

and

$$\int_0^1 t^{\lambda+1} I_\lambda(bct)\, \nu[c^2(1-t^2),\alpha]dt = \frac{1}{2a^2}(\frac{2a}{b})^{\alpha+1}\int_0^\infty (\frac{2a}{b})^u I_{u+\alpha+\lambda+1}(ab)du \tag{118.4.6}$$

Introducing b = 2a into (118.4.6) we have

$$\int_0^1 t^{\lambda+1} I_\lambda(at)\, \nu[\frac{a}{2}(1-t^2),\alpha]dt = \frac{1}{a}\int_0^\infty I_{u+\alpha+\lambda+1}(a)du \tag{119.4.6}$$

For α = -1/2 and considering that

$$I_{-1/2}(at) = \sqrt{\frac{2}{\pi at}}\cosh(at) \tag{120.4.6}$$

$$I_{1/2}(at) = \sqrt{\frac{2}{\pi at}}\sinh(at) \tag{121.4.6}$$

it follows from (119.4.6) that

$$\int_0^1 \cosh(at)\, \nu[\frac{a}{2}(1-t^2),-\frac{1}{2}]dt = \sqrt{\frac{\pi}{2a}}\int_0^\infty I_u(a)du \tag{122.4.6}$$

$$\int_0^1 t\sinh(at)\, \nu[\frac{a}{2}(1-t^2),-\frac{1}{2}]dt = \sqrt{\frac{\pi}{2a}}\int_0^\infty I_{u+1}(a)du \tag{123.4.6}$$

where improper integrals of the modified Bessel functions with respect to the order are [4]

$$\int_0^\infty I_u(a)du = \frac{e^a}{2} - \int_0^\infty \frac{e^{-a\cosh y}}{\pi^2+y^2}dy \tag{124.4.6}$$

and

$$\int_0^\infty I_{u+1}(a)du = \int_1^\infty I_x(a)dx = \frac{e^a}{2} - \frac{1}{\pi}\int_0^\pi e^{a\cos y}\sin y\frac{dy}{y} + \int_0^\infty \frac{e^{-y-a\cosh y}}{\pi^2+y^2}dy \tag{125.4.6}$$

The equivalent form of (122.4.6) is derived by introducing $x = 2t - 1$

$$\int_0^1 e^{2at}\nu[2at(1-t),-\frac{1}{2}]dt = \sqrt{\frac{\pi}{2a}}e^a\int_0^\infty I_u(a)du \tag{126.4.6}$$

and then

$$\int_0^1 e^{2at}\nu[2at(1-t),-\frac{1}{2}]dt = e^a\int_0^1 \cosh(at)\nu[\frac{a}{2}(1-t^2),-\frac{1}{2}]dt \tag{127.4.6}$$

Similarly as in (115.4.6), by combining (119.4.6) and (123.4.6) with α = λ = 0, we have

$$\int_0^1 t\sinh(at)\nu[2at(1-t),-\frac{1}{2}]dt = \sqrt{\frac{\pi a}{2}}\int_0^1 t I_0(at)\,\nu[\frac{a}{2}(1-t^2)]dt \tag{128.4.6}$$

The Riemann-Liouville fractional integrals [44, 13.1.88, 13.1.90]

$$\frac{1}{\Gamma(\rho)}\int_0^a t^{\lambda/2}H_\lambda(b\sqrt{t})(a-t)^{\rho-1}dt = (\frac{2}{b})^\rho a^{(\lambda+\rho)/2}H_{\lambda+\rho}(b\sqrt{a}) \tag{129.4.6}$$

$$\frac{1}{\Gamma(\rho)}\int_0^a t^{\lambda/2}L_\lambda(b\sqrt{t})(a-t)^{\rho-1}dt = (\frac{2}{b})^\rho a^{(\lambda+\rho)/2}L_{\lambda+\rho}(b\sqrt{a}) \tag{130.4.6}$$

$$\mathrm{Re}\,\lambda > -\frac{3}{2} \quad , \quad \mathrm{Re}\,\rho > 0$$

become using (87.4.6)

$$\int_0^1 t^{\lambda+1}H_\lambda(abt)\,\nu[a^2(1-t^2),\alpha]dt = \frac{1}{2a^2}(\frac{2a}{b})^{\alpha+1}\int_0^\infty (\frac{2a}{b})^u H_{u+\alpha+\lambda+1}(ab)du \tag{131.4.6}$$

$$\int_0^1 t^{\lambda+1}L_\lambda(abt)\,\nu[a^2(1-t^2),\alpha]dt = \frac{1}{2a^2}(\frac{2a}{b})^{\alpha+1}\int_0^\infty (\frac{2a}{b})^u L_{u+\alpha+\lambda+1}(ab)du \tag{132.4.6}$$

$$\mathrm{Re}\,\alpha > -1 \quad , \quad \mathrm{Re}\,\lambda > -\frac{3}{2}$$

and

$$\int_0^1 t^{\lambda+1}H_\lambda(at)\,\nu[\frac{a}{2}(1-t^2),\alpha]dt = \frac{1}{a}\int_0^\infty H_{u+\alpha+\lambda+1}(a)du \tag{133.4.6}$$

$$\int_0^1 t^{\lambda+1} L_\lambda(at)\, \nu[\frac{a}{2}(1-t^2),\alpha]dt = \frac{1}{a}\int_0^\infty L_{u+\alpha+\lambda+1}(a)du \qquad (134.4.6)$$

For α = -1/2 and introducing

$$H_{-1/2}(at) = \sqrt{\frac{2}{\pi at}}\sin(at) \qquad (135.4.6)$$

$$H_{1/2}(at) = \sqrt{\frac{2}{\pi at}}[1-\cos(at)] \qquad (136.4.6)$$

$$L_{-1/2}(at) = \sqrt{\frac{2}{\pi at}}\sinh(at) \qquad (137.4.6)$$

$$L_{1/2}(at) = \sqrt{\frac{2}{\pi at}}[\cosh(at)-1] \qquad (138.4.6)$$

we have

$$\int_0^1 \sin(at)\, \nu[\frac{a}{2}(1-t^2),-\frac{1}{2}]dt = \sqrt{\frac{\pi}{2a}}\int_0^\infty H_u(a)du \qquad (139.4.6)$$

$$\int_0^1 t\cos(at)\, \nu[\frac{a}{2}(1-t^2),-\frac{1}{2}]dt = \frac{1}{a}\nu(\frac{a}{2},\frac{1}{2}) - \sqrt{\frac{\pi}{2a}}\int_0^\infty H_{u+1}(a)du \qquad (140.4.6)$$

$$\int_0^1 \sinh(at)\, \nu[\frac{a}{2}(1-t^2),-\frac{1}{2}]dt = \sqrt{\frac{\pi}{2a}}\int_0^\infty L_u(a)du \qquad (141.4.6)$$

$$\int_0^1 t\cosh(at)\, \nu[\frac{a}{2}(1-t^2),-\frac{1}{2}]dt = \frac{1}{a}\nu(\frac{a}{2},\frac{1}{2}) + \sqrt{\frac{\pi}{2a}}\int_0^\infty L_{u+1}(a)du \qquad (142.4.6)$$

where

$$\int_0^1 t\, \nu[\frac{a}{2}(1-t^2),-\frac{1}{2}]dt = \frac{1}{a}\nu(\frac{a}{2},\frac{1}{2}) \qquad (143.4.6)$$

was used in (140.4.6) and (142.4.6).

If α = λ = 0, then (133.4.6) and (134.4.6) take the form

$$\int_0^1 t\,H_0(at)\,\nu[\frac{a}{2}(1-t^2),\alpha]dt = \frac{1}{a}\int_0^\infty H_{u+1}(a)du \tag{144.4.6}$$

$$\int_0^1 t\,L_0(at)\,\nu[\frac{a}{2}(1-t^2),\alpha]dt = \frac{1}{a}\int_0^\infty L_{u+1}(a)du \tag{145.4.6}$$

and therefore

$$\begin{aligned}&\sqrt{\frac{\pi a}{2}}\int_0^1 t\,H_0(at)\,\nu[\frac{a}{2}(1-t^2),\alpha]dt = \\ &\frac{1}{a}\nu(\frac{a}{2},\frac{1}{2}) - \int_0^1 t\cos(at)\,\nu[\frac{a}{2}(1-t^2),-\frac{1}{2}]dt\end{aligned} \tag{146.4.6}$$

$$\begin{aligned}&\sqrt{\frac{\pi a}{2}}\int_0^1 t\,L_0(at)\,\nu[\frac{a}{2}(1-t^2),\alpha]dt = \\ &\int_0^1 t\cosh(at)\,\nu[\frac{a}{2}(1-t^2),-\frac{1}{2}]dt - \frac{1}{a}\nu(\frac{a}{2},\frac{1}{2})\end{aligned} \tag{147.4.6}$$

Only one example of the two-sided Laplace transform of the Volterra function will be presented. Starting with [96]

$$L_{II}\{f(e^t);s\} = \int_{-\infty}^{\infty} e^{-st} f(e^t)dt = L_{II}(s) \tag{148.4.6}$$

and introducing $x = e^t$

$$\begin{aligned}&L_{II}(s) = \int_0^\infty x^{-(s+1)} f(x)dx = \frac{1}{\Gamma(s+1)}\int_0^\infty f(x)(\int_0^\infty y^s e^{-xy}dy)dx = \\ &\frac{1}{\Gamma(s+1)}\int_0^\infty y^s(\int_0^\infty e^{-yx} f(x)dx)dy = \frac{1}{\Gamma(s+1)}\int_0^\infty y^s F(y)dy\end{aligned} \tag{149.4.6}$$

where the left-hand side integral is written in the form of a repeated integral and $F(y)$ denotes the Laplace transform of function $f(x)$.

In the case of the Volterra function $\nu(e^{-t},\alpha)$

$$\nu(e^{-x},\alpha)=\int_0^\infty \frac{(e^{-x})^{u+\alpha}}{\Gamma(u+\alpha+1)}du=\int_\alpha^\infty \frac{e^{-xu}}{\Gamma(u+1)}du \tag{150.4.6}$$

the Laplace transform is

$$L\{\nu(e^{-x},\alpha)\}=\int_0^\infty e^{-yx}(\int_\alpha^\infty \frac{e^{-xu}}{\Gamma(u+1)}du)dx=\int_\alpha^\infty \frac{1}{\Gamma(u+1)}(\int_0^\infty e^{-(y+u)x}dx)du= \\ \int_\alpha^\infty \frac{1}{(y+u)\Gamma(u+1)}du \tag{151.4.6}$$

and using (149.4.6)

$$L_{II}(s)=\frac{1}{\Gamma(s+1)}\int_0^\infty y^s(\int_\alpha^\infty \frac{1}{(y+u)\Gamma(u+1)}du)dy= \\ \frac{1}{\Gamma(s+1)}\int_\alpha^\infty \frac{1}{\Gamma(u+1)}(\int_0^\infty \frac{y^s}{y+u}dy)du \tag{152.4.6}$$

But

$$\int_0^\infty \frac{y^s}{y+u}dy=B(s+1,-s)u^s=\Gamma(s+1)\Gamma(-s)u^s \\ \Gamma(-s)=-s\Gamma(1-s) \tag{153.4.6}$$

and from (148.4.6) the final result can be written as

$$\int_{-\infty}^\infty e^{-st}\nu(e^{-e^t},\alpha)dt=-\frac{\Gamma(1-s)}{s}\int_\alpha^\infty \frac{u^s}{\Gamma(u+1)}du \\ -1<\mathrm{Re}\,s<0 \quad , \quad \mathrm{Re}\,\alpha\ge 0 \tag{154.4.6}$$

Volterra functions and improper integrals of gamma and digamma functions are associated with the factorial transformation [48,49]

$$T\{f(t)\}=\int_0^\infty B(t+1,s)f(t)dt=\Gamma(s)\int_0^\infty \frac{\Gamma(t+1)}{\Gamma(t+s+1)}f(t)dt=T(s) \tag{155.4.6}$$

Evidently for the reciprocal gamma function

$$T\{\frac{1}{\Gamma(t+1)}\} = \Gamma(s)\int_0^\infty \frac{1}{\Gamma(t+s+1)}dt = \Gamma(s)\,\nu(1,s) \tag{156.4.6}$$

The factorial transform is related to the Laplace transform by taking into account that [88, 1.5.8]

$$L\{(1-e^{-t})^{\lambda}\} = B(\lambda+1,s) \qquad \mathrm{Re}\,\lambda > -1 \quad , \quad \mathrm{Re}\,s > 0 \tag{157.4.6}$$

and therefore

$$F(s) = L\{\int_0^\infty (1-e^{-t})^u f(u)du\} = \int_0^\infty B(u+1,s)f(u)du = T(s) \tag{158.4.6}$$

For example if $f(u) = u^n/\Gamma(u+1)$

$$L\{\int_0^\infty \frac{(1-e^{-t})^u u^n}{\Gamma(u+1)}du\} = \Gamma(s)n!\mu(1,n,s) \qquad n = 0,1,2,3,\ldots \tag{159.4.6}$$

Repeated differentiation of (158.4.6) gives

$$\frac{d^nT(s)}{ds^n} = (-1)^n\int_0^\infty f(u)\{\int_0^\infty t^n e^{-st}(1-e^{-t})^u dt\}du \qquad n = 0,1,2,3,\ldots \tag{160.4.6}$$

For n = 1, the inner integral in (160.4.6) is known [88, 1.5.9]

$$\int_0^\infty te^{-st}(1-e^{-t})^u dt = B(u+1,s)[\psi(u+s+1)-\psi(s)] \tag{161.4.6}$$

and therefore

$$\int_0^\infty B(u+1,s)\psi(u+s+1)f(u)du = \psi(s)T(s) - \frac{dT(s)}{ds} \tag{162.4.6}$$

In a particular case of $f(u) = 1/\Gamma(u+1)$, the integral in (162.4.6) is [3]

$$\Gamma(s)\int_0^\infty \frac{\psi(u+s+1)}{\Gamma(t+s+1)}du = \frac{\Gamma(s)}{\Gamma(s+1)} = \frac{1}{s} \tag{163.4.6}$$

and

$$\frac{dT(s)}{ds} - \psi(s)T(s) = -\frac{1}{s} \tag{164.4.6}$$

where $T(s)$ from (156.4.6) is a solution for (164.4.6). Thus, the linear nonhomogeneous first order differential equation

$$y'(x) - \psi(x)y(x) = -\frac{1}{x} \tag{165.4.6}$$

has a general solution

$$y(x) = \Gamma(x)[C + \nu(1,x)] \tag{166.4.6}$$

where C denotes the integration constant.

Repeated differentiation of (165.4.6) leads to the second order differential equation

$$y''(x) - [\psi(x)^2 + \psi'(x)]y(x) = \frac{1}{x^2} - \frac{1}{x}\psi(x) \tag{167.4.6}$$

which has the solution

$$y(x) = \Gamma(x)[C_1 + C_2\int^x \frac{1}{\Gamma(u)^2}du + \nu(1,x)] \tag{168.4.6}$$

where C_1 and C_2 are integration constants.

Closely related to the factorial transform is the integral transformation where the reciprocal gamma function serves as the kernel [27]

$$G\{f(t)\} = \int_0^\infty \frac{f(t)}{\Gamma(t+s+1)}dt = G(s) \tag{169.4.6}$$

where

$$T\{f(t)\} = \Gamma(s)G\{\Gamma(t+1)f(t)\} \tag{170.4.6}$$

and

$$G\{\Gamma(t+1)\} = \frac{1}{\Gamma(s)}\int_0^\infty B(t+1,s)dt = \frac{1}{\Gamma(s)}T\{1\} \tag{171.4.6}$$

Considering that

$$\frac{d}{dt}\{\frac{1}{\Gamma(t+s+1)}\} = \frac{d}{ds}\{\frac{1}{\Gamma(t+s+1)}\} \tag{172.4.6}$$

and using (169.4.6) it follows that [27]

$$G\{tf(t)\} = G(s-1) - sG(s) \tag{173.4.6}$$

$$G\{f'(t)\} = -\frac{f(0)}{\Gamma(s+1)} - G'(s) \tag{174.4.6}$$

$$G\{t\,f'(t)\} = sG'(s) - G'(s-1) \tag{175.4.6}$$

Integral transforms $G(s)$ are frequently expressed in terms of the Volterra functions

$$G\{a^t\} = a^{-s}\nu(a,s) \tag{176.4.6}$$

$$G\{1\} = \nu(1,s) = \mu(1,0,s) \tag{177.4.6}$$

$$G\{t\} = \mu(1,1,s) \tag{178.4.6}$$

$$G\{t^2\} = 2\mu(1,2,s) \tag{179.4.6}$$

$$G\{t^n\} = n!\,\mu(1,n,s) \tag{180.4.6}$$

If an entire function $f(t)$ is represented by the Taylor series

$$f(t) = \sum_{n=0}^{\infty} a_n t^n \tag{181.4.6}$$

then using (180.4.6) we have

$$G\{f(t)\} = \sum_{n=0}^{\infty} a_n n! \mu(1,n,s) \tag{182.4.6}$$

and for example

$$G\{e^t\} = \sum_{n=0}^{\infty} \mu(1,n,s) = e^{-s}\nu(e,s) \tag{183.4.6}$$

$$G\{e^{-t^2}\} = \sum_{n=0}^{\infty} (-1)^n \mu(1,2n,s) \tag{184.4.6}$$

$$G\{J_0(2\sqrt{t})\} = \sum_{n=0}^{\infty} \frac{(-1)^n}{n!} \mu(1,n,s) \tag{185.4.6}$$

$$G\{I_0(2\sqrt{t})\} = \sum_{n=0}^{\infty} \frac{1}{n!} \mu(1,n,s) \tag{186.4.6}$$

Considering that

$$a^t = \sum_{n=0}^{\infty} \frac{1}{n!} (t \ln a)^n \tag{187.4.6}$$

we have from (176.4.6) and (182.4.6)

$$G\{a^t\} = \sum_{n=0}^{\infty} (\ln a)^n \mu(1,n,s) = a^{-s}\nu(a,s) \tag{188.4.6}$$

the series of the Volterra functions already presented in (30.2.2).

Application of the operational rule (173.4.6) to (183.4.6) gives

$$G\{te^t\} = e^{-s}[e\nu(e,s-1) - s\nu(e,s)] = e^{-s}\mu(e,1,s) \tag{189.4.6}$$

and by repeating this operation with the aid of (2.2.2)

$$G\{t^2 e^t\} = e^{-s}[e\mu(e,1,s-1) - s\mu(e,1,s)] = 2e^{-s}\mu(e,2,s) \tag{190.4.6}$$

$$G\{t^3 e^t\} = 2e^{-s}[e\mu(e,2,s-1) - s\mu(e,2,s)] = 2 \cdot 3 e^{-s}\mu(e,3,s) \tag{191.4.6}$$

$$G\{t^n e^t\} = n! e^{-s} \mu(e,n,s) \qquad n = 0,1,2,3,\ldots \tag{192.4.6}$$

Evidently, (192.4.6) comes directly from a definition of the generalized Volterra function (2.2.1).

4.7. Volterra Functions and the Two-Dimensional Laplace Transformation

In the case of functions of two real variables x and y, an analogy with the Laplace transformation leads to the extension

$$F(p,q) = \int_0^\infty \int_0^\infty e^{-px} e^{-qy} f(x,y) dx dy \tag{1.4.7}$$

where p and q are complex variables. $F(p,q)$ is called *the two-dimensional Laplace transform* of the original function $f(x,y)$ which is defined in the positive quadrant of xy-plane. If the double integral (1.4.7) converges absolutely for some pair p_o and q_o then $F(p,q)$ exists for $\mathrm{Re}(p-p_o) > 0$ and $\mathrm{Re}(q-q_o) > 0$. The transform converges uniformly for $\mathrm{Re}\, p \to \infty$ and $\mathrm{Re}\, q \to \infty$. $F(p,q)$ can be differentiated infinitely with respect to p and q, being an analytic function in the domain of convergence inside a closed contour, whether finite or extending to infinity in some directions. The value of the double integral (1.4.7) is independent of the order of integration, thus the double integral may be replaced by the iterated integral.

The direct and inverse two-dimensional Laplace transforms will be denoted as

$$L_{xy}\{f(x,y)\} = F(p,q) \tag{2.4.7}$$

$$L_{xy}^{-1}\{F(p,q)\} = f(x,y) \tag{3.4.7}$$

For more details about the existance and uniqueness theorems, asymptotic properties, operational rules and applications of the two-dimensional Laplace transformation see the literature [36,56,94,96,98,113] .

Since our knowledge about two-dimensional Laplace transforms of the Volterra functions or transforms expressed by the Volterra functions is limited to four transforms derived by Colombo [27]

$$L_{xy}\{\nu(x,y)\} = \frac{1}{p \ln p (q + \ln p)} \tag{4.4.7}$$

$$L_{xy}\{\mu(x,y,\alpha)\} = \frac{1}{p^{\alpha+1}\ln p[q+\ln(\ln p)]} \quad (5.4.7)$$

$$L_{xy}\{\mu(x,\beta,y)\} = \frac{1}{p(\ln p)^{\beta+1}(q+\ln p)} \quad (6.4.7)$$

$$L_{xy}\{\frac{1}{\Gamma(x+y+1)}\} = \frac{1}{q-p}\{\nu(e^{-p}) - \nu(e^{-q})\} \quad (7.4.7)$$

it is wortwhile to evaluate and present a more complete list of them. In this section we will deal mainly with the Volterra function ν(*x*,*y*) and the obtained results can be easily extended to the generalized Volterra functions.

The first three transforms of the Volterra functions (4.4.7) - (6.4.7) follow directly from (1.4.7) and the transform of the reciprocal gamma function (7.4.7) is evaluated using (31.4.1). As expected, a similarity exists between the ordinary and the two-dimensional Laplace transformations considering that in both cases the inverse powers of the logarithmic functions are transforms of the Volterra functions and transforms of the reciprocal gamma functions are associated with the Volterra functions. A more general form of these and other transforms can be presented by applying the following operational rules

$$L_{xy}\{f(ax,by)\} = \frac{1}{ab}F(\frac{p}{a},\frac{q}{b}) \quad (8.4.7)$$
$$a>0 \quad , \quad b>0$$

$$L_{xy}\{e^{-ax-by}f(x,y)\} = F(p+a,q+b) \quad (9.4.7)$$

$$L_{xy}\{f(x+y)\} = \frac{1}{q-p}\{F(p)-F(q)\} \quad (10.4.7)$$

The last rule is especially useful because it gives immediately a large number of transforms which are directly based on the ordinary Laplace transforms of the Volterra functions of composite variables

$$L_{xy}\{\nu(x+y,\alpha)\} = \frac{1}{q-p}\{\frac{1}{p^{\alpha+1}\ln p} - \frac{1}{q^{\alpha+1}\ln q}\} \quad (11.4.7)$$

$$L_{xy}\{\nu(a\sqrt{x+y},\alpha)\} = \frac{\sqrt{\pi}a}{q-p}\{\frac{1}{p^{3/2}}\nu(\frac{a^2}{4p},\frac{\alpha-1}{2}) - \frac{1}{q^{3/2}}\nu(\frac{a^2}{4q},\frac{\alpha-1}{2})\} \quad (12.4.7)$$

$$L_{xy}\{\frac{1}{\sqrt{x+y}}\nu(a\sqrt{x+y},\alpha)\} = \frac{2\sqrt{\pi}}{q-p}\{\frac{1}{\sqrt{p}}\nu(\frac{a^2}{4p},\frac{\alpha}{2}) - \frac{1}{\sqrt{q}}\nu(\frac{a^2}{4q},\frac{\alpha}{2})\} \quad (13.4.7)$$

$$L_{xy}\{\nu[a(1-e^{-b(x+y)}),\alpha]\}=$$

$$\frac{1}{b(q-p)}\{\frac{\Gamma(\frac{p}{b})}{a^{p/b}}\nu(a,\alpha+\frac{p}{b})-\frac{\Gamma(\frac{q}{b})}{a^{q/b}}\nu(a,\alpha+\frac{q}{b})\} \quad (14.4.7)$$

$$L_{xy}\{\nu(e^{-(x+y)},\alpha)\}=\int_0^\infty \frac{1}{(u+p+\alpha)(u+q+\alpha)\Gamma(u+\alpha+1)}du \quad (15.4.7)$$

There are many general formulas of the two-dimensional Laplace transformation when p and q can be replaced by functions of $\ln p$ and $\ln q$. These formulas which include the Volterra functions are

$$L_{xy}\{\int_0^x \nu(x-\xi,y)f(\xi)d\xi\}=\frac{F(p)}{p\ln p(q+\ln p)} \quad (16.4.7)$$

$$L_{xy}\{\int_0^\infty \nu(y,\xi+\alpha)f(x,\xi)d\xi\}=\frac{F(p,\ln q)}{q^{\alpha+1}\ln q} \quad (17.4.7)$$

$$L_{xy}\{\int_0^\infty \mu(y,\beta,\xi+\alpha)f(x,\xi)d\xi\}=\frac{F(p,\ln q)}{q^{\alpha+1}(\ln q)^{\beta+1}} \quad (18.4.7)$$

$$L_{xy}\{\int_0^\infty \mu(y,\beta\xi-1,\alpha\xi-1)f(x,\xi)d\xi\}=F\{p,\ln[q^\alpha(\ln q)^\beta]\} \quad (19.4.7)$$

$$L_{xy}\{\int_0^\infty \mu(y,\beta\xi-1,\alpha\xi)f(x,\xi)d\xi\}=\frac{1}{q}F\{p,\ln[q^\alpha(\ln q)^\beta]\} \quad (20.4.7)$$

$$L_{xy}\{\int_0^\infty\int_0^\infty \nu(x,\xi)\nu(y,\eta)f(\xi,\eta)d\xi d\eta\}=\frac{F(\ln p,\ln q)}{pq\ln p\ln q} \quad (21.4.7)$$

and these with the reciprocal gamma functions are

$$L_{xy}\{\frac{y^{\alpha x-1}}{\Gamma(\alpha x)}f(x)\}=F(p+\alpha\ln q) \quad (22.4.7)$$

$$L_{xy}\{\int_0^x \frac{y^{\alpha\xi-1}}{\Gamma(\alpha\xi)} f(\xi)d\xi = \frac{1}{p}F(p+\alpha\ln q) \qquad (23.4.7)$$

$$L_{xy}\{\int_0^x \frac{(x-\xi)^{\alpha\xi-1}}{\Gamma(\alpha\xi)} f(\xi,y)d\xi = F(p+\alpha\ln q,q) \qquad (24.4.7)$$

$$L_{xy}\{\int_0^\infty \frac{x^{\alpha\xi-1}}{\Gamma(\alpha\xi)} f(\xi,y)d\xi\} = F(\alpha\ln p,q) \qquad (25.4.7)$$

$$L_{xy}\{\int_0^x\int_0^y \frac{(x-\xi)^{\alpha\xi-1}(y-\eta)^{\beta\eta-1}}{\Gamma(\alpha\xi)\Gamma(\beta\eta)} f(\xi,\eta)d\xi d\eta\} = F(p+\alpha\ln p,q+\beta\ln q) \qquad (26.4.7)$$

$$L_{xy}\{\int_0^\infty\int_0^\infty \frac{x^{\alpha\xi-1}y^{\beta\eta-1}}{\Gamma(\alpha\xi)\Gamma(\beta\eta)} f(\xi,\eta)d\xi d\eta\} = F(\alpha\ln p,\beta\ln q) \qquad (27.4.7)$$

$$\alpha \neq 0 \quad , \quad \beta \neq 0$$

and

$$L_{xy}\{\frac{y^{\alpha x}}{\Gamma(\alpha x+1)} f(x)\} = \frac{1}{q}F(p+\alpha\ln q) \qquad (28.4.7)$$

$$L_{xy}\{\int_0^x \frac{y^{\alpha\xi}}{\Gamma(\alpha\xi+1)} f(\xi)d\xi = \frac{1}{pq}F(p+\alpha\ln q) \qquad (29.4.7)$$

$$L_{xy}\{\int_0^x \frac{(x-\xi)^{\alpha\xi}}{\Gamma(\alpha\xi+1)} f(\xi,y)d\xi\} = \frac{1}{pq}F(p+\alpha\ln q,q) \qquad (30.4.7)$$

$$L_{xy}\{\int_0^\infty \frac{x^{\alpha\xi}}{\Gamma(\alpha\xi+1)} f(\xi,y)d\xi = \frac{1}{p}F(\alpha\ln p,q) \qquad (31.4.7)$$

$$L_{xy}\{\int_0^x\int_0^y \frac{(x-\xi)^{\alpha\xi}(y-\eta)^{\beta\eta}}{\Gamma(\alpha\xi+1)\Gamma(\beta\eta+1)} f(\xi,\eta)d\xi d\eta\} = \frac{1}{pq}F(p+\alpha\ln p, q+\beta\ln q) \qquad (32.4.7)$$

$$L_{xy}\{\int_0^\infty\int_0^\infty \frac{x^{\alpha\xi}y^{\beta\eta}}{\Gamma(\alpha\xi+1)\Gamma(\beta\eta+1)} f(\xi,\eta)d\xi d\eta\} = \frac{1}{pq}F(\alpha\ln p, \beta\ln q) \qquad (33.4.7)$$

$$\alpha \neq 0 \quad , \quad \beta \neq 0$$

These formulas are used in a variety of cases starting with integrals of the Volterra functions

$$L_{xy}\{\int_0^\infty \nu(x,\xi)d\xi\} = \frac{1}{pq(\ln p)^2} \qquad (34.4.7)$$

$$L_{xy}\{\int_0^\infty \mu(y,\beta\xi-1,\alpha\xi-1)d\xi\} = \frac{1}{p\ln[q^\alpha(\ln q)^\beta]} \qquad (35.4.7)$$

$$L_{xy}\{\int_0^\infty \mu(y,\beta\xi-1,\alpha\xi)d\xi\} = \frac{1}{pq\ln[q^\alpha(\ln q)^\beta]} \qquad (36.4.7)$$

$$L_{xy}\{\int_0^\infty \mu(y,\beta\xi-1,\alpha\xi-1)\nu(x,\xi)d\xi\} = \frac{1}{p\ln p\ln[pq^\alpha(\ln q)^\beta]} \qquad (37.4.7)$$

$$L_{xy}\{\int_0^\infty \mu(y,\beta\xi-1,\alpha\xi)\nu(x,\xi)d\xi\} = \frac{1}{pq\ln p\ln[pq^\alpha(\ln q)^\beta]} \qquad (38.4.7)$$

$$L_{xy}\{\int_0^\infty\int_0^\infty \nu(x,\xi)\nu(y,\eta)d\xi d\eta\} = \frac{1}{pq(\ln p\ln q)^2} \qquad (39.4.7)$$

$$L_{xy}\{\int_0^\infty\int_0^\infty \nu(x,\xi)\nu(y,\eta)\,\nu(\xi,\eta)d\xi d\eta\} = \frac{1}{pq\ln q(\ln p)^2\ln(\ln p^q)} \qquad (40.4.7)$$

Combinations of the Volterra functions and the reciprocal gamma functions form the next group of two-dimensional Laplace transforms

$$L_{xy}\{\frac{y^{\beta x-1}}{\Gamma(\beta x)}\nu(x,\alpha)\}=\frac{1}{(p+\ln q^{\beta})^{\alpha+1}\ln(p+\ln q^{\beta})} \tag{41.4.7}$$

$$L_{xy}\{\frac{y^{\beta x}}{\Gamma(\beta x+1)}\nu(x,\alpha)\}=\frac{1}{q(p+\ln q^{\beta})^{\alpha+1}\ln(p+\ln q^{\beta})} \tag{42.4.7}$$

$$L_{xy}\{\int_0^x\frac{\nu(x-\xi,y)}{\Gamma(\xi+1)}d\xi\}=\frac{\nu(e^{-p})}{p\ln p(q+\ln p)} \tag{43.4.7}$$

$$L_{xy}\{\int_0^\infty\frac{x^{\beta\xi-1}}{\Gamma(\beta\xi)}\nu(\xi,y)d\xi\}=\frac{1}{\ln p^{\beta}\ln(\ln p^{\beta})[q+\ln(\ln p^{\beta})]} \tag{44.4.7}$$

$$L_{xy}\{\int_0^\infty\frac{x^{\beta\xi}}{\Gamma(\beta\xi+1)}\nu(\xi,y)d\xi\}=\frac{1}{p\ln p^{\beta}\ln(\ln p^{\beta})[q+\ln(\ln p^{\beta})]} \tag{45.4.7}$$

and

$$L_{xy}\{\int_0^\infty\frac{\nu(y,\xi+\alpha)}{\Gamma(x+\xi+1)}d\xi\}=\frac{\nu(e^{-p})-\nu(\frac{1}{q})}{q^{\alpha+1}\ln q(\ln q-p)} \tag{46.4.7}$$

$$L_{xy}\{\int_0^\infty\int_0^\infty\frac{\nu(x,\xi)\nu(y,\eta)}{\Gamma(\xi+\eta+1)}d\xi d\eta\}=\frac{\nu(\frac{1}{p})-\nu(\frac{1}{q})}{pq\ln p\ln q(\ln q/p)} \tag{47.4.7}$$

$$L_{xy}\{\int_0^\infty\int_0^\infty\frac{x^{\alpha\xi-1}y^{\beta\eta-1}}{\Gamma(\alpha\xi)\Gamma(\beta\eta)}\nu(\xi,\eta)d\xi d\eta\}=\frac{1}{\ln p^{\alpha}\ln(\ln p^{\alpha})\ln[q^{\beta}\ln(\ln p^{\alpha})]} \tag{48.4.7}$$

$$L_{xy}\{\int_0^\infty\int_0^\infty\frac{x^{\alpha\xi}y^{\beta\eta}}{\Gamma(\alpha\xi+1)\Gamma(\beta\eta+1)}\nu(\xi,\eta)d\xi d\eta\}=\frac{1}{pq\ln p^{\alpha}\ln(\ln p^{\alpha})\ln[q^{\beta}\ln(\ln p^{\alpha})]} \tag{49.4.7}$$

If the Laplace transforms $F(s)$ of functions $f(t)$ are known and the Laplace transforms of functions $g(u,w)$ and $h(u,w)$ can be written in the form

$$\int_0^\infty e^{-pu} g(u,w)du = A(p)e^{-B(p)w} \tag{50.4.7}$$

$$\int_0^\infty e^{-qw} h(u,v)dw = C(q)e^{-D(q)u} \tag{51.4.7}$$

then the following two-dimensional Laplace transforms of products of functions or infinite integrals of these products can be evaluated from

$$L_{xy}\{g(x,y)f(y)\} = A(p)F[q + B(p)] \tag{52.4.7}$$

$$L_{xy}\{h(x,y)f(x)\} = C(q)F[p + D(q)] \tag{53.4.7}$$

$$L_{xy}\{\int_0^\infty g(x,\xi)h(y,\xi)f(\xi)d\xi\} = A(p)C(q)F[B(p)+D(q)] \tag{54.4.7}$$

$$L_{xy}\{\int_0^\infty\int_0^\infty g(x,\xi)h(y,\eta)f(\xi,\eta)d\xi d\eta\} = A(p)C(q)F[B(p),D(q)] \tag{55.4.7}$$

Obviously, the last formula desires a knowledge of $F(p,q)$, the two-dimensional Laplace transform of $f(x,y)$. There is a number of elementary and special functions of two variables x and y that satisfy (50.4.7) or (51.4.7) and therefore it is possible to obtain many transforms of products which contain the Volterra functions. We will systematically consider them starting with [88, 1.5.28, 1.5.30]

$$L_y\{\frac{1}{\sqrt{y}} e^{-x^2/4y}\} = \sqrt{\frac{\pi}{q}} e^{-x\sqrt{q}} \tag{56.4.7}$$

$$L_y\{\frac{1}{y^{3/2}} e^{-x^2/4y}\} = \frac{2\sqrt{\pi}}{x} e^{-x\sqrt{q}} \tag{57.4.7}$$

This leads together with other operational rules to

$$L_{xy}\{\frac{1}{\sqrt{y}} e^{-x^2/4y} \nu(x,\alpha)\} = \frac{\sqrt{\pi}}{\sqrt{q}(p+\sqrt{q})^{\alpha+1}\ln(p+\sqrt{q})} \tag{58.4.7}$$

$$L_{xy}\{\frac{x}{y^{3/2}}e^{-x^2/4y}\nu(x,\alpha)\} = \frac{2\sqrt{\pi}}{(p+\sqrt{q})^{\alpha+1}\ln(p+\sqrt{q})} \quad (59.4.7)$$

$$L_{xy}\{\int_0^x \frac{e^{-\xi^2/4y}}{\sqrt{y}}\nu(x-\xi,\alpha)d\xi\} = \frac{\sqrt{\pi}}{\sqrt{q}p^{\alpha+1}(p+\sqrt{q})\ln p} \quad (60.4.7)$$

$$L_{xy}\{x\int_0^y \frac{e^{-x^2/4\xi}}{\xi^{3/2}}\nu(y-\xi,\alpha)d\xi\} = \frac{2\sqrt{\pi}}{q^{\alpha+1}(p+\sqrt{q})\ln q} \quad (61.4.7)$$

and to transforms of integrals of the Volterra functions

$$L_{xy}\{\int_0^\infty \frac{e^{-\xi^2/4x}}{\sqrt{x}}\nu(\xi,y)d\xi\} = \frac{\sqrt{\pi}}{p(q+\ln\sqrt{p})\ln\sqrt{p}} \quad (62.4.7)$$

$$L_{xy}\{\int_0^\infty \frac{e^{-\xi^2/4y}}{\sqrt{y}}\nu(x+\xi,\alpha)d\xi\} = \frac{\sqrt{\pi}}{\sqrt{q}(\sqrt{q}-p)}\{\frac{1}{p^{\alpha+1}\ln p} - \frac{1}{q^{(\alpha+1)/2}\ln\sqrt{q}}\} \quad (63.4.7)$$

$$L_{xy}\{\int_0^\infty \frac{e^{-(x+\xi)^2/4y}}{\sqrt{y}}\nu(\xi,\alpha)d\xi\} = \frac{\sqrt{\pi}}{q^{1+\alpha/2}(p+\sqrt{q})\ln\sqrt{q}} \quad (64.4.7)$$

$$L_{xy}\{\int_0^\infty \frac{e^{-\xi^2/4x}e^{-\xi^2/4y}}{\sqrt{xy}}\nu(\xi,\alpha)d\xi\} = \frac{\pi}{\sqrt{pq}(\sqrt{p}+\sqrt{q})^{\alpha+1}\ln(\sqrt{p}+\sqrt{q})} \quad (65.4.7)$$

$$L_{xy}\{\int_0^\infty \int_0^\infty \frac{\xi e^{-\xi^2/4x}}{x^{3/2}}\nu(\xi,y)d\xi\} = \frac{2\sqrt{\pi}}{\sqrt{p}\ln\sqrt{p}(q+\ln\sqrt{p})} \quad (66.4.7)$$

$$L_{xy}\{\int_0^\infty \frac{(x+\xi)e^{-(x+\xi)^2/4y}}{y^{3/2}}\nu(\xi,\alpha)d\xi\} = \frac{2\sqrt{\pi}}{q^{(\alpha+1)/2}(p+\sqrt{q})\ln\sqrt{q}} \quad (67.4.7)$$

In the case of double infinite integrals of the Volterra functions we have

$$L_{xy}\{\int_0^\infty\int_0^\infty \frac{e^{-\xi^2/4x}e^{-\eta^2/4y}}{\sqrt{xy}}\nu(\xi,\eta)d\xi d\eta\} = \frac{\pi}{p\sqrt{q}\ln\sqrt{p}(\sqrt{q}+\ln\sqrt{p})} \tag{68.4.7}$$

$$L_{xy}\{\int_0^\infty\int_0^\infty \frac{\xi\eta e^{-\xi^2/4x}e^{-\eta^2/4y}}{(xy)^{3/2}}\nu(\xi,\eta)d\xi d\eta\} = \frac{4\pi}{\sqrt{p}\ln\sqrt{p}(\sqrt{q}+\ln\sqrt{p})} \tag{69.4.7}$$

$$L_{xy}\{\int_0^\infty\int_0^\infty \frac{e^{-\xi^2/4x}e^{-\eta^2/4y}}{\sqrt{xy}}\nu(\xi+\eta)d\xi d\eta\} = \frac{\pi}{\sqrt{pq}(\sqrt{q}-\sqrt{p})}\{\frac{1}{\sqrt{p}\ln\sqrt{q}} - \frac{1}{\sqrt{q}\ln\sqrt{q}}\} \tag{70.4.7}$$

$$L_{xy}\{\int_0^\infty\int_0^\infty \frac{\xi\eta e^{-\xi^2/4x}e^{-\eta^2/4y}}{(xy)^{3/2}}\nu(\xi+\eta)d\xi d\eta\} = \frac{4\pi}{\sqrt{q}-\sqrt{p}}\{\frac{1}{\sqrt{p}\ln\sqrt{p}} - \frac{1}{\sqrt{q}\ln\sqrt{q}}\} \tag{71.4.7}$$

Trigonometric functions satisfy [88, 1.7.74, 1.7.80]

$$L_x\{\sin(2\sqrt{ayx})\} = \frac{\sqrt{\pi a y}}{p^{3/2}}e^{-ay/p} \tag{72.4.7}$$

$$L_x\{\frac{1}{\sqrt{x}}\cos(2\sqrt{ayx})\} = \sqrt{\frac{\pi}{p}}e^{-ay/p} \tag{73.4.7}$$

and therefore from (52.4.7)-(55.4.7) it follows that

$$L_{xy}\{\frac{1}{\sqrt{x}}\sin(2\sqrt{axy})\nu(x,\alpha)\} = \frac{\sqrt{\pi a}}{q^{3/2}(p+\frac{a}{q})^{\alpha+1}\ln(p+\frac{a}{q})} \tag{74.4.7}$$

$$L_{xy}\{\frac{1}{\sqrt{y}}\cos(2\sqrt{axy})\nu(x,\alpha)\} = \frac{\sqrt{\pi}}{\sqrt{q}(p+\frac{a}{q})^{\alpha+1}\ln(p+\frac{a}{q})} \tag{75.4.7}$$

$$L_{xy}\{\int_0^\infty \frac{1}{\sqrt{\xi}}\sin(2\sqrt{ax\xi})\nu(\xi,y)d\xi\} = \frac{\sqrt{\pi}}{\sqrt{ap}\ln(\frac{a}{p})[q+\ln(\frac{a}{p})]} \tag{76.4.7}$$

$$L_{xy}\{\int_0^\infty \frac{1}{\sqrt{x}}\cos(2\sqrt{ax\xi})\nu(\xi,y)d\xi\} = \frac{\sqrt{\pi p}}{a\ln(\frac{a}{p})[q+\ln(\frac{a}{p})]} \quad (77.4.7)$$

$$L_{xy}\{\int_0^\infty\int_0^\infty \frac{1}{\sqrt{\xi\eta}}\sin(2\sqrt{ax\xi})\sin(2\sqrt{by\eta})\nu(\xi,\eta)d\xi d\eta\} = \frac{\pi\sqrt{b}}{\sqrt{ap}q^{3/2}\ln(\frac{a}{p})[\frac{b}{q}+\ln(\frac{a}{p})]} \quad (78.4.7)$$

$$L_{xy}\{\int_0^\infty\int_0^\infty \frac{1}{\sqrt{xy}}\cos(2\sqrt{ax\xi})\cos(2\sqrt{by\eta})\nu(\xi,\eta)d\xi d\eta\} = \frac{\pi\sqrt{p}}{a\sqrt{q}\ln(\frac{a}{p})[\frac{b}{q}+\ln(\frac{a}{p})]} \quad (79.4.7)$$

$$L_{xy}\{\int_0^\infty\int_0^\infty \frac{1}{\sqrt{y\xi}}\sin(2\sqrt{ax\xi})\cos(2\sqrt{by\eta})\nu(\xi,\eta)d\xi d\eta\} = \frac{\pi}{\sqrt{apq}\ln(\frac{a}{p})[\frac{b}{q}+\ln(\frac{a}{p})]} \quad (80.4.7)$$

Similarly, the two-dimensional Laplace transforms of the Volterra functions with the hyperbolic functions are

$$L_{xy}\{\frac{1}{\sqrt{x}}\sinh(2\sqrt{axy})\nu(x,\alpha)\} = \frac{\sqrt{\pi a}}{q^{3/2}(p-\frac{a}{q})^{\alpha+1}\ln(p-\frac{a}{q})} \quad (81.4.7)$$

$$L_{xy}\{\frac{1}{\sqrt{y}}\cosh(2\sqrt{axy})\nu(x,\alpha)\} = \frac{\sqrt{\pi}}{\sqrt{q}(p-\frac{a}{q})^{\alpha+1}\ln(p-\frac{a}{q})} \quad (82.4.7)$$

$$L_{xy}\{\frac{1}{\sqrt{y}}\sinh^2(2\sqrt{axy})\nu(x,\alpha)\} = \sqrt{\frac{\pi}{4q}}\{\frac{1}{(p-\frac{a}{q})^{\alpha+1}\ln(p-\frac{a}{q})} - \frac{1}{p^{\alpha+1}\ln p}\} \quad (83.4.7)$$

$$L_{xy}\{\frac{1}{\sqrt{y}}\cosh^2(2\sqrt{axy})\nu(x,\alpha)\} = \sqrt{\frac{\pi}{4q}}\{\frac{1}{(p-\frac{a}{q})^{\alpha+1}\ln(p-\frac{a}{q})} + \frac{1}{p^{\alpha+1}\ln p}\} \quad (84.4.7)$$

The two-dimensional Laplace transforms of products of the Volterra functions and special functions (error functions, Hermite polynomials, parabolic cylinder functions and Bessel functions) are evaluated in the same way as elementary functions based on the exponential form of Laplace transforms [88, 1.11.20, 1.17.66,67, 85, 102, 16.2.5]

$$L_y\{erfc(\frac{x}{2\sqrt{y}})\} = \frac{1}{q}e^{-x\sqrt{q}} \quad (85.4.7)$$

$$L_y\{e^{-ax}erfc(\frac{x}{2\sqrt{y}} - a\sqrt{y}) + e^{ax}erfc(\frac{x}{2\sqrt{y}} + a\sqrt{y})\} = \frac{2}{q}e^{-x\sqrt{q+a^2}} \quad (86.4.7)$$

$$L_y\{\frac{e^{-a^2x^2/4y}}{y^{(n+1)/2}}He_n(\frac{ax}{\sqrt{2y}})\} = 2^{n/2}\sqrt{\pi}q^{(n-1)/2}e^{-a\sqrt{q}} \quad (87.4.7)$$

$$L_y\{\frac{e^{-x^2/4y}}{y^{\lambda+1}}D_{2\lambda+1}(\frac{x}{\sqrt{y}})\} = 2^{\lambda+1/2}\sqrt{\pi}q^{\lambda}e^{-a\sqrt{2q}} \quad (88.4.7)$$

In the case of the error functions we have

$$L_{xy}\{erf(\frac{x}{2\sqrt{y}})\nu(x,\alpha\} = \frac{1}{q}\{\frac{1}{p^{\alpha+1}\ln p} - \frac{1}{(p+\sqrt{q})^{\alpha+1}\ln(p+\sqrt{q})}\} \quad (89.4.7)$$

$$L_{xy}\{erfc(\frac{x}{2\sqrt{y}})\nu(x,\alpha\} = \frac{1}{q(p+\sqrt{q})^{\alpha+1}\ln(p+\sqrt{q})} \quad (90.4.7)$$

$$L_{xy}\{[e^{-ax}erfc(\frac{x}{2\sqrt{y}}-a\sqrt{y})+e^{ax}erfc(\frac{x}{2\sqrt{y}}+a\sqrt{y})]\nu(x,\alpha\}=$$
$$\frac{2}{q(p+\sqrt{q+a^2})^{\alpha+1}\ln(p+\sqrt{q+a^2})} \qquad (91.4.7)$$

$$L_{xy}\{\int_0^\infty\int_0^\infty erfc(\frac{\xi+\eta}{2\sqrt{x}})\nu(\xi,\eta)d\xi d\eta\}=\frac{1}{p^{3/2}q(\sqrt{p}+\ln\sqrt{p})\ln\sqrt{p}} \qquad (92.4.7)$$

and for the Hermite polynomials

$$L_{xy}\{\frac{e^{-a^2x^2/4y}}{y^{(n+1)/2}}He_n(\frac{ax}{\sqrt{2y}})\nu(x,\alpha)\}=\frac{2^{n/2}\sqrt{\pi}q^{(n-1)/2}}{(p+a\sqrt{q})^{\alpha+1}\ln(p+a\sqrt{q})} \qquad (93.4.7)$$

$$L_{xy}\{\int_0^\infty\frac{e^{-a^2\xi^2/4y}}{y^{(n+1)/2}}He_n(\frac{a\xi}{\sqrt{2y}})\nu(\xi,x)d\xi\}=$$
$$\frac{2^{n/2}\sqrt{\pi}q^{(n-1)/2}}{a\sqrt{q}\ln(a\sqrt{q})[p+\ln(a\sqrt{q})]} \qquad (94.4.7)$$

$$L_{xy}\{\int_0^\infty\int_0^\infty\frac{e^{-a^2\xi^2/4x}e^{-b^2\eta^2/4y}}{x^{(n+1)/2}y^{(m+1)/2}}He_n(\frac{a\xi}{\sqrt{2x}})He_m(\frac{b\eta}{\sqrt{2y}})\nu(\xi,\eta)d\xi d\eta\}=$$
$$\frac{2^{(m+n)/2}\pi p^{(n-1)/2}q^{(m-1)/2}}{a\sqrt{p}\ln(a\sqrt{p})[b\sqrt{q}+\ln(a\sqrt{p})]} \qquad (95.4.7)$$

The next group of transforms are those of the parabolic cylinder functions

$$L_{xy}\{\frac{e^{-x^2/4y}}{y^{\lambda+1}}D_{2\lambda+1}(\frac{x}{\sqrt{y}})\nu(x,\alpha)\}=\frac{2^{\lambda+1/2}\sqrt{\pi}q^{\lambda}}{(p+\sqrt{2q})^{\alpha+1}\ln(p+\sqrt{2q})} \qquad (96.4.7)$$

$$L_{xy}\{\int_0^\infty\frac{e^{-(x+\xi)^2/4y}}{y^{\lambda+1}}D_{2\lambda+1}(\frac{x+\xi}{\sqrt{y}})\nu(\xi,x)d\xi\}=$$
$$\frac{2^{\lambda}\sqrt{\pi}q^{\lambda-1/2}}{\ln(\sqrt{2q})[p+\sqrt{2q}+\ln\sqrt{2q}]} \qquad (97.4.7)$$

$$L_{xy}\{\int_0^\infty\int_0^\infty \frac{e^{-\xi^2/4x}e^{-\eta^2/4y}}{x^{\lambda+1}y^{\sigma+1}} D_{2\lambda+1}(\frac{\xi}{\sqrt{x}}) D_{2\sigma+1}(\frac{\eta}{\sqrt{y}}) \nu(\xi,\eta)d\xi d\eta\} = \frac{2^{\lambda+\sigma+1/2}\pi p^{\lambda-1/2}q^{\sigma}}{\ln(\sqrt{2p})[\sqrt{2q}+\ln(\sqrt{2p})]} \quad (98.4.7)$$

There is a large number of two-dimensional Laplace transforms which are associated with the Bessel functions. They are based on [102, 10.2.8, 10.2.10, 12.2.1. 12.2.10]

$$L_y\{\frac{1}{\sqrt{y}} J_1(2\sqrt{axy})\} = \frac{1}{\sqrt{ax}}(1-e^{-ax/q}) \quad (99.4.7)$$

$$L_y\{\frac{1}{\sqrt{y}} I_1(2\sqrt{axy})\} = \frac{1}{\sqrt{ax}}(e^{ax/q}-1) \quad (100.4.7)$$

$$L_y\{y^{\lambda/2} J_\lambda(2\sqrt{axy})\} = \frac{(ax)^{\lambda/2}e^{-ax/q}}{q^{\lambda+1}} \quad (101.4.7)$$

$$L_y\{y^{\lambda/2} I_\lambda(2\sqrt{axy})\} = \frac{(ax)^{\lambda/2}e^{ax/q}}{q^{\lambda+1}} \quad (102.4.7)$$

$\mathrm{Re}\,\lambda > -1$

and therefore from (52.4.7) - (53.4.7)

$$L_{xy}\{\sqrt{\frac{x}{y}} J_1(2\sqrt{axy})\nu(x,\alpha)\} = \frac{1}{\sqrt{a}}\{\frac{1}{p^{\alpha+1}\ln p} - \frac{1}{(p+\frac{a}{q})^{\alpha+1}\ln(p+\frac{a}{q})}\} \quad (103.4.7)$$

$$L_{xy}\{\sqrt{\frac{x}{y}} I_1(2\sqrt{axy})\nu(x,\alpha)\} = \frac{1}{\sqrt{a}}\{\frac{1}{(p-\frac{a}{q})^{\alpha+1}\ln(p-\frac{a}{q})} - \frac{1}{p^{\alpha+1}\ln p}\} \quad (104.4.7)$$

$$L_{xy}\{(\frac{y}{x})^{\lambda/2} J_\lambda(2\sqrt{axy})\nu(x,\alpha)\} = \frac{a^{\lambda/2}}{q^{\lambda+1}(p+\frac{a}{q})^{\alpha+1}\ln(p+\frac{a}{q})} \quad (105.4.7)$$

$$L_{xy}\{(\frac{y}{x})^{\lambda/2} I_\lambda(2\sqrt{axy})\nu(x,\alpha)\} = \frac{a^{\lambda/2}}{q^{\lambda+1}(p-\frac{a}{q})^{\alpha+1}\ln(p-\frac{a}{q})} \qquad (106.4.7)$$

and using [88, 2.5.77-78, 102, 13.2.5-7]

$$L_{xy}\{J_0(2\sqrt{axy})\ln(\frac{y}{ax}) + \pi Y_0(2\sqrt{axy})]\nu(x,\alpha)\} = \frac{2\ln(1/q)}{q(p+\frac{a}{q})^{\alpha+1}\ln(p+\frac{a}{q})} \qquad (107.4.7)$$

$$L_{xy}\{K_0(2\sqrt{axy}) - \ln(\frac{y}{ax})I_0(2\sqrt{axy})]\nu(x,\alpha)\} = \frac{\ln q}{q(p-\frac{a}{q})^{\alpha+1}\ln(p-\frac{a}{q})} \qquad (108.4.7)$$

$$L_{xy}\{\sqrt{\frac{x}{y}}K_{1/3}(\frac{2x^{3/2}}{\sqrt{y}})\,\nu(x,\alpha)\} = \frac{\pi}{\sqrt{3}q^{2/3}(p+3q^{1/3})^{\alpha+1}\ln(p+3q^{1/3})} \qquad (109.4.7)$$

$$L_{xy}\{(\frac{x}{y})^{3/2}K_{1/3}(\frac{2x^{3/2}}{\sqrt{y}})\,\nu(x,\alpha)\} = \frac{\pi}{\sqrt{3}(p+3q^{1/3})^{\alpha+1}\ln(p+3q^{1/3})} \qquad (110.4.7)$$

$$L_{xy}\{(\frac{x}{y})K_{2/3}(\frac{2x^{3/2}}{\sqrt{y}})\,\nu(x,\alpha)\} = \frac{\pi}{\sqrt{3}q^{1/3}(p+3q^{1/3})^{\alpha+1}\ln(p+3q^{1/3})} \qquad (111.4.7)$$

In the case of infinite integrals of the Bessel functions, (54.4.7) - (55.4.7) give

$$L_{xy}\{\int_0^\infty J_0(2\sqrt{ay\xi})\nu(x+\xi,\alpha)d\xi\} = \frac{1}{(a-pq)}\{\frac{1}{p^{\alpha+1}\ln p} - \frac{1}{(\frac{a}{q})^{\alpha+1}\ln(\frac{a}{q})}\} \tag{112.4.7}$$

$$L_{xy}\{\int_0^\infty (\frac{x}{\xi})^{\lambda/2} J_\lambda(2\sqrt{ax\xi})\nu(\xi,y)d\xi\} = \frac{1}{ap^\lambda \ln(\frac{a}{p})[q+\ln(\frac{a}{p})]} \tag{113.4.7}$$

$$L_{xy}\{\int_0^\infty J_0(2\sqrt{ax\xi})J_0(2\sqrt{by\eta})\nu(\xi,\alpha)d\xi\} = \frac{1}{pq(\frac{a}{p}+\frac{b}{q})^{\alpha+1}\ln(\frac{a}{p}+\frac{b}{q})} \tag{114.4.7}$$

$$L_{xy}\{\int_0^\infty\int_0^\infty J_\lambda(2\sqrt{ax\xi})J_\sigma(2\sqrt{by\eta})\nu(\xi,\eta)d\xi d\eta\} = \frac{a^{\lambda/2}b^{\sigma/2}}{ap^\lambda q^{\sigma+1}\ln(\frac{a}{p})[\frac{b}{q}+\ln(\frac{a}{p})]} \tag{115.4.7}$$

Now, we will consider the two-dimensional Laplace transforms of the reciprocal gamma functions with elementary and special functions which are expressed by the Volterra functions

$$L_{xy}\{\frac{y^{\beta x-1}}{\Gamma(\beta x)\Gamma(x+\alpha+1)}\} = e^{\alpha p}q^{\alpha\beta}\nu(\frac{e^{-p}}{q^\beta},\alpha) \tag{116.4.7}$$

$$L_{xy}\{\frac{y^{\beta x}}{\Gamma(\beta x+1)\Gamma(x+\alpha+1)}\} = e^{\alpha p}q^{\alpha\beta-1}\nu(\frac{e^{-p}}{q^\beta},\alpha) \tag{117.4.7}$$

$$L_{xy}\{\int_0^x \frac{y^\xi}{\Gamma(\xi+1)^2}d\xi\} = \frac{1}{pq}\nu(\frac{e^{-p}}{q}) \tag{118.4.7}$$

$$L_{xy}\{\int_0^\infty \frac{x^{\alpha\xi}}{\Gamma(\alpha\xi+1)\Gamma(\xi+y+1)}d\xi\} = \frac{1}{p(q-\alpha\ln p)}\{\nu(\frac{1}{p^\alpha})-\nu(e^{-q})\} \quad (119.4.7)$$

$$L_{xy}\{\frac{y^{\beta x}}{\Gamma(\beta x+1)}\int_0^\infty \frac{x^{\xi}}{\Gamma(\xi+1)\Gamma(\xi+\alpha+1)}d\xi\} = \frac{e^{\alpha p}q^{\alpha\beta}}{p+\ln q^\beta}\nu(\frac{e^{-p}}{q^\beta},\alpha) \quad (120.4.7)$$

$$L_{xy}\{\int_0^\infty\int_0^\infty \frac{x^{\alpha\xi}y^{\beta\eta}}{\Gamma(\alpha\xi+1)\Gamma(\beta\eta+1)\Gamma(\xi+\eta+1)}d\xi d\eta\} = \frac{1}{pq\ln(q^\beta/p^\alpha)}\{\nu(\frac{1}{p^\alpha})-\nu(\frac{1}{q^\beta})\} \quad (121.4.7)$$

Examples of the exponential, trigonometric and hyperbolic functions are presented below

$$L_{xy}\{\frac{e^{-x^2/4y}}{\sqrt{y}\Gamma(x+\alpha+1)}\} = \sqrt{\frac{\pi}{q}}e^{\alpha(p+\sqrt{q})}\nu(e^{-(p+\sqrt{q})},\alpha) \quad (122.4.7)$$

$$L_{xy}\{\frac{e^{-x^2/4y}}{y^{3/2}\Gamma(x+\alpha+1)}\} = 2\sqrt{\pi}e^{\alpha(p+\sqrt{q})}\nu(e^{-(p+\sqrt{q})},\alpha) \quad (123.4.7)$$

$$L_{xy}\{\frac{\sin(2\sqrt{axy})}{\sqrt{x}\Gamma(x+\alpha+1)}\} = \frac{\sqrt{\pi a}}{q^{3/2}}e^{\alpha(p+a/q)}\nu(e^{-(p+a/q)},\alpha) \quad (124.4.7)$$

$$L_{xy}\{\frac{\cos(2\sqrt{axy})}{\sqrt{y}\Gamma(x+\alpha+1)}\} = \sqrt{\frac{\pi}{q}}e^{\alpha(p+a/q)}\nu(e^{-(p+a/q)},\alpha) \quad (125.4.7)$$

$$L_{xy}\{\frac{\sinh(2\sqrt{axy})}{\sqrt{x}\Gamma(x+\alpha+1)}\} = \frac{\sqrt{\pi a}}{q^{3/2}}e^{\alpha(p-a/q)}\nu(e^{-(p-a/q)},\alpha) \quad (126.4.7)$$

$$L_{xy}\{\frac{\cosh(2\sqrt{axy})}{\sqrt{y}\Gamma(x+\alpha+1)}\} = \sqrt{\frac{\pi}{q}}e^{\alpha(p-a/q)}\nu(e^{-(p-a/q)},\alpha) \quad (127.4.7)$$

The two-dimensional Laplace transforms of infinite integrals of the exponential and trigonometric functions are

$$L_{xy}\{\int_0^\infty \frac{e^{-\xi^2/4y}}{\sqrt{y}\Gamma(\xi+x+1)}d\xi\} = \frac{\sqrt{\pi}}{q-p\sqrt{q}}\{\nu(e^{-p})-\nu(e^{-\sqrt{q}})\} \tag{128.4.7}$$

$$L_{xy}\{\int_0^\infty \frac{\xi e^{-\xi^2/4y}}{y^{3/2}\Gamma(\xi+x+1)}d\xi\} = \frac{2\sqrt{\pi}}{\sqrt{q}-p}\{\nu(e^{-p})-\nu(e^{-\sqrt{q}})\} \tag{129.4.7}$$

$$L_{xy}\{\int_0^\infty \frac{\sin(2\sqrt{ax\xi})}{\sqrt{\xi}\Gamma(\xi+y+1)}d\xi\} = \frac{\sqrt{\pi a}}{\sqrt{p}(pq-a)}\{\nu(e^{-a/p})-\nu(e^{-q})\} \tag{130.4.7}$$

$$L_{xy}\{\int_0^\infty \frac{\cos(2\sqrt{ax\xi})}{\sqrt{x}\Gamma(\xi+y+1)}d\xi\} = \frac{\sqrt{\pi p}}{(pq-a)}\{\nu(e^{-a/p})-\nu(e^{-q})\} \tag{131.4.7}$$

and

$$L_{xy}\{\int_0^\infty\int_0^\infty \frac{e^{-\xi^2/4x}e^{-\eta^2/4y}}{\sqrt{xy}\Gamma(\xi+\eta+1)}d\xi d\eta\} = \frac{\pi}{q\sqrt{p}-p\sqrt{q}}\{\nu(e^{-\sqrt{p}})-\nu(e^{-\sqrt{q}})\} \tag{132.4.7}$$

$$L_{xy}\{\int_0^\infty\int_0^\infty \frac{\xi\eta e^{-\xi^2/4x}e^{-\eta^2/4y}}{(xy)^{3/2}\Gamma(\xi+\eta+1)}d\xi d\eta\} = \frac{\pi}{\sqrt{q}-\sqrt{p}}\{\nu(e^{-\sqrt{p}})-\nu(e^{-\sqrt{q}})\} \tag{133.4.7}$$

$$L_{xy}\{\int_0^\infty\int_0^\infty \frac{\sin(2\sqrt{ax\xi})\sin(2\sqrt{by\eta})}{\sqrt{\xi\eta}\Gamma(\xi+\eta+1)}d\xi d\eta\} = \frac{\pi\sqrt{ab}}{\sqrt{pq}(bp-aq)}\{\nu(e^{-a/p})-\nu(e^{-b/q})\} \tag{134.4.7}$$

$$L_{xy}\{\int_0^\infty\int_0^\infty \frac{\cos(2\sqrt{ax\xi})\cos(2\sqrt{by\eta})}{\sqrt{xy}\Gamma(\xi+\eta+1)}d\xi d\eta\} = \frac{\pi\sqrt{pq}}{(bp-aq)}\{\nu(e^{-a/p})-\nu(e^{-b/q})\} \tag{135.4.7}$$

$$L_{xy}\{\int_0^\infty\int_0^\infty \frac{\sin(2\sqrt{ax\xi})\cos(2\sqrt{by\eta})}{\sqrt{\xi y}\Gamma(\xi+\eta+1)}d\xi d\eta\} = \frac{\pi\sqrt{aq}}{\sqrt{p}(bp-aq)}\{\nu(e^{-a/p})-\nu(e^{-b/q})\} \quad (136.4.7)$$

In the case of special functions, the two-dimensional Laplace transforms of the error and parabolic cylinder functions with the reciprocal gamma functions are

$$L_{xy}\{erfc(\frac{x}{2\sqrt{y}})\frac{1}{\Gamma(x+\alpha+1)}\} = \frac{1}{q}e^{\alpha(p+\sqrt{q})}\nu(e^{-(p+\sqrt{q})},\alpha) \quad (137.4.7)$$

$$L_{xy}\{\int_0^\infty erfc(\frac{\xi}{2\sqrt{y}})\frac{d\xi}{\Gamma(\xi+x+1)}\} = \frac{1}{q(\sqrt{q}-p)}\{\nu(e^{-p})-\nu(e^{-\sqrt{q}})\} \quad (138.4.7)$$

$$L_{xy}\{\int_0^\infty\int_0^\infty erfc(\frac{\xi}{2\sqrt{x}})erfc(\frac{\eta}{2\sqrt{y}})\frac{d\xi d\eta}{\Gamma(\xi+\eta+1)}\} = \frac{1}{pq(\sqrt{q}-\sqrt{p})}\{\nu(e^{-\sqrt{p}})-\nu(e^{-\sqrt{q}})\} \quad (139.4.7)$$

$$L_{xy}\{\frac{e^{-x^2/4y}}{y^{\lambda+1}}D_{2\lambda+1}(\frac{x}{\sqrt{y}})\frac{1}{\Gamma(x+\alpha+1)}\} = 2^{\lambda+1/2}\sqrt{\pi}\,q^{\lambda}e^{\alpha(p+\sqrt{2q})}\nu(e^{-(p+\sqrt{2q})},\alpha) \quad (140.4.7)$$

$$L_{xy}\{\int_0^\infty \frac{e^{-\xi^2/4y}}{y^{\lambda+1}}D_{2\lambda+1}(\frac{\xi}{\sqrt{y}})\frac{d\xi}{\Gamma(\xi+x+1)}\} = \frac{2^{\lambda+1/2}\sqrt{\pi}q^{\lambda}}{\sqrt{2q}-p}\{\nu(e^{-p})-\nu(e^{-\sqrt{2q}})\} \quad (141.4.7)$$

$$L_{xy}\{\int_0^\infty\int_0^\infty \frac{e^{-\xi^2/4x}e^{-\eta^2/4y}}{x^{\lambda+1}y^{\sigma+1}\Gamma(\xi+\eta+1)}D_{2\lambda+1}(\frac{\xi}{\sqrt{x}})D_{2\lambda+1}(\frac{\eta}{\sqrt{y}})d\xi d\eta\} = \frac{2^{\lambda+\sigma+1/2}\pi p^{\lambda}q^{\sigma}}{\sqrt{q}-\sqrt{p}}\{\nu(e^{-\sqrt{2p}})-\nu(e^{-\sqrt{2q}})\} \quad (142.4.7)$$

Finally in the case of the Bessel functions

$$L_{xy}\{\sqrt{\frac{x}{y}}J_1(2\sqrt{axy})\frac{1}{\Gamma(x+\alpha+1)}\}= \frac{1}{\sqrt{a}}\{e^{\alpha p}\nu(e^{-p},\alpha)-e^{\alpha(p+a/q)}\nu(e^{-(p+a/q)},\alpha)\} \quad (143.4.7)$$

$$L_{xy}\{(\frac{x}{y})^{\lambda/2}J_\lambda(2\sqrt{axy})\frac{1}{\Gamma(x+\alpha+1)}\}= \frac{a^{\lambda/2}}{q^{\lambda+1}}e^{\alpha(p+a/q)}\nu(e^{-(p+a/q)},\alpha) \quad (144.4.7)$$

$$L_{xy}\{\int_0^\infty(\frac{x}{\xi})^{\lambda/2}\frac{J_\lambda(2\sqrt{ax\xi})}{\Gamma(\xi+y+1)}d\xi\}= \frac{a^{\lambda/2}p}{q^{\lambda+1}(pq-a)}\{\nu(e^{-a/p})-\nu(e^{-q})\} \quad (145.4.7)$$

$$L_{xy}\{\int_0^\infty\int_0^\infty(\frac{x}{\xi})^{\lambda/2}(\frac{y}{\eta})^{\sigma/2}\frac{J_\lambda(2\sqrt{ax\xi})J_\lambda(2\sqrt{by\eta})}{\Gamma(\xi+\eta+1)}d\xi d\eta\}= \frac{a^{\lambda/2}b^{\sigma/2}}{p^\lambda q^\sigma(bp-aq)}\{\nu(e^{-a/p})-\nu(e^{-b/q})\} \quad (146.4.7)$$

Using (70.4.5) and (66.4.3) it is possible to obtain the two-dimensional Laplace transform of infinite integral with respect to the order of the Bessel function

$$L_{xy}\{\int_\alpha^\infty y^{\xi/2}\frac{J_\xi(2\sqrt{axy})}{\Gamma(\xi+1)}d\xi\}=\frac{\sqrt{\pi a}}{\sqrt{q}(pq+a)^{3/2}}\nu(\frac{a}{4q^2},\frac{\alpha-1}{2}) \quad (147.4.7)$$

The Laplace transform of the Ramanujan and Touchard integral representations of the Volterra functions (1.4.3)-(4.4.3), together with the operational rule (10.4.7) leads to

$$L_{xy}\{\frac{1}{(x+y)(\pi^2+[\ln(x+y)]^2)}\}=\frac{1}{q-p}\{e^p-e^q+\nu(q)-\nu(p)\} \quad (148.4.7)$$

$$L_{xy}\{\frac{\ln(x+y)}{\sqrt{x+y}(\pi^2+[\ln(x+y)]^2)}\}= \frac{\pi}{q-p}\{e^q-e^p+\nu(p,-\frac{1}{2})-\nu(q,-\frac{1}{2})\} \tag{149.4.7}$$

$$L_{xy}\{\tan^{-1}[\frac{\ln(x+y)}{\pi}]\}= \frac{\pi}{q-p}\{\frac{1}{p}[e^p-\frac{1}{2}-\nu(p)]-\frac{1}{q}[e^q-\frac{1}{2}-\nu(q)]\} \tag{150.4.7}$$

$$L_{xy}\{\tan^{-1}[\frac{\pi}{\ln(x+y)}]\}= \frac{\pi}{q-p}\{\frac{1}{p}[\nu(p)-2\sinh p]-\frac{1}{q}[\nu(q)-2\sinh q]\} \tag{151.4.7}$$

Similarly as in the case of the one-dimensional Laplace transforms, only the differences of the exponential and the Volterra functions in (148.4.7) - (151.4.7) are the two-dimensional Laplace transforms. These functions are not the Laplace transforms when considered separately.

Finally, using (31.4.1) and (10.4.7)

$$L_{xy}\{\nu(\alpha,x+y)\}= \frac{\pi}{q-p}\{\frac{1}{p}[\nu(\alpha)-\nu(\alpha e^{-p})]-\frac{1}{q}[\nu(\alpha)-\nu(\alpha e^{-q})]\} \tag{152.4.7}$$

The convolution of two functions $f(x,y)$ and $g(x,y)$ is defined by

$$(x,y)**g(x,y)=\int_0^x\int_0^y f(\xi,\eta)g(x-\xi,y-\eta)d\xi d\eta= \int_0^x\int_0^y f(x-\xi,y-\eta)g(\xi,\eta)d\xi d\eta \tag{153.4.7}$$

and *the product theorem of the two-dimensional Laplace transformation* is

$$L_{xy}\{f(x,y)**g(x,y)\}=F(p,q)\cdot G(p,q) \tag{154.4.7}$$

where $L_{xy}\{f(x,y)\}=F(p,q)$ and $L_{xy}\{g(x,y)\}=G(p,q)$.

Frequently, the convolutions performed with respect to x or y only are met and then (154.4.7) becomes

$$L_{xy}\{f(x,y)*g(x)\} = F(p,q)\cdot G(p) \tag{155.4.7}$$

$$L_{xy}\{f(x,y)*h(x)\} = F(p,q)\cdot H(q) \tag{156.4.7}$$

where $L_x\{g(x)\} = G(p)$ and $L_y\{h(y)\} = H(q)$.

The two-dimensional Laplace transform of the Volterra function $\nu(x,y+\alpha+1)$ is

$$L_{xy}\{\nu(x,y+\alpha+1)\} = \frac{1}{p^{\alpha+2}\ln p(q+\ln p)} = \frac{1}{p(q+\ln p)}\cdot\frac{1}{p^{\alpha+1}\ln p} \tag{157.4.7}$$

and therefore from (155.4.7)

$$\nu(x,y+\alpha+1) = L_{xy}{}^{-1}\{\frac{1}{p(q+\ln p)}\} * L_{xy}{}^{-1}\{\frac{1}{p^{\alpha+1}\ln p}\} \tag{158.4.7}$$

Using

$$L_{xy}{}^{-1}\{\frac{1}{p^{\lambda+1}(q+\ln p)^{\sigma+1}}\} = \frac{x^{y+\lambda}y^{\sigma}}{\Gamma(\sigma+1)\Gamma(y+\lambda+1)} \tag{159.4.7}$$

with $\lambda = 0$ and $\sigma = 0$

$$\Gamma(y+1)\nu(x,y+\alpha+1) = \int_0^x (x-u)^y \nu(u,\alpha)du \tag{160.4.7}$$

which is equivalent to (5.4.2) and (28.4.4).

Similarly, it follows from

$$L_{xy}\{\mu(x,\beta,y+\alpha+1)\} = \frac{1}{p^{\alpha+2}(\ln p)^{\beta+1}(q+\ln p)} = \frac{1}{p(q+\ln p)}\cdot\frac{1}{p^{\alpha+1}(\ln p)^{\beta+1}} = \frac{1}{p\ln p}\cdot\frac{1}{p^{\alpha+1}(\ln p)^{\beta}(q+\ln p)} \tag{161.4.7}$$

that

$$\mu(x,\beta,y+\alpha+1)=\frac{x^y}{\Gamma(y+1)}*\mu(x,\beta,\alpha)=\nu(x)*\mu(x,\beta-1,y+\alpha) \tag{162.4.7}$$

In particular for $\beta = 1$, (162.4.7) becomes

$$\mu(x,1,y+\alpha+1)=\nu(x)*\nu(x,y+\alpha) \tag{163.4.7}$$

Both convolutions are equivalent to (11.4.2) and (12.4.2).

In the case of positive integer powers, the convolution powers of a function may be defined in the following way

$$\begin{aligned}
&f(x,y)^{**1}=f(x,y)\\
&f(x,y)^{**2}=f(x,y)**f(x,y)\\
&f(x,y)^{**(n+1)}=f(x,y)**[f(x,y)^{**n}]\\
&n=1,2,3,\ldots
\end{aligned} \tag{164.4.7}$$

and more generally for any $\lambda > 0$, by

$$f(x,y)^{**\lambda}=L_{xy}{}^{-1}\{[F(p,q)]^{\lambda}\} \tag{165.4.7}$$

Applying (164.4.7) to $\nu(x,y+\alpha)$ with $n = 1$

$$\begin{aligned}
&L_{xy}\{\nu(x,y+\alpha)**\nu(x,y+\alpha)\}=\frac{1}{[p^{\alpha+1}\ln p(q+\ln p)]^2}=\\
&\frac{1}{p(q+\ln p)}\cdot\frac{1}{p^{2\alpha+1}(\ln p)^2(q+\ln p)}=\frac{1}{p^{2\alpha+1}(\ln p)^2}\cdot\frac{1}{p(q+\ln p)^2}
\end{aligned} \tag{166.4.7}$$

The inverse of (166.4.7) gives the second power of convolution of the Volterra function in the equivalent forms

$$\begin{aligned}
&\nu(x,y+\alpha)^{**2}=\nu(x,y+\alpha)**\nu(x,y+\alpha)=\\
&\nu(x,y+\alpha_1)**\nu(x,y+\alpha_2)=\frac{x^y}{\Gamma(y+1)}**\mu(x,1,y+2\alpha)=\\
&\frac{x^y}{\Gamma(y)}*\mu(x,1,2\alpha)\\
&\alpha_1+\alpha_2=2\alpha
\end{aligned} \tag{167.4.7}$$

This result can be generalized to

$$L_{xy}\{\nu(x,y+\alpha)^{**(n+1)}\} = \frac{1}{[p^{\alpha+1}\ln p(q+\ln p)]^{n+1}} = \frac{1}{p(q+\ln p)^n}\cdot\frac{1}{p^{\alpha(n+1)+n}(\ln p)^{n+1}(q+\ln p)} = \frac{1}{p^n(q+\ln p)^{n+1}}\cdot\frac{1}{p^{\alpha(n+1)+1}(\ln p)^{n+1}} \tag{168.4.7}$$

Using (159.4.7) and (161.4.7), the inverse of (168.4.7) is

$$n!\nu(x,y+\alpha)^{**(n+1)} = \frac{ny^{n-1}x^y}{\Gamma(y+1)}**\mu(x,n,y+\alpha n+\alpha+n-1) = \frac{y^n x^{y+n-1}}{\Gamma(y+n)}*\mu[x,n,y+\alpha(n+1)] \quad n=1,2,3,\ldots \tag{169.4.7}$$

Similarly from

$$L_{xy}\{\nu(x,y+\alpha)^{**(\lambda+1)}\} = \frac{1}{[p^{\alpha+1}\ln p(q+\ln p)]^{\lambda+1}} = \frac{1}{p(q+\ln p)^\lambda}\cdot\frac{1}{p^{\alpha(\lambda+1)+\lambda}(\ln p)^{\lambda+1}(q+\ln p)} = \frac{1}{p^\lambda(q+\ln p)^{\lambda+1}}\cdot\frac{1}{p^{\alpha(\lambda+1)+1}(\ln p)^{\lambda+1}} \tag{170.4.7}$$

we have

$$\Gamma(\lambda+1)\nu(x,y+\alpha)^{**(\lambda+1)} = \frac{\lambda y^{\lambda-1}x^y}{\Gamma(y+1)}**\mu(x,\lambda,y+\alpha\lambda+\alpha+\lambda-1) = \frac{y^\lambda x^{y+\lambda-1}}{\Gamma(y+\lambda)}*\mu[x,\lambda,\alpha(\lambda+1)] \quad \mathrm{Re}\,\lambda>-1 \;,\; \mathrm{Re}[\alpha(\lambda+1)]>-1 \tag{171.4.7}$$

or in the explicit form

$$\int_0^x\int_0^y \frac{(x-\xi)^{(y-\eta)}(y-\eta)^{\lambda-1}}{\Gamma(y-\eta+1)}\mu[\xi,\lambda,\eta+\alpha\lambda+\alpha+\lambda-1]d\xi d\eta = \frac{y^{\lambda}}{\lambda\Gamma(y+\lambda)}\int_0^x (x-u)^{y+\lambda-1}\mu[u,\lambda,\alpha(\lambda+1)]du \quad (172.4.7)$$

The function $f(x,y)$ multiplied by powers of x and y can be expressed in the two-dimensional Laplace transformation by the partial differentiation of $F(p,q)$ by powers of p and q

$$L_{xy}\{(-x)^m(-y)^n f(x,y)\} = \frac{\partial^{m+n}F(p,q)}{\partial p^m \partial q^m} \quad (173.4.7)$$

$$m,n = 0,1,2,3,\ldots$$

Applying this operational rule to $\nu(x,y+\alpha)$ we have

$$L_{xy}\{y\nu(x,y+\alpha)\} = -\frac{\partial}{\partial q}\left[\frac{1}{p^{\alpha+1}\ln p(q+\ln p)}\right] = \frac{1}{p^{\alpha+1}\ln p(q+\ln p)^2} \quad (174.4.7)$$

The inverse of (174.4.7) gives

$$\Gamma(y)\nu(x,y+\alpha) = \int_0^x (x-u)^{y-1}\nu(u,\alpha)du \quad (175.4.7)$$

which is the equivalent form of (90.4.6).

Similarly from

$$L_{xy}\{x\nu(x,y+\alpha)\} = -\frac{\partial}{\partial p}\left[\frac{1}{p^{\alpha+1}\ln p(q+\ln p)}\right] = \frac{\alpha+1}{p^{\alpha+2}\ln p(q+\ln p)} + \frac{1}{p^{\alpha+2}(\ln p)^2(q+\ln p)} + \frac{1}{p^{\alpha+2}\ln p(q+\ln p)^2} \quad (176.4.7)$$

it is possible to obtain

$$x\nu(x,y+\alpha) = \frac{(y+\alpha+1)x^y}{\Gamma(y+1)} * \nu(x,\alpha) + \nu(x,y+\alpha) * \nu(x) \quad (177.4.7)$$

and finally from

$$L_{xy}\{xy\nu(x,y+\alpha)\} = \frac{\partial^2}{\partial p \partial q}[\frac{1}{p^{\alpha+1}\ln p(q+\ln p)}] \qquad (178.4.7)$$

we have

$$xy\nu(x,y+\alpha) = \frac{x^y}{\Gamma(y)} * [(y+\alpha+1)\nu(x,\alpha) + \mu(x,1,\alpha)] \qquad (179.4.7)$$

By comparing (177.4.7) with (179.4.7) we obtain once again (162.4.7), which is valid for $\beta = 1$

$$\nu(x,y+\alpha) * \nu(x) = \frac{x^y}{\Gamma(y+1)} * \mu(x,1,\alpha) \qquad (180.4.7)$$

Sometimes, there is an advantage to put $q = p$ in (1.4.7) and then we have

$$F(p,p) = \int_0^\infty \int_0^\infty e^{-p(x+y)} f(x,y) dx dy \qquad (181.4.7)$$

but by introducing a new variable, $\xi = x + y$

$$\int_0^\infty e^{-px} \int_0^x f(\xi, x-\xi) d\xi dx = F(p,p) \qquad (182.4.7)$$

Thus, in this case the two-dimensional Laplace transformation is reduced to the ordinary Laplace transform of the integral

$$L\{\int_0^t f(u,t-u) du\} = F(s,s) \qquad (183.4.7)$$

and other operational rules of this type are

$$L\{\int_0^t f(t-u,au) du\} = \frac{1}{a} F(s,as) \qquad (184.4.7)$$

$$a > 0$$

$$L\{\int_0^\infty f(t,udu\} = F(s,0) \qquad (185.4.7)$$

For example using (4.4.7) - (7.4.7) and (183.4.7) - (184.4.7) we have

$$L\{\int_0^t \nu(u,t-u)du\} = \frac{1}{s\ln s(s+\ln s)} \tag{186.4.7}$$

$$L\{\int_0^t \mu(u,\beta,t-u)du\} = \frac{1}{s(\ln s)^{\beta+1}(s+\ln s)} \tag{187.4.7}$$

$$L\{\int_0^t \nu(t-u,au)du\} = \frac{1}{s\ln s[1+\ln(s^a)]} \tag{188.4.7}$$

$$L\{\int_0^t \frac{1}{\Gamma(t+au+1)}du\} = \frac{1}{as}\{\nu(e^{-s/(a+1)})-\nu(e^{-s}) \tag{189.4.7}$$

Inroducing $q = 0$ in (4.4.7) together with (185.4.7) gives

$$L\{\int_0^\infty \nu(t,u)du\} = \frac{1}{s(\ln s)^2} \tag{190.4.7}$$

but the inverse of (190.4.7) is the improper integral of the Volterra function with respect to the parameter

$$\int_0^\infty \nu(t,u)du = \mu(t,1) \tag{191.4.7}$$

and in a similar way it is possible to obtain

$$\int_0^\infty \mu(t,\beta,u)du = \mu(t,\beta+1) \tag{192.4.7}$$

4.8. Volterra Functions and Other Two-Dimensional Integral Transformations

In an analogy with the ordinary sine and cosine Fourier, Mellin, Hankel and other integral transformations, the corresponding two-dimensional integral transformations were introduced [34-36,94,98,100]. They can be written in the general form as

$$F(p,q)=\int_0^\infty\int_0^\infty K_1(x,p)K_2(y,q)f(x,y)dxdy \tag{1.4.8}$$

where $K_1(x,p)$ and $K_2(y,q)$ are the kernels of one-dimensional integral transformations. Unfortunately, the two-dimensional integral transformations are considerably less developed than the two-dimensional Laplace transformation and two-dimensional transforms of the Volterra functions are unknown in the literature. Considering $F(p,q)$ from (1.4.8) we will limit ourselves mainly to *the two-dimensional Mellin* and *Laplace-Hankel transformations*.

In the first case both kernels are represented by the same function

$$M(p,q)=M_{xy}\{f(x,y)\}=\int_0^\infty\int_0^\infty x^{p-1}y^{q-1}f(x,y)dxdy \tag{2.4.8}$$

where p and q are complex variables.

In the second case, in the two-dimensional Laplace-Hankel transformation, the kernels are different

$$H_\lambda(p,q)=H_{xy}\{f(x,y);\lambda\}=\int_0^\infty\int_0^\infty e^{-px}yJ_\lambda(qy)f(x,y)dxdy \tag{3.4.8}$$

where $\lambda > -1/2$, p is real and positive and $\mathrm{Re}(q-q_0) > 0$. This is *the two-dimensional mixed transformation* which was introduced by Huguette Delavault [34,35] who considered solutions of the heat transfer problems.

If a function $f(x+y)$ depends on x and y only in the combination $x+y$, then an evaluation of the two-dimensional Mellin transform is reduced to that of the ordinary Mellin transform of the function $f(t)$. The change of independent variables from x and y to $t = x+y$ and $u = y$ gives

$$\int_0^\infty\int_0^\infty x^{p-1}y^{q-1}f(x+y)dxdy=\int_0^\infty\left[\int_0^t(t-u)^{p-1}u^{q-1}du\right]f(t)dt \tag{4.4.8}$$

But the inner integral can be recognized as the beta function and therefore

$$\begin{aligned}M(p,q)=M_{xy}\{f(x+y)\}&=B(p,q)\int_0^\infty t^{p+q-1}f(t)dt=\\ B(p,q)M\{f(t);p+q\}&=B(p,q)M(p+q)\end{aligned} \tag{5.4.8}$$

Evidently, this operational rule can serve to obtain some of the Mellin transforms of the Volterra functions. For example from (6.4.6), (8.4.6), (12.4.6) - (14.4.6) and (5.4.8) we have

$$M_{xy}\{\nu(e^{-(x+y)})\}=\Gamma(p)\Gamma(q)\Gamma(1-p-q)\mu(1,-p-q)$$
$$0<\operatorname{Re}(p+q)<1 \quad , \quad \operatorname{Re}p>0 \quad , \quad \operatorname{Re}q>0 \tag{6.4.8}$$

$$M_{xy}\{\nu(ae^{-b(x+y)})\}=\frac{\Gamma(p)\Gamma(q)\Gamma(1-p-q)}{b^{p+q}}\mu(a,-p-q)$$
$$a>0 \quad , \quad b>0 \tag{7.4.8}$$
$$0<\operatorname{Re}(p+q)<1 \quad , \quad \operatorname{Re}p>0 \quad , \quad \operatorname{Re}q>0$$

$$M_{xy}\{(x+y)^{\lambda}\nu(e^{-(x+y)^{\rho}})\}=$$
$$\frac{1}{\rho}B(p,q)\Gamma(\frac{p+q+\lambda}{\rho})\Gamma(1-\frac{p+q+\lambda}{\rho})\mu(1,-\frac{p+q+\lambda}{\rho})$$
$$\rho>0 \quad , \quad 0<\operatorname{Re}(\frac{p+q+\lambda}{\rho})<1 \tag{8.4.8}$$
$$\operatorname{Re}p>0 \quad , \operatorname{Re}q>0$$

$$M_{xy}\{(x+y)^{\lambda}\nu(e^{-1/(x+y)^{\rho}})\}=$$
$$\frac{1}{\rho}B(p,q)\Gamma(-\frac{p+q+\lambda}{\rho})\Gamma(1+\frac{p+q+\lambda}{\rho})\mu(1,\frac{p+q+\lambda}{\rho}) \tag{9.4.8}$$
$$\rho>0 \quad , \quad -1<\operatorname{Re}(\frac{p+q+\lambda}{\rho})<0$$

$$M_{xy}\{\frac{1}{x+y}\nu(e^{-1/(x+y)})\}=\Gamma(p)\Gamma(q)\Gamma(1-p-q)\mu(1,p+q-1)$$
$$0<\operatorname{Re}(p+q)<1 \quad , \quad \operatorname{Re}p>0 \quad , \quad \operatorname{Re}q>0 \tag{10.4.8}$$

From (25.4.6) and (5.4.8) it follows that

$$\int_0^{\infty}\int_0^{\infty}x^{p-1}y^{q-1}(x+y)^n e^{-(x+y)^2}\nu(x+y)dxdy=$$

$$B(p,q)\int_0^{\infty}t^{p+q+n-1}e^{-t^2}\nu(t)dt=\frac{1}{2}B(p,q)\int_0^{\infty}\frac{\Gamma(\frac{u+p+q+n}{2})}{\Gamma(u+1)}du \tag{11.4.8}$$
$$n=0,1,2,3,\ldots$$

For $p=q=1$, (11.4.8) reduces to

$$\int_0^\infty\int_0^\infty (x+y)^n e^{-(x+y)^2} \nu(x+y)dxdy = \int_0^\infty t^{n+1} e^{-t^2} \nu(t)dt =$$

$$\frac{1}{2}\int_0^\infty \frac{\Gamma(\frac{u+n}{2}+1)}{\Gamma(u+1)} du \tag{12.4.8}$$

$$n = 0,1,2,3,\ldots$$

Like (27.4.6), by using the duplication formula of gamma function, these integrals are expressed in terms of the Volterra functions. The first four integrals are presented below

$$\int_0^\infty t e^{-t^2} \nu(t)dt = \frac{\sqrt{\pi}}{2} \nu(\frac{1}{4}, -\frac{1}{2}) \tag{13.4.8}$$

$$\int_0^\infty t^2 e^{-t^2} \nu(t)dt = \frac{\sqrt{\pi}}{2} [\nu(\frac{1}{4}) + 2\mu(\frac{1}{4}, 1)] \tag{14.4.8}$$

$$\int_0^\infty t^3 e^{-t^2} \nu(t)dt = \frac{\sqrt{\pi}}{2} [\nu(\frac{1}{4}, -\frac{1}{2}) + \mu(\frac{1}{4}, 1, -\frac{1}{2})] \tag{15.4.8}$$

$$\int_0^\infty t^4 e^{-t^2} \nu(t)dt = \frac{\sqrt{\pi}}{2} [\frac{3}{2}\nu(\frac{1}{4}) + 4\mu(\frac{1}{4}, 1) + 4\mu(\frac{1}{4}, 2)] \tag{16.4.8}$$

Consider again (11.4.8), but in a general form by replacing n with λ. Then for $\lambda = 1\text{-}p\text{-}q$, the double integral is

$$\int_0^\infty\int_0^\infty x^{p-1} y^{q-1} (x+y)^{1-p-q} e^{-(x+y)^2} \nu(x+y)dxdy = \sqrt{\pi} B(p,q) \nu(\frac{1}{4}) \tag{17.4.8}$$

and for $p = q = (1\text{-}\lambda)/2$

$$\int_0^\infty\int_0^\infty (xy)^{-(1+\lambda)/2}(x+y)^\lambda e^{-(x+y)^2}\nu(x+y)dxdy = 2^\lambda \pi \frac{\Gamma(\frac{1-\lambda}{2})}{\Gamma(1-\frac{\lambda}{2})}\nu(\frac{1}{4}) \tag{18.4.8}$$

This reduces for $\lambda = -1$ and $\lambda = 0$ to

$$\int_0^\infty\int_0^\infty \frac{e^{-(x+y)^2}}{x+y}\nu(x+y)dxdy = \sqrt{\pi}\nu(\frac{1}{4}) \tag{19.4.8}$$

$$\int_0^\infty\int_0^\infty (xy)^{-1/2}e^{-(x+y)^2}\nu(x+y)dxdy = \pi^{3/2}\nu(\frac{1}{4}) \tag{20.4.8}$$

If $f(x+y)$ in (5.4.8) is replaced by $f(x+y)\nu(x+y+\alpha)$ then we have

$$M_{xy}\{f(x+y)\nu(x+y+\alpha)\} = \int_0^\infty \frac{B(p,q)}{\Gamma(u+\alpha+1)}[\int_0^\infty t^{u+p+q+\alpha-1}f(t)dt]du = B(p,q)\int_0^\infty \frac{M(u+p+q+\alpha)}{\Gamma(u+\alpha+1)}du \tag{21.4.8}$$

and this result is similar to that given in (22.4.6) for the one-dimensional Mellin transformation. For example from

$$M\{erfc(at)\} = \frac{\Gamma(\frac{s+1}{2})}{\sqrt{\pi}a^s s} \qquad a>0 \quad , \quad \operatorname{Re} s>0 \tag{22.4.9}$$

it follows that

$$M_{xy}\{erfc[a(x+y)]\nu(x+y+\alpha)\} = \frac{B(p,q)}{\sqrt{\pi}a^{p+q+\alpha}}\int_0^\infty \frac{\Gamma(\frac{u+p+q+\alpha+1}{2})}{a^u(u+p+q+\alpha)\Gamma(u+\alpha+1)}du \tag{23.4.8}$$

In particular, by applying the duplication formula of gamma function and (1.2.2) we have

$$\int_0^\infty\int_0^\infty erfc[a(x+y)]\nu(x+y+\alpha)dxdy = \frac{1}{a^4}\nu(\frac{1}{4a^2},\frac{\alpha}{2}) - \frac{2}{a^2}\nu(\frac{1}{4a^2},\frac{\alpha}{2}+1) \tag{24.4.8}$$

$$\int_0^\infty\int_0^\infty xy\,erfc[a(x+y)]\nu(x+y+\alpha)dxdy = \frac{1}{24a^6}\nu(\frac{1}{4a^2},\frac{\alpha}{2}-1) + \frac{1}{2a^2}\nu(\frac{1}{4a^2},\frac{\alpha}{2}+1) - 2\nu(\frac{1}{4a^2},\frac{\alpha}{2}+2) \tag{25.4.8}$$

As can be expected, the two-dimensional Mellin transforms of the reciprocal gamma functions are represented by the Volterra functions

$$M_{xy}\{\frac{1}{\Gamma(x+y+\alpha+1)}\} = \Gamma(p)\Gamma(q)\mu(1,p+q-1,\alpha) \tag{26.4.8}$$

$$M_{xy}\{\frac{a^{x+y+\alpha}(x+y)^{\lambda}}{\Gamma(x+y+\alpha+1)}\} = B(p,q))\Gamma(p+q+\lambda)\mu(a,p+q+\lambda-1,\alpha)$$

$$a>0 \quad , \quad \operatorname{Re}\alpha > -1 \quad , \quad \operatorname{Re}\lambda \geq 0 \tag{27.4.8}$$

$$\operatorname{Re}p > 0 \quad , \quad \operatorname{Re}q > 0$$

Double integrals based on (26.4.8) have a simple form for $p = q$, $p+q = 2n+1$ and $p+q = 2n+2$ where n is zero or positive integer

$$\int_0^\infty\int_0^\infty \frac{(xy)^{n-1/2}}{\Gamma(x+y+\alpha+1)}dxdy = \frac{\pi}{2^{4n}}[\frac{(2n)!}{n!}]^2\mu(1,2n,\alpha) \tag{28.4.8}$$

$$\int_0^\infty\int_0^\infty \frac{(xy)^{n}}{\Gamma(x+y+\alpha+1)}dxdy = (n!)^2\mu(1,2n+1,\alpha) \tag{29.4.8}$$

$$n = 0,1,2,3,...$$

and they can be expressed in terms of different $\nu(1,\alpha-n)$ functions , for example

$$\int_0^\infty\int_0^\infty \frac{1}{\sqrt{xy}\,\Gamma(x+y+\alpha+1)}dxdy = \pi\mu(1,0,\alpha) = \pi\nu(1,\alpha) \tag{30.4.8}$$

$$\int_0^\infty\int_0^\infty \frac{1}{\Gamma(x+y+\alpha+1)}dxdy = \mu(1,1,\alpha) = \nu(1,\alpha-1)-\alpha\nu(1,\alpha) \tag{31.4.8}$$

$$\int_0^\infty\int_0^\infty \frac{\sqrt{xy}}{\Gamma(x+y+\alpha+1)}dxdy = \frac{\pi}{4}\mu(1,2,\alpha) = \frac{\pi}{8}[\nu(1,\alpha-2)-(2\alpha-1)\,\nu(1,\alpha-1)+\alpha^2\nu(1,\alpha)] \tag{32.4.8}$$

Applying the operational rule of the two-dimensional Mellin transformation [56]

$$M_{xy}\{\frac{1}{xy}f(\frac{1}{x},\frac{1}{y})\} = M(1-p,1-q) \tag{33.4.8}$$

to (26.4.8) we have

$$M_{xy}\{\frac{1}{xy\Gamma(\frac{1}{x}+\frac{1}{y}+\alpha+1)}\} = \Gamma(1-p)\Gamma(1-q)\mu(1,1-p-q,\alpha) \tag{34.4.8}$$

$$\operatorname{Re}\alpha > -1 \quad , \quad \operatorname{Re}p < 1 \quad , \quad \operatorname{Re}q < 1$$

In particular, for $p = q = 0$ and $p = q = \pm 1/2$, the double integrals of reciprocal gamma functions are

$$\int_0^\infty\int_0^\infty \frac{1}{(xy)^2\Gamma(\frac{1}{x}+\frac{1}{y}+\alpha+1)}dxdy = \mu(1,1,\alpha) \tag{35.4.8}$$

$$\int_0^\infty\int_0^\infty \frac{1}{(xy)^{3/2}\Gamma(\frac{1}{x}+\frac{1}{y}+\alpha+1)}dxdy = \pi\nu(1,\alpha) \tag{36.4.8}$$

$$\int_0^\infty\int_0^\infty \frac{1}{(xy)^{5/2}\Gamma(\frac{1}{x}+\frac{1}{y}+\alpha+1)}dxdy = \frac{\pi}{4}\mu(1,2,\alpha) \tag{37.4.8}$$

If also $\alpha = 0$, then (35.4.8)-(37.4.8) become

$$\int_0^\infty\int_0^\infty \frac{1}{xy(x+y)\Gamma(\frac{x+y}{xy})} dxdy = \mu(1,1) = \nu(1,-1) = 2.807770... \tag{38.4.8}$$

$$\int_0^\infty\int_0^\infty \frac{1}{\sqrt{xy}(x+y)\Gamma(\frac{x+y}{xy})} dxdy = \pi\nu(1) = 7.120528... \tag{39.4.8}$$

$$\int_0^\infty\int_0^\infty \frac{1}{(xy)^{3/2}(x+y)\Gamma(\frac{x+y}{xy})} dxdy = \frac{\pi}{4}\mu(1,2) = \frac{\pi}{8}\mu(1,1,-1) = \\ \frac{\pi}{8}[\nu(1,-2)+\nu(1,-1)] = 2.133070... \tag{40.4.8}$$

and for $p = q = 0$ and $\alpha = 1$

$$\int_0^\infty\int_0^\infty \frac{1}{(x+y)(x+y+xy)\Gamma(\frac{x+y}{xy})} dxdy = \nu(1) - \nu(1,1) = 1.085142... \tag{41.4.8}$$

Applying (33.4.8) to (27.4.8), but with a = 1, gives

$$M_{xy}\{\frac{(\frac{1}{x}+\frac{1}{y})^{\lambda}}{xy\Gamma(\frac{1}{x}+\frac{1}{y}+\alpha+1)}\} = \\ B(1-p,1-q)\Gamma(2-p-q+\lambda)\mu(1,1-p-q+\lambda,\alpha) \tag{42.4.8}$$

which for $p = q = \lambda$ becomes

$$\int_0^\infty\int_0^\infty \frac{(x+y)^{p}}{(xy)^2\Gamma(\frac{1}{x}+\frac{1}{y}+\alpha+1)} dxdy = \frac{\Gamma(1-p)^2\Gamma(2-p)}{\Gamma(2-2p)}\mu(1,1-p,\alpha) \tag{43.4.8}$$

If $p = 0$, then (43.4.8) reduces to (35.4.8) and for $p = \pm 1/2$ and $\alpha = 0$

$$\int_0^\infty\int_0^\infty \frac{1}{xy\sqrt{x+y}\,\Gamma(\frac{x+y}{xy})}dxdy = \frac{\pi^{3/2}}{2}\mu(1,\frac{1}{2}) = 7.321551... \tag{44.4.8}$$

$$\int_0^\infty\int_0^\infty \frac{1}{xy(x+y)^{3/2}\Gamma(\frac{x+y}{xy})}dxdy = \frac{3\pi^{3/2}}{32}\mu(1,\frac{3}{2}) = 1.474758... \tag{45.4.8}$$

Now, we will consider the Laplace-Hankel transformation

$$H_\lambda(p,q) = H_{xy}\{f(x,y);\lambda\} = \int_0^\infty\int_0^\infty e^{-px} yJ_\lambda(qy)f(x,y)dxdy \tag{46.4.8}$$

in the special case of $f(x,y) = g(y)\nu(x,y+\alpha)$.

If the ordinary Laplace transform

$$L\{g(y)yJ_\lambda(qy)\} = F(s,q) \tag{47.4.8}$$

is available, then the Laplace-Hankel transform is given by

$$H_\lambda(p,q) = H_{xy}\{g(y)\nu(x,y+\alpha);\lambda\} = \frac{F(\ln p,q)}{p^{\alpha+1}\ln p} \tag{48.4.8}$$
$$\operatorname{Re}\alpha > -1 \quad , \quad \operatorname{Re}p > 1 \quad , \quad \operatorname{Re}q > 0$$

For example [88, 1.14.2]

$$L\{yJ_\lambda(qy)\} = \frac{q^\lambda[s+\lambda\sqrt{s^2+q^2}]}{(s^2+q^2)^{3/2}[s+\sqrt{s^2+q^2}]^\lambda} \tag{49.4.8}$$
$$\operatorname{Re}\lambda > -2$$

and therefore the Laplace-Hankel transform of the Volterra function is

$$H_{xy}\{\nu(x,y+\alpha);\lambda\} =$$
$$\frac{q^\lambda[\ln p+\lambda\sqrt{(\ln p)^2+q^2}]}{p^{\alpha+1}\ln p[(\ln p)^2+q^2]^{3/2}[\ln p+\sqrt{(\ln p)^2+q^2}]^\lambda} \tag{50.4.8}$$

which for $\lambda = 0$ reduces to

$$H_0(p,q) = \int_0^\infty \int_0^\infty e^{-px} y J_0(qy) \nu(x, y+\alpha) dx dy = \frac{q^\lambda}{p^{\alpha+1}[(\ln p)^2 + q^2]^{3/2}} \tag{51.4.8}$$

Similarly from [88, 1.14.1, 1.14.5]

$$H_{xy}\{\frac{1}{y}\nu(x, y+\alpha); \lambda\} = \frac{q^\lambda}{p^{\alpha+1} \ln p \sqrt{(\ln p)^2 + q^2}\,[\ln p + \sqrt{(\ln p)^2 + q^2}\,]^\lambda} \tag{52.4.8}$$

$$\operatorname{Re}\lambda > -1$$

$$H_{xy}\{\frac{1}{y^2}\nu(x, y+\alpha); \lambda\} = \frac{q^\lambda}{\lambda p^{\alpha+1} \ln p [\ln p + \sqrt{(\ln p)^2 + q^2}\,]^\lambda} \tag{53.4.8}$$

$$\operatorname{Re}\lambda > 0$$

Applying the operational rules, the evaluated here Laplace-Hankel transforms can be used to obtain

$$H_{xy}\{e^{-ax} g(y) \nu(x, y+\alpha); \lambda\} = \frac{F[\ln(p+a), q]}{(p+a)^{\alpha+1} \ln(p+a)} \tag{54.4.80}$$

$$a > 0$$

$$H_{xy}\{x^n g(y) \nu(x, y+\alpha); \lambda\} = (-1)^n \frac{\partial^n}{\partial p^n}[H_\lambda(p,q)] \tag{55.4.8}$$

$$n = 0, 1, 2, 3, \ldots$$

Delavault showed that if $x^{1/2} f(x,y)$ is absolutely integrable with respect to x, the two-dimensional Laplace transform $F(p,q)$ and the Laplace-Hankel transform $H_\lambda(p,q)$ satisfy the following relation [34, 56]

$$F(p,q) = \int_0^\infty \frac{u^{\lambda+1} H_\lambda(p,u)}{\sqrt{q^2+u^2}\,[q + \sqrt{q^2+u^2}\,]^\lambda} du \tag{56.4.8}$$

If both $F(p,q)$ and $H_\lambda(p,q)$ are known, the relation (56.4.8) is useful in evaluation of rather complicated algebraic integrals.

In the case of the Volterra function $\nu(x,y+\alpha)$, the two-dimensional Laplace transform is

$$F(p,q) = L_{xy}\{\nu(x,y+\alpha)\} = \frac{1}{p^{\alpha+1}\ln p(q+\ln p)} \tag{57.4.8}$$

and introducing $H_\lambda(p,q)$ from (50.4.8) into (56.4.8) we have

$$\int_0^\infty \frac{u^{2\lambda+1}[\ln p+\lambda\sqrt{(\ln p)^2+u^2}\,]}{\sqrt{q^2+u^2}[(\ln p)^2+u^2]^{3/2}\{[q+\sqrt{q^2+u^2}\,][\ln p+\sqrt{(\ln p)^2+u^2}\,]\}^\lambda}du = \frac{1}{q+\ln p} \tag{58.4.8}$$

After changing the variable of integration, it is possible to bring the integral (58.4.8) to the form

$$\int_0^\infty \frac{t^\lambda(1+\lambda\sqrt{1+t})}{\sqrt{a+t}(1+t)^{3/2}[(1+\sqrt{1+t})(\sqrt{a}+\sqrt{a+t})]^\lambda}dt = \frac{2}{\sqrt{a}+1} \tag{59.4.8}$$

$$a>0$$

For $\lambda = 0$, repeated differentiation of the integrand in (59.4.8) with regard to the parameter a, gives

$$\int_0^\infty \frac{1}{(1+t)^{1/2}(a+t)^{n+3/2}}dt = (-1)^n\frac{2^{2n+2}(n+1)!}{[2(n+1)]!}\frac{d^n}{da^n}\left[\frac{1}{\sqrt{a}(\sqrt{a}+1)}\right] \tag{60.4.8}$$

$$\int_0^\infty \frac{1}{(1+t)^{3/2}(a+t)^{n+1/2}}dt = (-1)^n\frac{2^{2n+1}n!}{(2n)!}\frac{d^n}{da^n}\left[\frac{1}{\sqrt{a}+1}\right] \tag{61.4.8}$$

$$n=0,1,2,3,\ldots$$

Other algebraic integrals of this type can be evaluated in a similar way from (56.4.8) by using (55.4.8) and

$$L_{xy}\{x^n f(x,y)\} = (-1)^n\frac{\partial^n F(p,q)}{\partial p^n} \tag{62.4.8}$$

$$n=0,1,2,3,\ldots$$

The Laplace-Hankel transformation is a particular case of the two-dimensional mixed Laplace transformation with a general kernel $K(y,q)$

$$T(p,q) = \int_0^\infty \int_0^\infty e^{-px} K(y,q) f(x,y) dx dy \tag{63.4.8}$$

If ordinary Laplace transforms of

$$L\{f(y,q)\} = L\{g(y)K(y,q)\} = F(s,q) \tag{64.4.8}$$

exist and are known, then the mixed Laplace transforms of $g(y)\nu(x,y+\alpha)$ follow directly from

$$T(p,q) = \int_0^\infty \int_0^\infty e^{-px} f(y,q) \nu(x,y+\alpha) dx dy = \frac{F(\ln p, q)}{p^{\alpha+1} \ln p} \tag{65.4.8}$$

$$\operatorname{Re}\alpha > -1 \quad , \quad \operatorname{Re} p > 1$$

For the Hankel kernel such expression was already given in (48.4.8) and for ordinary Laplace transform in (53.4.3). Introduction of other integral kernels will also lead to evaluation of integrals of the Volterra functions or to establishment of integral equalities. In the case of the Mellin kernel $K(y,q) = y^{q-1}$

$$\int_0^\infty \int_0^\infty e^{-px} y^{q-1} \nu(x,y+\alpha) dx dy = \frac{\Gamma(q)}{p^{\alpha+1} (\ln p)^{q+1}} \tag{66.4.8}$$

$$\operatorname{Re}\alpha > -1 \quad , \quad \operatorname{Re} p > 1 \quad , \quad \operatorname{Re} q > -1$$

The inverse Laplace transform of (66.4.8) gives

$$\int_0^\infty y^{q-1} \nu(x,y+\alpha) dy = \Gamma(q) \mu(x,q,\alpha) \tag{67.4.8}$$

and by introducing $q = 1$

$$\int_0^\infty \nu(x,y+\alpha) dy = \mu(x,1,\alpha) \tag{68.4.8}$$

It is possible to obtain these integrals also from (53.4.3).

If kernels of the Fourier sine and cosine transformations are considered, then

$$\int_0^\infty\int_0^\infty e^{-px}\sin(qy)\nu(x,y+\alpha)dxdy = \frac{q}{p^{\alpha+1}\ln p[(\ln p)^2+q^2]} \tag{69.4.8}$$

$$\int_0^\infty\int_0^\infty e^{-px}\cos(qy)\nu(x,y+\alpha)dxdy = \frac{1}{p^{\alpha+1}[(\ln p)^2+q^2]} \tag{70.4.8}$$

The inversion performed with the aid of (17.4.1) and (18.4.1) leads to the equality of integrals

$$\int_0^\infty \sin(qy)\nu(x,y+\alpha)dy = \int_0^x Ts(u,\frac{q}{\pi})\nu(x-u,\alpha-1)du \tag{71.4.8}$$

$$\int_0^\infty \cos(qy)\nu(x,y+\alpha)dy = \int_0^x Tc(u,\frac{q}{\pi})\nu(x-u,\alpha-1)du \tag{72.4.8}$$

where the auxiliary Volterra functions $Tc(z,\alpha)$ and $Ts(z,\alpha)$ are defined in (13.2.2) and (14.2.2) respectively.

The generalized Stieltjes kernel $K(y,q) = 1/(q+y)^\rho$ when introduced into (65.4.8) leads to [102, 2.7]

$$\int_0^\infty\int_0^\infty \frac{e^{-px}}{(q+y)^\rho}\nu(x,y+\alpha)dxdy = \frac{\Gamma(1-\rho,q\ln p)}{p^{\alpha-q+1}(\ln p)^{2-\rho}} \tag{73.4.8}$$

which for ρ = 1/2, 3/2, 1 and 2 is [102, 2.5, 2.6, 1.4, 1.6]

$$\int_0^\infty\int_0^\infty \frac{e^{-px}}{(q+y)^{1/2}}\nu(x,y+\alpha)dxdy = \frac{\sqrt{\pi}}{p^{\alpha-q+1}(\ln p)^{3/2}}erfc(\sqrt{q\ln p}) \tag{74.4.8}$$

$$\int_0^\infty\int_0^\infty \frac{e^{-px}}{(q+y)^{3/2}}\nu(x,y+\alpha)dxdy =$$
$$\frac{2}{p^{\alpha+1}\ln p}\{\frac{1}{\sqrt{q}} - p^q\sqrt{\pi\ln p}\,erfc(\sqrt{q\ln p})\} \tag{75.4.8}$$

$$\int_0^\infty\int_0^\infty \frac{e^{-px}}{(q+y)}\nu(x,y+\alpha)dxdy = -\frac{1}{p^{\alpha-q+1}\ln p}Ei(-q\ln p) \tag{76.4.8}$$

$$\int_0^\infty\int_0^\infty \frac{e^{-px}}{(q+y)^2}\nu(x,y+\alpha)dxdy = \frac{1}{p^{\alpha+1}}\{\frac{1}{q\ln p} + p^q Ei(-q\ln p)\} \tag{77.4.8}$$

Considering that [88.2.6.1]

$$L^{-1}\{\frac{\ln p}{p}\} = -\ln(e^\gamma x) \quad \gamma = 0.577215... \tag{78.4.8}$$

and denoting the generalized Stieltjes transform of the Volterra function $\nu(x,y+\alpha)$ as

$$S_\rho(x,\alpha) = \int_0^\infty \frac{\nu(x,y+\alpha)}{(q+y)^\rho}dy \tag{79.4.8}$$

the convolution theorem as applied to (74.4.8), (75.4.8) and (76.4.8),(77.4.8) gives

$$\nu(x,y+\alpha) = \sqrt{q}[\frac{1}{2}S_{3/2}(x,\alpha) + S_{1/2}(x,\alpha-1) * \ln(e^\gamma x)] \tag{80.4.8}$$

$$\nu(x,y+\alpha) = q[S_2(x,\alpha) + S_1(x,\alpha-1) * \ln(e^\gamma x)] \tag{81.4.8}$$

Finally, consider once again the relation (65.4.8), but this time with the Laplace kernel $K(y,q) = e^{-qy}$. Thus, the two-dimensional Laplace transformation is formally treated as the "mixed" Laplace-Laplace transformation.

Since

$$L\{e^{-qy}f(y)\} = F(s+q) \quad \mathrm{Re}\, q > 0 \tag{82.4.8}$$

it follows that the two-dimensional Laplace transforms of $f(y)\nu(x,y+\alpha)$ and of $f(y)\mu(x,\beta, y+\alpha)$ are given by

$$L_{xy}\{f(y)\nu(x,y+\alpha)\} = \int_0^\infty\int_0^\infty e^{-px}e^{-qy}f(y)\nu(x,y+\alpha)dxdy = \frac{F(q+\ln p)}{p^{\alpha+1}\ln p} \tag{83.4.8}$$

$$L_{xy}\{f(y)\mu(x,\beta,y+\alpha)\}=\int_0^\infty\int_0^\infty e^{-px}e^{-qy}f(y)\mu(x,\beta,y+\alpha)dxdy=$$
$$\frac{F(q+\ln p)}{p^{\alpha+1}(\ln p)^{\beta+1}} \tag{84.4.8}$$

For example, in the simplest case, $L\{y^{\lambda}\}=\Gamma(\lambda+1)/s^{\lambda+1}$, we have immediately from (83.4.8) that

$$L_{xy}\{y^{\lambda}\nu(x,y+\alpha)\}=\frac{\Gamma(\lambda+1)}{p^{\alpha+1}\ln p(q+\ln p)^{\lambda+1}}$$
$$\operatorname{Re}\alpha>-1 \quad , \quad \operatorname{Re}\lambda>-1 \tag{85.4.8}$$
$$\operatorname{Re}p>1 \quad , \quad \operatorname{Re}q>0$$

which reduces to (157.4.7) or (57.4.8) for $\lambda = 0$ and to (174.4.7) for $\lambda = 1$.

Using (83.4.8), in the case of trigonometric functions, then we have

$$L_{xy}\{\sin(ay)\nu(x,y+\alpha)\}=\frac{a}{p^{\alpha+1}\ln p[a^2+(q+\ln p)^2]} \tag{86.4.8}$$

$$L_{xy}\{\cos(ay)\nu(x,y+\alpha)\}=\frac{q+\ln p}{p^{\alpha+1}\ln p[a^2+(q+\ln p)^2]} \tag{87.4.8}$$

Applying the same procedure as for the evaluation of (80.4.8) and (81.4.8), the ordinary Laplace transforms of the products of the Volterra functions and trigonometric functions are

$$L\{\sin(ay)\nu(x,y+\alpha)\}=e^{q}\nu(x,\alpha-1)*Ts(e^{-q}x,\frac{a}{\pi}) \tag{88.4.8}$$

$$L\{\cos(ay)\nu(x,y+\alpha)\}=e^{q}\nu(x,\alpha-1)*Tc(e^{-q}x,\frac{a}{\pi}) \tag{89.4.8}$$

In the last example, the two-dimensional transforms of products of the Volterra functions are presented

$$L_{xy}\{\nu(y,\lambda)\,\nu(x,y+\alpha)\}=$$
$$\frac{1}{p^{\alpha+1}\ln p(q+\ln p)^{\lambda+1}\ln(q+\ln p)} \tag{90.4.8}$$

$$L_{xy}\{\mu(y,\delta,\lambda)\,\nu(x,y+\alpha)\} = \frac{1}{p^{\alpha+1}\ln p(q+\ln p)^{\lambda+1}[\ln(q+\ln p)]^{\delta+1}} \tag{91.4.8}$$

$$L_{xy}\{\nu(y,\lambda)\mu(x,\beta,y+\alpha)\} = \frac{1}{p^{\alpha+1}(\ln p)^{\beta+1}(q+\ln p)^{\lambda+1}\ln(q+\ln p)} \tag{92.4.8}$$

$$L_{xy}\{\mu(y,\delta,\lambda)\mu(x,\beta,y+\alpha)\} = \frac{1}{p^{\alpha+1}(\ln p)^{\beta+1}(q+\ln p)^{\lambda+1}[\ln(q+\ln p)]^{\delta+1}} \tag{93.4.8}$$

Chapter 5

5. INTEGRALS, SERIES AND LIMITS ASSOCIATED WITH VOLTERRA FUNCTIONS

A systematic description of procedures and techniques used in the evaluation of integrals which are not classified as integral transforms is given here. The presented material is mainly based on the Apelblat work [3] in which integrals of the Volterra, gamma and polygamma functions were discussed.

However, many new integrals, in addition to these which are known in the literature [3,8,43], will be mentioned. In some cases, the evaluated integrals are represented by the infinite series of the Volterra functions. Since numerous integrals of the Volterra functions or those that are associated with them were already presented in different contexts in many places of this book (e.g. (56.4.6) - (64.4.6) and (13.4.8) - (16.4.8)), some repetitions are therefore unavoidable. Double integrals of the Volterra functions were discussed in section 4.8 and will not be repeated here.

The Hadamard finite parts of divergent integrals [17,104] which are associated with the Volterra functions were considered by Butzer [17] and they also will not be discussed here.

5.1. INDEFINITE AND DEFINITE INTEGRALS OF THE VOLTERRA FUNCTIONS

Integrals of the type

$$I(t,n) = \int_0^t x^n \nu(x,\alpha)dx \qquad n = 0,1,2,3,... \tag{1.5.1}$$

can be evaluated successively from

$$I(t,n) = L^{-1}\{(-1)^n \frac{1}{s}\frac{d^n}{ds^n}(\frac{1}{s^{\alpha+1}\ln s})\} \tag{2.5.1}$$

For n = 0,1 and 2, using the inverse Laplace transforms and (2.2.2) we have

$$\int_0^t \nu(x,\alpha)dx = L^{-1}\{\frac{1}{s^{\alpha+2}\ln s}\} = \nu(t,\alpha+1) \tag{3.5.1}$$

$$\int_0^t x\,\nu(x,\alpha)dx = L^{-1}\{\frac{\alpha+1}{s^{\alpha+3}\ln s}+\frac{1}{s^{\alpha+3}(\ln s)^2}\} = \\ (\alpha+1)\,\nu(t,\alpha+2)+\mu(t,1,\alpha+2) = t\nu(t,\alpha+1)-\nu(t,\alpha+2) \tag{4.5.1}$$

and

$$\int_0^t x^2\nu(x,\alpha)dx = L^{-1}\{\frac{(\alpha+1)(\alpha+2)}{s^{\alpha+4}\ln s}+\frac{2\alpha+3}{s^{\alpha+4}(\ln s)^2}+\frac{1}{s^{\alpha+4}(\ln s)^3}\} = \\ (\alpha+1)(\alpha+2)\,\nu(t,\alpha+3)+(2\alpha+3)\mu(t,1,\alpha+3)+\mu(t,2,\alpha+3) = \\ t^2\nu(t,\alpha+1)-2t\nu(t,\alpha+2)-(\alpha-2)\nu(t,\alpha+3) \tag{5.5.1}$$

For α = 0, the equivalent forms of these integrals were already given in (59.4.6) - (62.4.6).

From

$$L\{e^{at}\nu(t,\alpha)\} = \frac{1}{(s-a)^{\alpha+1}\ln(s-a)} \\ a>0 \tag{6.5.1}$$

and expanding the Laplace transform into the series

$$L\{\int_0^t e^{ax}\nu(x,\alpha)dx\} = \frac{1}{s(s-a)^{\alpha+1}\ln(s-a)} = \\ \sum_{n=0}^{\infty}\frac{(-1)^n a^n}{(s-a)^{\alpha+n+2}\ln(s-a)} \tag{7.5.1}$$

then, the term-by-term inversion of (7.5.1) gives

$$\int_0^t e^{ax}\nu(x,\alpha)dx = e^{at}\sum_{n=0}^{\infty}(-1)^n a^n\nu(t,\alpha+n+1) \tag{8.5.1}$$

and similarly

$$\int_0^t e^{-ax}\nu(x,\alpha)dx = e^{-at}\sum_{n=0}^{\infty} a^n \nu(t,\alpha+n+1) \tag{9.5.1}$$

After simple manipulations with (8.5.1) and (9.5.1) it is possible to obtain

$$\int_0^t \sinh(ax)\nu(x,\alpha)dx = \\ \sinh(at)\sum_{n=0}^{\infty} a^{2n}\nu(t,\alpha+2n+1) - \cosh(at)\sum_{n=0}^{\infty} a^{2n+1}\nu(t,\alpha+2n+2) \tag{10.5.1}$$

$$\int_0^t \cosh(ax)\nu(x,\alpha)dx = \\ \cosh(at)\sum_{n=0}^{\infty} a^{2n}\nu(t,\alpha+2n+1) - \sinh(at)\sum_{n=0}^{\infty} a^{2n+1}\nu(t,\alpha+2n+2) \tag{11.5.1}$$

and

$$\int_0^t \sinh[a(t-x)]\nu(x,\alpha)dx = \sum_{n=0}^{\infty} a^{2n+1}\nu(t,\alpha+2n+2) \tag{12.5.1}$$

$$\int_0^t \cosh[a(t-x)]\nu(x,\alpha)dx = \sum_{n=0}^{\infty} a^{2n}\nu(t,\alpha+2n+1) \tag{13.5.1}$$

Replacing the parameter a by ia in (10.5.1) - (13.5.1) leads to

$$\int_0^t \sin(ax)\nu(x,\alpha)dx = \sin(at)\sum_{n=0}^{\infty}(-1)^n a^{2n}\nu(t,\alpha+2n+1) - \\ \cos(at)\sum_{n=0}^{\infty}(-1)^n a^{2n+1}\nu(t,\alpha+2n+2) \tag{14.5.1}$$

$$\int_0^t \cos(ax)\nu(x,\alpha)dx = \cos(at)\sum_{n=0}^{\infty}(-1)^n a^{2n}\nu(t,\alpha+2n+1) + \\ \sin(at)\sum_{n=0}^{\infty}(-1)^n a^{2n+1}\nu(t,\alpha+2n+2) \tag{15.5.1}$$

and

$$\int_0^t \sin[a(t-x)]\nu(x,\alpha)dx = \sum_{n=0}^{\infty}(-1)^n a^{2n+1}\nu(t,\alpha+2n+2) \tag{16.5.1}$$

$$\int_0^t \cos[a(t-x)]\nu(x,\alpha)dx = \sum_{n=0}^{\infty}(-1)^n a^{2n}\nu(t,\alpha+2n+1) \tag{17.5.1}$$

Differentiation of (8.5.1) with respect to the parameter a and using (8.5.1) gives

$$\int_0^t (t-x)e^{ax}\nu(x,\alpha)dx = e^{at}\sum_{n=1}^{\infty}(-1)^{n+1} n a^{n-1}\nu(t,\alpha+n+1) \tag{18.5.1}$$

Similarly from (9.5.1)

$$\int_0^t (t-x)e^{-ax}\nu(x,\alpha)dx = e^{-at}\sum_{n=1}^{\infty} n a^{n-1}\nu(t,\alpha+n+1) \tag{19.5.1}$$

Combining (18.5.1) with (19.5.1) we obtain

$$\int_0^t (t-x)\sinh(ax)\nu(x,\alpha)dx = \sinh(at)\sum_{n=1}^{\infty}(2n-1)a^{2n-2}\nu(t,\alpha+2n) - \cosh(at)\sum_{n=1}^{\infty} 2n a^{2n-1}\nu(t,\alpha+2n+1) \tag{20.5.1}$$

$$\int_0^t (t-x)\cosh(ax)\nu(x,\alpha)dx = \cosh(at)\sum_{n=1}^{\infty}(2n-1)a^{2n-2}\nu(t,\alpha+2n) - \sinh(at)\sum_{n=1}^{\infty} 2n a^{2n-1}\nu(t,\alpha+2n+1) \tag{21.5.1}$$

The next differentiation of (8.5.1) and (9.5.1) leads to

$$\int_0^t (t-x)^2 e^{ax}\nu(x,\alpha)dx = e^{at}\sum_{n=2}^{\infty}(-1)^n n(n-1)a^{n-2}\nu(t,\alpha+n+1) \tag{22.5.1}$$

$$\int_0^t (t-x)^2 e^{-ax} \nu(x,\alpha)dx = e^{-at} \sum_{n=2}^{\infty} (-1)^n n(n-1)a^{n-2} \nu(t,\alpha+n+1) \tag{23.5.1}$$

These results can be generalized

$$\int_0^t (t-x)^k e^{ax} \nu(x,\alpha)dx = e^{at} \sum_{n=k}^{\infty} (-1)^{n+k} \nu(t,\alpha+n+1) \frac{d^k}{da^k}(a^n) \tag{24.5.1}$$

$$\int_0^t (t-x)^k e^{-ax} \nu(x,\alpha)dx = e^{-at} \sum_{n=k}^{\infty} \nu(t,\alpha+n+1) \frac{d^k}{da^k}(a^n) \tag{25.5.1}$$

$$a > 0 \quad , \quad k = 0,1,2,3,\ldots$$

If the integral (18.5.1) is written as the convolution integral

$$(te^{-at}) * \nu(t,\alpha) = \int_0^t (t-x)e^{-a(t-x)} \nu(x,\alpha)dx =$$

$$\sum_{n=1}^{\infty} (-1)^{n+1} n a^{n-1} \nu(t,\alpha+n+1) \tag{26.5.1}$$

then the Laplace transform of both sides of (26.5.1)

$$\frac{1}{(s+a)^2} \cdot \frac{1}{s^{\alpha+1} \ln s} = \sum_{n=1}^{\infty} (-1)^{n+1} n a^{n-1} \frac{1}{s^{\alpha+n+2} \ln s} \tag{27.5.1}$$

gives the sum of the well-known numerical series

$$\sum_{n=1}^{\infty} (-1)^{n+1} n \rho^{n-1} = \frac{1}{(1+\rho)^2} \tag{28.5.1}$$

$$\rho = a/s < 1$$

Following the same procedure, we have from (24.5.1) and (25.5.1)

$$\sum_{n=k}^{\infty} (-1)^{n+k} \frac{d^k}{d\rho^k}(\rho^n) = \frac{k!}{(1+\rho)^{k+1}} \tag{29.5.1}$$

$$\sum_{n=k}^{\infty} \frac{d^k}{d\rho^k}(\rho^n) = \frac{k!}{(1-\rho)^{k+1}} \qquad (30.5.1)$$
$$0 < \rho < 1 \quad , \quad k = 0,1,2,3,...$$

and

$$\sum_{n=k}^{\infty} [(-1)^{n+k} \pm 1] \frac{d^k}{d\rho^k}(\rho^n) = \frac{k!}{(1-\rho^2)^{k+1}} [(1-\rho)^{k+1} \pm (1-\rho)^{k+1}] \qquad (31.5.1)$$

For example, from (31.5.1) for k = 2,3 and 4 with $|\rho| < 1$, the following power series can be derived

$$\sum_{n=1}^{\infty} n(2n-1)\rho^{n-1} = 1\cdot 1 + 2\cdot 3\rho + 3\cdot 5\rho^2 + 4\cdot 7\rho^3 + ... = \frac{1+3\rho}{(1-\rho)^3} \qquad (32.5.1)$$

$$\sum_{n=1}^{\infty} n(2n-1)(2n+1)\rho^{n-1} = 1\cdot 1\cdot 3 + 2\cdot 3\cdot 5\rho + 3\cdot 5\cdot 7\rho^2 + 4\cdot 7\cdot 9\rho^3 + ... = \frac{3(1+6\rho+\rho^2)}{(1-\rho)^4} \qquad (33.5.1)$$

$$\sum_{n=1}^{\infty} n(n+1)(2n-1)(2n+1)\rho^{n-1} = 1\cdot 1\cdot 2\cdot 3 + 2\cdot 3\cdot 3\cdot 5\rho + 3\cdot 4\cdot 5\cdot 7\rho^2 + 4\cdot 5\cdot 7\cdot 9\rho^3 + ... = \frac{6(1+10\rho+5\rho^2)}{(1-\rho)^5} \qquad (34.5.1)$$

and

$$\sum_{n=1}^{\infty} n(2n+1)\rho^{n-1} = 1\cdot 3 + 2\cdot 5\rho + 3\cdot 7\rho^2 + 4\cdot 9\rho^3 + ... = \frac{3+\rho}{(1-\rho)^3} \qquad (35.5.1)$$

$$\sum_{n=1}^{\infty} n(n+1)(2n+1)\rho^{n-1} = 1\cdot 2\cdot 3 + 2\cdot 3\cdot 5\rho + 3\cdot 4\cdot 7\rho^2 + 4\cdot 5\cdot 9\rho^3 + ... = \frac{6(1+\rho)}{(1-\rho)^4} \qquad (36.5.1)$$

$$\sum_{n=1}^{\infty} n(n+1)(2n+1)(2n+3)\rho^{n-1} =$$

$$1\cdot2\cdot3\cdot5+2\cdot3\cdot5\cdot7\rho+3\cdot4\cdot7\cdot9\rho^2+4\cdot5\cdot9\cdot11\rho^3+...= \tag{37.5.1}$$

$$\frac{6(5+10\rho+\rho^2)}{(1-\rho)^5}$$

5.2. Convolution Integrals

Various convolution integrals were already considered in the previous section where they were represented by infinite series of the Volterra functions. However, in many cases, this type of integrals can be evaluated in the closed form if the inverse Laplace transforms are available

$$f(t)*\nu(t,\alpha)=\int_0^t f(t-x)\,\nu(x,\alpha)dx=L^{-1}\{\frac{F(s)}{s^{\alpha+1}\ln s}\} \tag{1.5.2}$$

$$f(t)*\mu(t,\beta,\alpha)=\int_0^t f(t-x)\mu(x,\beta,\alpha)dx=L^{-1}\{\frac{F(s)}{s^{\alpha+1}(\ln s)^{\beta+1}}\} \tag{2.5.2}$$

$$\mathrm{Re}\,\alpha>-1 \quad , \quad \mathrm{Re}\,\beta>-1 \quad , \quad \mathrm{Re}\,s>1$$

where $F(s) = L\{f(t)\}$. To a great extent, these expressions are used here starting with the Volterra functions

$$\int_0^t \nu(t-x,\alpha')\,\nu(x,\alpha)dx=L^{-1}\{\frac{1}{s^{\alpha+\alpha'+2}(\ln s)^2}\}=\mu(t,1,\alpha+\alpha'+1) \tag{3.5.2}$$

$$\int_0^t \mu(t-x,\beta',\alpha')\mu(x,\beta,\alpha)dx=L^{-1}\{\frac{1}{s^{\alpha+\alpha'+2}(\ln s)^{\beta+\beta'+2}}\}= \tag{4.5.2}$$

$$\mu(t,\beta+\beta'+1,\alpha+\alpha'+1)$$

and they are equivalent to (10.4.2) and (11.4.2).

The next simple case is $f(t) = t^\lambda$, $\mathrm{Re}\lambda > -1$ with $F(s) = \Gamma(\lambda+1)/s^{\lambda+1}$ and therefore

$$\int_0^t (t-x)^\lambda \nu(x,\alpha)dx=L^{-1}\{\frac{1}{s^{\alpha+\lambda+2}\ln s}\}=\Gamma(\lambda+1)\nu(t,\alpha+\lambda+1) \tag{5.5.2}$$

$$\int_0^t (t-x)^\lambda \mu(x,\beta,\alpha)dx = L^{-1}\{\frac{1}{s^{\alpha+\lambda+2}(\ln s)^{\beta+1}}\} = \Gamma(\lambda+1)\mu(t,\beta,\alpha+\lambda+1) \tag{6.5.2}$$

These results are related to (28.4.4), (56.4.6) - (62.4.6), (191.4.7) and (192.4.7). Since [88, 1.6.37]

$$L\{t^\lambda \ln t\} = \frac{\Gamma(\lambda+1)}{s^{\lambda+1}}[\psi(\lambda+1) - \ln s] \qquad \mathrm{Re}\,\lambda > -1 \quad , \quad \mathrm{Re}\,s > 1 \tag{7.5.2}$$

it follows from (1.5.2) and (2.5.2) that

$$\int_0^t (t-x)^\lambda \ln(t-x)\nu(x,\alpha)dx = L^{-1}\{\frac{\Gamma(\lambda+1)[\psi(\lambda+1)-\ln s]}{s^{\alpha+\lambda+2}\ln s}\} = \Gamma(\lambda+1)\{\psi(\lambda+1)\,\nu(t,\alpha+\lambda+1) - \frac{t^{\alpha+\lambda+1}}{\Gamma(\alpha+\lambda+2)}\} \tag{8.5.2}$$

$$\int_0^t (t-x)^\lambda \ln(t-x)\mu(x,\beta,\alpha)dx = L^{-1}\{\frac{\Gamma(\lambda+1)[\psi(\lambda+1)-\ln s]}{s^{\alpha+\lambda+2}(\ln s)^{\beta+1}}\} = \Gamma(\lambda+1)\{\psi(\lambda+1)\mu(t,\beta,\alpha+\lambda+1) - \mu(t,\beta-1,\alpha+\lambda+1)\} \tag{9.5.2}$$

The convolution integral (8.5.2) was already given in (99.4.6). The cases when λ is zero or positive integer

$$J_n(t,\alpha) = \int_0^t (t-x)^n \ln(t-x)\nu(x,\alpha)dx = n!\{\psi(n+1)\,\nu(t,\alpha+n+1) - \frac{t^{\alpha+n+1}}{\Gamma(\alpha+n+2)}\}$$
$$\psi(n+1) = -\gamma + \sum_{k=1}^{n}\frac{1}{k} \qquad n = 0,1,2,3,... \tag{10.5.2}$$

are of interest because the convolution integrals of the type

$$I_n(t,\alpha)=\int_0^t \ln(t-x)x^n \nu(x,\alpha)dx \qquad (11.5.2)$$
$$n=0,1,2,3,..$$

can be expressed by these in (10.5.2)

$$I_n(t,\alpha)=\sum_{k=0}^{n}(-1)^k\binom{n}{k}t^{n-k}J_k(t,\alpha) \qquad (12.5.2)$$

Many Laplace transforms which are similar to that in (7.5.2) but with higher powers of logarithms are known [69,88] and they can be obtained from the recurrence formula established by Kölbig [69]

$$L\{t^{\lambda}(\ln t)^n\}=\frac{\Gamma(\lambda+1)}{s^{\lambda+1}}\{\frac{d}{d\lambda}R_{n-1}(\lambda,s)-R_1(\lambda,s)R_{n-1}(\lambda,s)\}$$
$$R_0(\lambda,s)=1 \quad , \quad R_1(\lambda,s)=[\psi(\lambda+1)-\ln s] \qquad (13.5.2)$$
$$\text{Re}\,\lambda>-1 \quad , \quad n=1,2,3,...$$

For example for n = 2

$$L\{t^{\lambda}(\ln t)^2\}=\frac{\Gamma(\lambda+1)}{s^{\lambda+1}}\{\psi'(\lambda+1)+[\psi(\lambda+1)-\ln s]^2\}=$$
$$\frac{\Gamma(\lambda+1)}{s^{\lambda+1}}\{\zeta(2,\lambda+1)+[\psi(\lambda+1)-\ln s]^2\} \qquad (14.5.2)$$

where derivatives of the psi function are related to the generalized zeta function in the following way [1, 6.4.2]

$$\psi^{(k)}(\lambda+1)=(-1)^{k+1}k!\zeta(k+1,\lambda+1)$$
$$k=1,2,3,... \qquad (15.5.2)$$

Using (1.5.2), (2.5.2) and (14.5.2) we have

$$\int_0^t (t-x)^{\lambda}[\ln(t-x)]^2\nu(x,\alpha)dx=$$
$$\Gamma(\lambda+1)\{[\psi(\lambda+1)^2+\zeta(2,\lambda+1)]\,\nu(t,\alpha+\lambda+1)+ \qquad (16.5.2)$$
$$\frac{t^{\alpha+\lambda+1}}{\Gamma(\alpha+\lambda+2)}[\psi(\alpha+\lambda+1)-2\psi(\lambda+1)-\ln t]\}$$

$$\int_0^t (t-x)^{\lambda}[\ln(t-x)]^2 \mu(x,\beta,\alpha)dx =$$
$$\Gamma(\lambda+1)\{[\psi(\lambda+1)^2+\zeta(2,\lambda+1)]\mu(t,\beta,\alpha+\lambda+1)- \tag{17.5.2}$$
$$2\psi(\lambda+1)\mu(t,\beta-1,\alpha+\lambda+1)+\mu(t,\beta-2,\alpha+\lambda+1)\}$$

Evidently, from (16.5.2) for integer values of λ, the convolution integrals of the type $(\ln t)^2 * [t^n \nu(t,\alpha)]$, n = 1,2,3,... can be evaluated in a similar way as those in (12.5.2).

Since [88, 2.6.8, 2.6.24]

$$L^{-1}\{\frac{\ln s}{s^{\lambda+1}}\} = \frac{t^{\lambda}}{\Gamma(\lambda+1)}[\psi(\lambda+1)-\ln t] \tag{18.5.2}$$

$$L^{-1}\{\frac{(\ln s)^2}{s^{\lambda+1}}\} = \frac{t^{\lambda}}{\Gamma(\lambda+1)}\{[\psi(\lambda+1)-\ln t]^2-\psi^{(1)}(\lambda+1)\} \tag{19.5.2}$$

the product theorem of the Laplace transformation gives

$$\frac{t^{\lambda}}{\Gamma(\lambda+1)}[\psi(\lambda+1)-\ln t]*\nu(t,\alpha) = \frac{t^{\alpha+\lambda+1}}{\Gamma(\alpha+\lambda+2)} \tag{20.5.2}$$

$$\frac{t^{\lambda}}{\Gamma(\lambda+1)}[\psi(\lambda+1)-\ln t]*\mu(t,\beta,\alpha) = \mu(t,\beta-1,\alpha+\lambda+1) \tag{21.5.2}$$

and

$$\frac{t^{\lambda}}{\Gamma(\lambda+1)}\{[\psi(\lambda+1)-\ln t]^2-\psi^{(1)}(\lambda+1)\}*\nu(t,\alpha) =$$
$$\frac{t^{\alpha+\lambda+1}}{\Gamma(\alpha+\lambda+2)}[\psi(\alpha+\lambda+2)-\ln t] \tag{22.5.2}$$

$$\frac{t^{\lambda}}{\Gamma(\lambda+1)}\{[\psi(\lambda+1)-\ln t]^2-\psi^{(1)}(\lambda+1)\}*\mu(t,\beta,\alpha) =$$
$$\mu(t,\beta-2,\alpha+\lambda+1) \tag{23.5.2}$$

In particular for α = 0 and λ = 0 we have

$$\int_0^t [\gamma + \ln(t-x)]\nu(x)dx = -t \tag{24.5.2}$$

$$\int_0^t \{[\gamma + \ln(t-x)]^2 - \frac{\pi^2}{6}\}\nu(x)dx = t(1-\gamma-\ln t) \tag{25.5.2}$$

If the Laplace transforms in (1.5.2) are in the form $F(s) = G(s)\cdot\ln s$, then

$$f(t)*\nu(t,\alpha) = \int_0^t f(t-x)\,\nu(x,\alpha)dx = L^{-1}\{\frac{G(s)}{s^{\alpha+1}}\} \tag{26.5.2}$$

This case is associated with the sine and cosine integral functions, the exponential integrals and the Bessel functions. From [88, 2.6.17]

$$L\{f(t)\} = \frac{a\ln s}{s^2+a^2}$$
$$f(t) = \{\cos(at)Si(at) + \sin(at)[\ln a - Ci(at)]\} \tag{27.5.2}$$
$$a > 0$$

it follows that [88, 2.4.33]

$$f(t)*\nu(t,\alpha) = L^{-1}\{\frac{a}{s^{\alpha+1}(s^2+a^2)}\} = \frac{t^\alpha}{\Gamma(\alpha+1)} * \sin(at) =$$
$$\frac{at^{\alpha+2}}{\Gamma(\alpha+3)}\,{}_1F_2(1;\frac{\alpha+3}{2},\frac{\alpha+4}{2};-\frac{a^2t^2}{4}) \tag{28.5.2}$$

For $\alpha = 0$, (28.5.2) reduces to

$$f(t)*\nu(t,\alpha) = \frac{1}{a}[1-\cos(at)] \tag{29.5.2}$$

Similarly from [88, 2.6.18]

$$L\{f(t)\} = \frac{s\ln s}{s^2+a^2}$$
$$f(t) = \{\cos(at)[\ln a - Ci(at)] - \sin(at)Si(at)\} \tag{30.5.2}$$
$$a > 0$$

we have

$$f(t)*v(t,\alpha)=L^{-1}\{\frac{s}{s^{\alpha+1}(s^2+a^2)}\}=\frac{t^{\alpha}}{\Gamma(\alpha+1)}*\cos(at)=$$
$$\frac{t^{\alpha+1}}{\Gamma(\alpha+2)}{}_1F_2(1;\frac{\alpha+2}{2},\frac{\alpha+3}{2};-\frac{a^2t^2}{4}) \qquad (31.5.2)$$

and

$$f(t)*v(t,\alpha)=\frac{1}{a}\sin(at) \qquad (32.5.2)$$

Other examples are available from (25.5.2) in the case of $F(s) = s^{-\rho}e^{\pm a/s}\ln s$, because [88, 2.5.75-76]

$$L^{-1}\{\frac{e^{-a/s}}{s^{\rho}}\}=(\frac{t}{a})^{(\rho-1)/2}J_{\rho-1}(2\sqrt{at}) \qquad (33.5.2)$$

$$L^{-1}\{\frac{e^{a/s}}{s^{\rho}}\}=(\frac{t}{a})^{(\rho-1)/2}I_{\rho-1}(2\sqrt{at}) \qquad (34.5.2)$$
$$a>0 \quad , \quad \operatorname{Re}\rho>0$$

For $\rho = 1/2$ and $\rho = 3/2$ from [88, 2.5.80, 2.5.82] we have the sine and cosine integral functions

$$f(t)=-\frac{1}{\sqrt{\pi t}}\{\frac{1}{2}\ln(\frac{t}{a})\cos(2\sqrt{at})+$$
$$\cos(2\sqrt{at})Ci(4\sqrt{at})+\sin(2\sqrt{at})Si(4\sqrt{at})\} \qquad (35.5.2)$$
$$f(t)*v(t,\alpha)=L^{-1}\{\frac{e^{-a/s}}{s^{\alpha+3/2}}\}=(\frac{t}{a})^{\alpha/2+1/4}J_{\alpha+1/2}(2\sqrt{at})$$

$$f(t)*v(t)=\frac{1}{\sqrt{\pi a}}\sin(2\sqrt{at}) \qquad (36.5.2)$$

$$f(t) = -\frac{1}{\sqrt{\pi t}}\{\frac{1}{2}\ln(\frac{t}{a})\sin(2\sqrt{at}) +$$

$$\sin(2\sqrt{at})Ci(4\sqrt{at}) - \cos(2\sqrt{at})Si(4\sqrt{at})\} \tag{37.5.2}$$

$$f(t) * \nu(t,\alpha) = L^{-1}\{\frac{e^{-a/s}}{s^{\alpha+5/2}}\} = (\frac{t}{a})^{\alpha/2+3/4} J_{\alpha+3/2}(2\sqrt{at})$$

$$f(t) * \nu(t) = \frac{1}{2\sqrt{\pi}a^{3/2}}[\sin(2\sqrt{at}) - 2\sqrt{at}\cos(2\sqrt{at})] \tag{38.5.2}$$

Similarly from [88, 2.5.79, 2.5.81] the exponential integrals are presented

$$f(t) = -\frac{1}{2\sqrt{\pi t}}\{\ln(\frac{t}{a})\cosh(2\sqrt{at}) + e^{2\sqrt{at}} Ei(-4\sqrt{at}) +$$

$$e^{-2\sqrt{at}}\overline{Ei}(-4\sqrt{at})\} \tag{39.5.2}$$

$$f(t) * \nu(t,\alpha) = L^{-1}\{\frac{e^{a/s}}{s^{\alpha+3/2}}\} = (\frac{t}{a})^{\alpha/2+1/4} I_{\alpha+1/2}(2\sqrt{at})$$

$$f(t) * \nu(t) = \frac{1}{\sqrt{\pi t}}\sinh(2\sqrt{at}) \tag{40.5.2}$$

Convolution integrals with the Bessel functions appear for ρ = 1 and ρ = 2 [88, 2.5.77-78]

$$f(t) = -\frac{1}{2}\{\ln(\frac{t}{a})J_0(2\sqrt{at}) + \pi Y_0(2\sqrt{at})\}$$

$$f(t) * \nu(t,\alpha) = L^{-1}\{\frac{e^{-a/s}}{s^{\alpha+2}}\} = (\frac{t}{a})^{\alpha/2+1/2} J_{\alpha+1}(2\sqrt{at}) \tag{41.5.2}$$

$$f(t) = -\frac{1}{2}\sqrt{\frac{t}{a}}\{\ln(\frac{t}{a})J_1(2\sqrt{at}) + \pi Y_1(2\sqrt{at}) + \frac{1}{\sqrt{at}}J_0(2\sqrt{at})\}$$

$$f(t) * \nu(t,\alpha) = L^{-1}\{\frac{e^{-a/s}}{s^{\alpha+3}}\} = (\frac{t}{a})^{\alpha/2+1} J_{\alpha+2}(2\sqrt{at}) \tag{42.5.2}$$

and with the modified Bessel functions [88, 2.5.83-84]

$$f(t)=\{K_0(2\sqrt{at})-\frac{1}{2}\ln(\frac{t}{a})I_0(2\sqrt{at})\}$$
$$f(t)*\nu(t,\alpha)=L^{-1}\{\frac{e^{a/s}}{s^{\alpha+2}}\}=(\frac{t}{a})^{\alpha/2+1/2}I_{\alpha+1}(2\sqrt{at}) \tag{43.5.2}$$

$$f(t)=-\sqrt{\frac{t}{a}}\{\frac{1}{2}\ln(\frac{t}{a})I_1(2\sqrt{at})+\pi K_1(2\sqrt{at})-\frac{1}{2\sqrt{at}}I_0(2\sqrt{at})\}$$
$$f(t)*\nu(t,\alpha)=L^{-1}\{\frac{e^{a/s}}{s^{\alpha+3}}\}=(\frac{t}{a})^{\alpha/2+1}I_{\alpha+2}(2\sqrt{at}) \tag{44.5.2}$$

5.3. Integral Formulas

There is a large group of integrals in the form

$$\int_0^\infty K(t,x)f(x)dx=\Phi(t) \tag{1.5.3}$$

where integrands and $\Phi(t)$ contain the Volterra functions. These solutions of integral equations (e.g. (37.4.2) - (41.4.2), (27.4.6), (31.4.6), (34.4.6), (13.4.8) - (16.4.8)) are based on the substitution formulas of operational calculus of the type

$$L\{\int_0^\infty K(t,x)f(x)dx\}=\frac{1}{s^\rho}F(s^\lambda) \qquad \lambda>0 \tag{2.5.3}$$

Introduction of $f(t)=\mu(t,\beta,\alpha)$ and $F(s)=1/[s^{\alpha+1}(\ln s)^{\beta+1}]$ into (2.5.3) gives

$$L\{\int_0^\infty K(t,x)\mu(x,\beta,\alpha)dx\}=\frac{1}{\lambda^{\beta+1}s^{\rho+\lambda(\alpha+1)}(\ln s)^{\beta+1}} \tag{3.5.3}$$

and

$$\int_0^\infty K(t,x)\mu(x,\beta,\alpha)dx=\mu[t,\beta,\rho+\lambda(\alpha+1)-1]$$
$$t>0 \quad , \quad \alpha>-1 \quad , \quad \beta>-1 \quad , \quad \rho+\lambda(\alpha+1)>0 \tag{4.5.3}$$

where $\mu(t,\beta,\alpha)$ in (4.5.3) represents all Volterra functions because $\nu(t)=\mu(t,0,0)$, $\nu(t,\alpha)=\mu(t,0,\alpha)$ and $\mu(t,\beta)=\mu(t,\beta,0)$.

The first group of integrals with the *heat-transfer kernels,* $K(t,x) = x^n \exp(-x^2/4t)$, n = 0,1,2, ... is based on

$$\frac{d^n}{dt^n}\{t^{-1/2}\int_0^\infty e^{-x^2/4t} f(x)dx\} = L^{-1}\{s^{n-1/2}F(s^{1/2})\} \tag{5.5.3}$$

In applying rule (5.5.3) we find successively the first five integrals

$$\int_0^\infty e^{-x^2/4t}\mu(x,\beta,\alpha)dx = 2^{\beta+1}\sqrt{\pi t}\,\mu(t,\beta,\frac{\alpha}{2}) \tag{6.5.3}$$

$\alpha > -1 \quad , \quad \beta > -1$

$$\int_0^\infty x e^{-x^2/4t}\mu(x,\beta,\alpha)dx = 2^{\beta+2}\sqrt{\pi}\,t^{3/2}\mu(t,\beta,\frac{\alpha-1}{2}) \tag{7.5.3}$$

$\alpha > -1 \quad , \quad \beta > -1$

$$\int_0^\infty x^2 e^{-x^2/4t}\mu(x,\beta,\alpha)dx = 2^{\beta+2}\sqrt{\pi}\,t^{5/2}\{2\mu(t,\beta,\frac{\alpha}{2}-1)+ \frac{1}{t}\mu(t,\beta,\frac{\alpha}{2})\} \tag{8.5.3}$$

$\alpha > 0 \quad , \quad \beta > -1$

$$\int_0^\infty x^3 e^{-x^2/4t}\mu(x,\beta,\alpha)dx = 2^{\beta+3}\sqrt{\pi}\,t^{7/2}\{2\mu(t,\beta,\frac{\alpha+3}{2})+ \frac{3}{t}\mu(t,\beta,\frac{\alpha-1}{2})\} \tag{9.5.3}$$

$\alpha > 1 \quad , \quad \beta > -1$

$$\int_0^\infty x^4 e^{-x^2/4t}\mu(x,\beta,\alpha)dx = 2^{\beta+4}\sqrt{\pi}\,t^{9/2}\{2\mu(t,\beta,\frac{\alpha}{2}-2)+ \frac{3}{t}\mu(t,\beta,\frac{\alpha}{2}-1)\} \tag{10.5.3}$$

$\alpha > 2 \quad , \quad \beta > -1$

From operational formula with the parabolic cylinder function [88, 1.62]

$$\sqrt{\frac{2}{\pi}}(2t)^{-(\lambda+1)/2}\int_0^\infty e^{-x^2/8t}D_\lambda(\frac{x}{\sqrt{2t}})f(x)dx = \\ L^{-1}\{s^{(\lambda-1)/2}F(s^{1/2})\} \tag{11.5.3}$$

we have

$$\int_0^\infty e^{-x^2/8t}D_\lambda(\frac{x}{\sqrt{2t}})\mu(x,\beta,\alpha)dx = \\ 2^{\beta+\lambda/2+1}\sqrt{\pi}\,t^{(\lambda+1)/2}\mu(x,\beta,\frac{\alpha-\lambda}{2}) \\ \alpha > -1 \quad , \quad \beta > -1 \quad , \quad \alpha-\lambda > -2 \tag{12.5.3}$$

In particular for $\lambda = -\alpha$

$$\int_0^\infty e^{-x^2/8t}D_{-\alpha}(\frac{x}{\sqrt{2t}})\mu(x,\beta,\alpha)dx = \\ 2^{\beta-\alpha/2+1}\sqrt{\pi}\,t^{(1-\alpha)/2}\mu(x,\beta,\alpha) \tag{13.5.3}$$

For special values of λ, the parabolic cylinder function is reduced to other special functions. If $\lambda = -(n+1)$, n = 0,1,2,3, ... the complementary error function is involved [80, 8.1.2]

$$D_{-n-1}(u) = \frac{(-1)^n}{n!}\sqrt{\frac{\pi}{2}}\,e^{-u^2/4}\frac{d^n}{du^n}[e^{u^2/2}erfc(\frac{u}{\sqrt{2}})] \tag{14.5.3}$$

Using (12.5.3) and (14.5.3), the first three integrals are

$$\int_0^\infty erfc(\frac{x}{2\sqrt{t}})\mu(x,\beta,\alpha)dx = 2^{\beta+1}\mu(t,\beta,\frac{\alpha+1}{2}) \tag{15.5.3}$$

$$\int_0^\infty x\,erfc(\frac{x}{2\sqrt{t}})\mu(x,\beta,\alpha)dx = 2^{\beta+1}\{2t\mu(t,\beta,\frac{\alpha+1}{2}) - \mu(t,\beta,\frac{\alpha}{2}+1)\} \tag{16.5.3}$$

$$\int_0^\infty x^2 erfc(\frac{x}{2\sqrt{t}})\mu(x,\beta,\alpha)dx = 2^{\beta+2}\{\mu(t,\beta,\frac{\alpha+3}{2}) - t\mu(t,\beta,\frac{\alpha+1}{2}) +$$

$$2t^2\mu(t,\beta,\frac{\alpha-1}{2})\} \qquad (17.5.3)$$

$$\alpha > -1 \quad , \quad \beta > -1$$

For the parameter λ being a positive integer, the parabolic cylinder function becomes the Hermite polynomial

$$D_n(u) = 2^{-n/2} e^{-u^2/4} H_n(\frac{u}{\sqrt{2}}) \qquad (18.5.3)$$

and therefore

$$\int_0^\infty e^{-x^2/4t} H_n(\frac{x}{2\sqrt{t}})\mu(x,\beta,\alpha)dx = 2^{\beta+n+1}\sqrt{\pi}\, t^{(n+1)/2}\mu(t,\beta,\frac{\alpha-n}{2}) \qquad (19.5.3)$$

$$\alpha > -1 \quad , \quad \beta > -1 \quad , \quad \alpha - n > -2$$

Combination of the parabolic cylinder functions with $\lambda = \pm$ (n-1/2), n = 0,1,2,3, ... leads to the modified Bessel functions $K_{n/2-1/4}$ ($u^2/4$), for example [80, 8.1.2]

$$K_{1/4}(\frac{u^2}{4}) = \sqrt{2\pi}\, u^{-1/2} D_{-1/2}(u) \qquad (20.5.3)$$

$$K_{3/4}(\frac{u^2}{4}) = \sqrt{\frac{\pi}{2}}\, u^{-3/2}[D_{-3/2}(u) + 2D_{1/2}(u)] \qquad (21.5.3)$$

and using (12.5.3) we have

$$\int_0^\infty x^{1/2} e^{-x^2/8t} K_{1/4}(\frac{x^2}{8t})\mu(x,\beta,\alpha)dx = 2^{\beta+3/2}\pi t^{1/2}\mu(t,\beta,\frac{2\alpha+1}{4}) \qquad (22.5.3)$$

$$\alpha > -1 \quad , \quad \beta > -1$$

$$\int_0^\infty x^{3/2} e^{-x^2/8t} K_{3/4}(\frac{x^2}{8t}) \mu(x,\beta,\alpha) dx = 2^{\beta+1/2} \pi t^{1/2} [\mu(t,\beta,\frac{2\alpha+3}{4}) +$$

$$4t\mu(t,\beta,\frac{2\alpha-1}{4})] \qquad (23.5.3)$$

$$\alpha > -1 \quad , \quad \beta > -1$$

From the operational rules dealing with the Laplace transform inverses of the type $1/s^\rho F(s^{1/3})$, $\rho = 0$, 1/3 and 2/3

$$\frac{1}{3\pi t^{3/2}} \int_0^\infty x^{3/2} K_{1/3}(\frac{2x^{3/2}}{\sqrt{27t}}) f(x) dx = L^{-1}\{F(s^{1/3})\} \qquad (24.5.3)$$

$$\frac{1}{\sqrt{3}\pi t} \int_0^\infty x K_{2/3}(\frac{2x^{3/2}}{\sqrt{27t}}) f(x) dx = L^{-1}\{\frac{1}{s^{1/3}} F(s^{1/3})\} \qquad (25.5.3)$$

$$\frac{1}{\pi\sqrt{t}} \int_0^\infty x^{1/2} K_{1/3}(\frac{2x^{3/2}}{\sqrt{27t}}) f(x) dx = L^{-1}\{\frac{1}{s^{2/3}} F(s^{1/3})\} \qquad (26.5.3)$$

the corresponding integrals are

$$\int_0^\infty x^{3/2} K_{1/3}(\frac{2x^{3/2}}{\sqrt{27t}}) \mu(x,\beta,\alpha) dx = 3^{\beta+2} \pi t^{3/2} \mu(t,\beta,\frac{\alpha-2}{3}) \qquad (27.5.3)$$

$$\int_0^\infty x K_{2/3}(\frac{2x^{3/2}}{\sqrt{27t}}) \mu(x,\beta,\alpha) dx = 3^{\beta+3/2} \pi t \mu(t,\beta,\frac{\alpha-1}{3}) \qquad (28.5.3)$$

$$\int_0^\infty x^{1/2} K_{1/3}(\frac{2x^{3/2}}{\sqrt{27t}}) \mu(x,\beta,\alpha) dx = 3^{\beta+1} \pi t^{1/2} \mu(t,\beta,\frac{\alpha}{3}) \qquad (29.5.3)$$

Considering that the modified Bessel functions in (27.5.3) - (29.5.3) can be represented in terms of the Airy function $Ai(u)$ [1]

$$Ai(u) = \frac{1}{\pi}\sqrt{\frac{u}{3}} K_{1/3}(\frac{2}{3}u^{3/2}) \qquad (30.5.3)$$

$$Ai'(u) = -\frac{u}{\pi\sqrt{3}} K_{\pm 2/3}(\frac{2}{3}u^{3/2}) \tag{31.5.3}$$

the integrals in (27.5.3)-(29.5.3) can be written in the form

$$\int_0^\infty u\,Ai(u)\mu[(3t)^{1/3}u,\beta,\alpha]du = 3^{\beta+2/3}t^{2/3}\mu(t,\beta,\frac{\alpha-2}{3}) \tag{32.5.3}$$

$$\int_0^\infty Ai'(u)\mu[(3t)^{1/3}u,\beta,\alpha]du = -3^{\beta+1/3}t^{1/3}\mu(t,\beta,\frac{\alpha-1}{3}) \tag{33.5.3}$$

$$\int_0^\infty Ai(u)\mu[(3t)^{1/3}u,\beta,\alpha]du = 3^{\beta}\mu(t,\beta,\frac{\alpha}{3}) \tag{34.5.3}$$

Vallée generalized the result in (33.5.3) by deriving the integral with the nth derivative of the Airy function [112]

$$\int_0^\infty Ai^{(n)}(u)\mu[(3t)^{1/3}u,\beta,\alpha]du = (-1)^n 3^{\beta+n/3}t^{n/3}\mu(t,\beta,\frac{\alpha-n}{3}) \tag{35.5.3}$$

and also the integral with the square of the Airy function

$$\int_0^\infty Ai^2(u)\mu[(3t)^{1/3}u,\beta,\alpha]du = \frac{3^{\beta-1/6}}{2\sqrt{\pi}}t^{-1/6}\mu(\frac{t}{4},\beta,\frac{2\alpha+1}{6}) \tag{36.5.3}$$

In some cases infinite integrals of the Volterra functions can be expressed as convolution integrals. One example is presented here, others will be considered in Section 5.4. From

$$e^{a^2t}\int_0^\infty e^{ax}erfc(\frac{x}{2\sqrt{t}}+a\sqrt{t})f(x)dx = L^{-1}\{\frac{F(s^{1/2})}{s^{1/2}(s^{1/2}+a)}\} \tag{37.5.3}$$

$$a \geq 0$$

and

$$L^{-1}\{\frac{1}{s^{1/2}(s^{1/2}+a)}\} = e^{a^2t}erfc(a\sqrt{t}) \tag{38.5.3}$$

it follows that

$$\int_0^\infty e^{ax} erfc(\frac{x}{2\sqrt{t}} + a\sqrt{t})\mu(x,\beta,\alpha)dx =$$

$$2^{\beta+1}\int_0^t e^{-a^2(t-u)} erfc(a\sqrt{t-u})\mu(u,\beta,\frac{\alpha-1}{2})du = \tag{39.5.3}$$

$$2^{\beta+1}[e^{-a^2 t} erfc(a\sqrt{t}) * \mu(t,\beta,\frac{\alpha-1}{2})]$$

However, using the equivalent form of (30.5.3)

$$\frac{F(s^{1/2})}{s^{1/2}(s^{1/2}+a)} = \frac{1}{s-a^2}[F(s^{1/2}) - \frac{a}{s^{1/2}}F(s^{1/2})] \tag{40.5.3}$$

the integral in (39.5.3) can also be expressed as

$$\int_0^\infty e^{ax} erfc(\frac{x}{2\sqrt{t}} + a\sqrt{t})\mu(x,\beta,\alpha)dx =$$

$$2^{\beta+1}\int_0^t e^{-a^2(t-u)}[\mu(u,\beta,\frac{\alpha-1}{2}) - a\mu(u,\beta,\frac{\alpha}{2})]du = \tag{41.5.3}$$

$$2^{\beta+1}e^{-a^2 t} * [\mu(t,\beta,\frac{\alpha-1}{2}) - a\mu(t,\beta,\frac{\alpha}{2})]$$

5.4. Integrals with Respect to Parameters of the Volterra Functions

The substitution formulas of the Laplace transformation (53.4.3) and (54.4.3)

$$\int_0^\infty \nu(t,\alpha+x)f(x)dx = L^{-1}\{\frac{F(\ln s)}{s^{\alpha+1}\ln s}\} \tag{1.5.4}$$

$$\mathrm{Re}\,\alpha > -1 \quad , \quad \mathrm{Re}\,s > 1$$

$$\int_0^\infty \mu(t,\beta,\alpha+x)f(x)dx = L^{-1}\{\frac{F(\ln s)}{s^{\alpha+1}(\ln s)^{\beta+1}}\} \tag{2.5.4}$$

$$\mathrm{Re}\,\alpha > -1 \quad , \quad \mathrm{Re}\,\beta > -1 \quad , \quad \mathrm{Re}\,s > 1$$

can be used for evaluation of the left-hand side integrals, if the inverse Laplace transforms in (1.5.4) and (2.5.4) are available. If the Laplace transforms $F(s)$ in (1.5.4) or in (2.5.4) are expressed by elementary and special functions of s, then the corresponding integrals with respect to parameters can frequently be reduced to the convolution integrals of the Volterra functions.

In the simplest case, $f(t) = t^{\lambda-1}$ with $F(s) = \Gamma(\lambda)/s^{\lambda}$, it follows from (1.5.4) that

$$\int_0^\infty x^{\lambda-1}\nu(t,\alpha+x)dx = L^{-1}\{\frac{\Gamma(\lambda)}{s^{\alpha+1}(\ln s)^{\lambda+1}}\} = \Gamma(\lambda)\mu(t,\lambda,\alpha) \qquad (3.5.4)$$

$$\operatorname{Re}\alpha > -1 \quad , \quad \operatorname{Re}\lambda > 0$$

and similarly using (2.5.4)

$$\int_0^\infty x^{\lambda-1}\mu(t,\beta,\alpha+x)dx = L^{-1}\{\frac{F(\ln s)}{s^{\alpha+1}(\ln s)^{\beta+\lambda+1}}\} = \Gamma(\lambda)\mu(t,\beta+\lambda,\alpha) \qquad (4.5.4)$$

$$\operatorname{Re}\alpha > -1 \quad , \quad \operatorname{Re}\beta > -1 \quad , \quad \operatorname{Re}\lambda > 0$$

In particular for $\lambda = 1$

$$\int_0^\infty \nu(t,\alpha+x)dx = \mu(t,1,\alpha) \qquad (5.5.4)$$

$$\int_0^\infty \nu(t,x)dx = \int_0^\infty \frac{t^u}{\Gamma(u)}du = \mu(t,1) \qquad (6.5.4)$$

$$\int_0^\infty \nu(1,x)dx = \int_0^\infty \frac{1}{\Gamma(u)}du = \mu(1,1) \qquad (7.5.4)$$

and

$$\int_0^\infty \mu(t,\beta,\alpha+x)dx = \Gamma(\lambda)\mu(t,\beta+1,\alpha) \qquad (8.5.4)$$

$$\int_0^\infty \mu(1,1,x)dx = \frac{1}{2}\int_0^\infty \frac{u}{\Gamma(u)}du = \mu(1,2) \qquad (9.5.4)$$

The finite integrals of the Volterra functions with respect to parameters can be evaluated from (1.5.4) using the Laplace transforms of sectionally defined functions, for example from [88, 1.2.4]

$$L\{[u(t)-u(t-a)]t^n\}=\frac{n!}{s^{n+1}}-n!e^{-as}\sum_{k=0}^{n}\frac{a^k}{k!s^{n-k+1}} \tag{10.5.4}$$

we have

$$\int_0^a x^n \nu(t,\alpha+x)dx = n!\{\mu(t,n+1,\alpha)-\sum_{k=0}^{n}\frac{a^k}{k!}\mu(t,n-k+1,\alpha+a)\} \tag{11,5.4}$$

which for n = 0 and n = 1 is

$$\int_0^a \nu(t,\alpha+x)dx = \mu(t,1,\alpha)-\mu(t,1,\alpha+a) \tag{12.5.4}$$

$$\int_0^a x\nu(t,\alpha+x)dx = \mu(t,2,\alpha)-\mu(t,2,\alpha+a)-a\mu(t,1,\alpha+a) \tag{13.5.4}$$

In the case of exponential function, $f(t)$ = e^{-at}, a > 0 with $F(s)$ = $1/(s+a)$, (1.5.4) gives

$$\int_0^\infty e^{-ax}\nu(t,\alpha+x)dx = L^{-1}\{\frac{1}{s^{\alpha+1}\ln s(\ln s+a)}\} =$$

$$\frac{1}{a}L^{-1}\{\frac{1}{s^{\alpha+1}\ln s}-\frac{1}{s^{\alpha+1}\ln(se^a)}\}=\frac{1}{a}\{\nu(t,\alpha)-e^{\alpha a}\nu(e^{-a}t,\alpha)\} \tag{14.5.4}$$

$$\text{Re}\,\alpha > -1 \quad , \quad a>0$$

This result can be considered as the Laplace transform of the Volterra function $f(t)$ = $\nu(a,t+\alpha)$ which for $\alpha = 0$ was already given in (31.4.1). Similarly, it is possible to obtain

$$\int_0^\infty e^{ax}\nu(t,\alpha+x)dx = \frac{1}{a}\{e^{-\alpha a}\nu(e^a t,\alpha)-\nu(t,\alpha)\} \tag{15.5.4}$$

$$\int_0^\infty x^\lambda \nu(t,\alpha+x)dx = \Gamma(\lambda+1)e^{(\alpha-1)a}[\nu(t)*\mu(e^{-a}t,\lambda,\alpha-1)] \tag{16.5.4}$$

$$\int_0^\infty \cosh(ax)\nu(t,\alpha+x)dx = \frac{1}{2a}\{e^{-\alpha a}\nu(e^a t,\alpha) - e^{\alpha a}\nu(e^{-a}t,\alpha)\} \tag{17.5.4}$$

$$\int_0^b e^{-ax}\nu(t,\alpha+x)dx = \nu(e^{-a}t) * [\nu(t,\alpha-1) - e^{-ab}\nu(t,\alpha+b-1)] \tag{18.5.4}$$

In more complex cases [44, 4.5.10] and [88, 5.16]

$$L\{\frac{t^{\lambda-1}}{1-e^{-at}}\} = \frac{\Gamma(\lambda)}{a^\lambda}\zeta(\lambda,\frac{s}{a}) = \frac{\Gamma(\lambda)}{a^\lambda}\sum_{n=0}^\infty (n+\frac{s}{a})^{-\lambda} \tag{19.5.4}$$

$$L\{\frac{1}{1+e^{-at}}\} = \frac{1}{2}[\psi(\frac{s+1}{2}) - \psi(\frac{s}{2})] = \sum_{n=0}^\infty \frac{(-1)^n}{s+n} \tag{20.5.4}$$

we have

$$\int_0^\infty \frac{x^{\lambda-1}}{1-e^{-ax}}\nu(t,\alpha+x)dx = \frac{\Gamma(\lambda)}{a^\lambda}\nu(t) * \sum_{n=0}^\infty e^{\alpha a(n-1)}\mu(te^{-an},\lambda-1,\alpha-1) \tag{21.5.4}$$

$\mathrm{Re}\,\lambda > 0$

$$\int_0^\infty \frac{\nu(t,\alpha+x)}{1+e^{-ax}}dx = \nu(t) * \sum_{n=0}^\infty (-1)^n e^{(\alpha-1)n}\nu(te^{-n},\alpha-1) \tag{22.5.4}$$

From the Laplace transforms [88, 1.2.9], (46.4.1), [52, 3.353.2], [44, 4.2.11 and 4.2.24]

$$L\{\frac{1}{t+\lambda}\} = -e^{\lambda s}Ei(-\lambda s) \tag{23.5.4}$$

$$L\{\nu i(t,\lambda\} = -\frac{1}{s}Ei(-\lambda s) \tag{24.5.4}$$

$$L\{\frac{1}{(t+\lambda)^2}\} = \frac{1}{\lambda} + se^{\lambda s}Ei(-\lambda s) \tag{25.5.4}$$

$\lambda > 0$

$$L\{\frac{t^n}{t+\lambda}\} = (-1)^{n-1}\lambda^n e^{\lambda s}Ei(-\lambda s) + \sum_{k=1}^n \frac{(-\lambda)^{n-k}(k-1)!}{s^k} \tag{26.5.4}$$

$$L\{\frac{1+2\lambda t}{\sqrt{\sqrt{t}}}\} = \pi\frac{s+\lambda}{s^{3/2}} \tag{27.5.4}$$

using (1.5.4), the following integrals of the Volterra functions with rational functions can be derived

$$\int_0^\infty \frac{\nu(t,\alpha+x)}{\lambda+x}dx = \nu(t,\alpha-\lambda-1)*\nu i(t,\lambda) \tag{28.5.4}$$

$$\int_0^\infty \frac{\nu(t,\alpha+x)}{(\lambda+x)^2}dx = \frac{\nu(t,\alpha)}{\lambda} - \frac{t^{\alpha-\lambda-1}}{\Gamma(\alpha-\lambda)}*\nu i(t,\lambda) \tag{29.5.4}$$

$\alpha > \lambda$

$$\int_0^\infty \frac{x^n \nu(t,\alpha+x)}{\lambda+x}dx = (-\lambda)^n \nu(t,\alpha-\lambda-1)*\nu i(t,\lambda) + \sum_{k=1}^{n}(-\lambda)^{n-k}(k-1)!\mu(t,k,\alpha) \tag{30.5.4}$$

$$\int_0^\infty \frac{(1+2\lambda x)}{\sqrt{x}}\nu(t,\alpha+x)dx = \sqrt{\pi}[\mu(t,\frac{1}{2},\alpha)+\lambda\mu(t,\frac{3}{2},\alpha)] \tag{31.5.4}$$

Using the Laplace transforms of logarithmic functions [88, 1.6.4 and 1.6.6]

$$L\{u(t-a)\ln t\} = \frac{1}{s}[e^{-as}\ln a - Ei(-as)] \tag{32.5.4}$$

$$L\{\ln(t+a)\} = \frac{1}{s}[\ln a - e^{as}Ei(-as)] \tag{33.5.4}$$

we have

$$\int_a^\infty \ln x\nu(t,\alpha+x)dx = \ln a\mu(t,1,\alpha+a) + \mu(t,1,\alpha-1)*\nu i(t,a) \tag{34.5.4}$$

$$\int_a^\infty \ln(x+a)\nu(t,\alpha+x)dx = \ln a\mu(t,1,\alpha) + \mu(t,1,\alpha-a-1)*\nu i(t,a) \tag{35.5.4}$$

In the case of logarithmic functions with composed arguments [102, 4.3.1, 4.3.6, 5.4.4, 5.4.6]

$$L\{\ln(1+e^{t})\}=\frac{\ln 2}{2s}[\psi(\frac{s+1}{2})-\psi(\frac{s}{2})]=\frac{\ln 2}{s}\sum_{n=0}^{\infty}\frac{(-1)^{n}}{s+n} \tag{36.5.4}$$

$$L\{\frac{\ln(1-e^{-t})}{(1-e^{-t})}\}=\sum_{n=1}^{\infty}\frac{(-1)^{n}}{s+n}[1-\frac{1}{2}+\frac{1}{3}-\ldots\frac{(-1)^{n-1}}{n}] \tag{37.5.4}$$

$$L\{\ln[2\sin(\frac{at}{2})]\}=-s\sum_{n=0}^{\infty}\frac{1}{n(s^{2}+n^{2}a^{2})} \tag{38.5.4}$$

$$L\{\ln[2\cos(\frac{at}{2})]\}=s\sum_{n=0}^{\infty}\frac{(-1)^{n-1}}{n(s^{2}+n^{2}a^{2})} \tag{39.5.4}$$

use of (1.5.4) gives

$$\int_{0}^{\infty}\ln(1+e^{x})\nu(t,\alpha+x)dx=\ln 2\mu(t,1)*\sum_{n=1}^{\infty}(-1)^{n}e^{(\alpha-1)n}\nu(te^{-n},\alpha-1) \tag{40.5.4}$$

$$\begin{aligned}&\int_{0}^{\infty}\frac{\ln(1-e^{x})}{1-e^{x}}\nu(t,\alpha+x)dx=\\&\nu(t)*\sum_{n=1}^{\infty}\{(-1)^{n}e^{(\alpha-1)n}[1-\frac{1}{2}+\frac{1}{3}-\ldots\frac{(-1)^{n-1}}{n}]\nu(te^{-n},\alpha-1)\}\end{aligned} \tag{41.5.4}$$

$$\int_{0}^{\infty}\ln[2\sin(\frac{ax}{2})]\nu(t,\alpha+x)dx=-\frac{t^{\alpha-1}}{a\Gamma(\lambda)}*\sum_{n=1}^{\infty}\frac{1}{n^{2}}Ts(t,\frac{an}{\pi}) \tag{42.5.4}$$

$$\int_{0}^{\infty}\ln[2\cos(\frac{ax}{2})]\nu(t,\alpha+x)dx=\frac{t^{\alpha-1}}{a\Gamma(\lambda)}*\sum_{n=1}^{\infty}\frac{(-1)^{n-1}}{n^{2}}Ts(t,\frac{an}{\pi}) \tag{43.5.4}$$

Integrals with trigonometric functions can be expressed in terms of the auxiliary Volterra functions which are defined in (13.2.2) and (14.2.2). For $f(t)$ = cos(at) with $F(s) = s/(s^2+a^2)$ and $\alpha = 0$, we have

$$\int_0^\infty \cos(ax)\nu(t,x)dx = L^{-1}\{\frac{1}{s[(\ln s)^2 + a^2]}\} \tag{44.5.4}$$

and using (18.4.1)

$$\int_0^\infty \cos(ax)\nu(t,x)dx = \frac{1}{a}Ts(t,\frac{a}{\pi}) \tag{45.5.4}$$

Similarly from $f(t) = \sin(at)$ and $F(s) = 1/(s^2+a^2)$ it follows that

$$\begin{gathered}\int_0^\infty \sin(ax)\nu(t,x)dx = L^{-1}\{\frac{1}{s\ln s[(\ln s)^2 + a^2]}\} = \\ \frac{1}{a}L^{-1}\{\frac{1}{s\ln s} - \frac{1}{s[(\ln s)^2 + a^2]}\}\end{gathered} \tag{46.5.4}$$

and using (17.4.1)

$$\int_0^\infty \sin(ax)\nu(t,x)dx = \frac{1}{a}[\nu(t) - Tc(t,\frac{a}{\pi})] \tag{47.5.4}$$

If $\alpha > 0$, then the integrals with trigonometric functions can only be expressed as convolution integrals

$$\begin{gathered}\int_0^\infty \cos(ax)\nu(t,\alpha+x)dx = L^{-1}\{\frac{1}{s^{\alpha+1}[(\ln s)^2 + a^2]}\} = \\ L^{-1}\{\frac{1}{s^\alpha}\cdot\frac{1}{s[(\ln s)^2 + a^2]}\} = \frac{t^{\alpha-1}}{a\Gamma(\alpha)} * Ts(t,\frac{a}{\pi})\end{gathered} \tag{48.5.4}$$

and

$$\begin{gathered}\int_0^\infty \sin(ax)\nu(t,\alpha+x)dx = \frac{1}{a}L^{-1}\{\frac{1}{s^{\alpha+1}\ln s} - \frac{\ln s}{s^{\alpha+1}[(\ln s)^2 + a^2]}\} = \\ \frac{1}{a}[\nu(t,\alpha) - \frac{t^{\alpha-1}}{a\Gamma(\alpha)} * Tc(t,\frac{a}{\pi})]\end{gathered} \tag{49.5.4}$$

Convolution integrals also appear when the function $\nu(t,\alpha+x)$ is replaced by $\mu(t,\beta,\alpha+x)$, for example

$$\int_0^\infty e^{-ax}\mu(t,\beta,\alpha+x)dx = L^{-1}\{\frac{1}{s\ln(se^a)}\cdot\frac{1}{s^\alpha(\ln s)^{\beta+1}}\} = \nu(e^{-a}t)*\mu(t,\beta,\alpha-1) \tag{50.5.4}$$

and

$$\int_0^\infty \sin(ax)\mu(t,\beta,\alpha+x)dx = L^{-1}\{\frac{a}{s[(\ln s)^2+a^2]}\cdot\frac{1}{s^\alpha(\ln s)^{\beta+1}}\} = Ts(t,\frac{a}{\pi})*\mu(t,\beta,\alpha-1) \tag{51.5.4}$$

$$\int_0^\infty \cos(ax)\mu(t,\beta,\alpha+x)dx = L^{-1}\{\frac{\ln s}{s[(\ln s)^2+a^2]}\cdot\frac{1}{s^\alpha(\ln s)^{\beta+1}}\} = Tc(t,\frac{a}{\pi})*\mu(t,\beta,\alpha-1) = \frac{1}{a}Ts(t,\frac{a}{\pi})*\mu(t,\beta-1,\alpha-1) \tag{52.5.4}$$

In some cases, the Laplace transforms are in such forms that it is possible to obtain integrals of the Volterra functions with special functions. For example, from [102, 30.3.2.2, 88, 17.6.2, 44, 4.12.30]

$$L\{t^{\lambda-1}{}_1F_1(b;\lambda;ax)\} = \Gamma(\lambda)\frac{s^{b-\lambda}}{(s-a)^b} \qquad \operatorname{Re}\lambda > 0 \tag{53.5.4}$$

$$L\{erf(\sqrt{at})\} = \frac{\sqrt{a}}{s\sqrt{s+a}} \tag{54.5.4}$$

$$L\{\Gamma(\lambda,at)\} = \frac{\Gamma(\lambda)}{s}[1-\frac{a^\lambda}{(s+a)^\lambda}] \qquad Re\lambda > -1 \ , \ \operatorname{Re}(s-a) > 0 \tag{55.5.4}$$

using (1.5.4) we have

$$\int_0^\infty x^{\lambda-1}{}_1F_1(b;\lambda;at)\nu(t,\alpha+x)dx = \Gamma(\lambda)\mu(t,\lambda-b,\alpha-1)*\mu(te^a,b-1) \tag{56.5.4}$$

$$\int_0^\infty erf(\sqrt{at})\nu(t,\alpha+x)dx = \sqrt{a}\mu(t,1,\alpha-1) * \mu(te^{-a},-\frac{1}{2}) \tag{57.5.4}$$

$$\int_0^\infty \Gamma(\lambda,ax)\nu(t,\alpha+x)dx = \\ \Gamma(\lambda)[\mu(t,1,\alpha)-a^{(\alpha-1)a}\mu(t,1) * \mu(te^{-a},\lambda-1,\alpha-1) \tag{58.5.4}$$

The Mittag-Leffler type functions are of special interest because they represent many elementary and special functions (see Appendix A). From (A-7)

$$L\{E_\alpha(at^\rho)\} = L\{\sum_{n=0}^\infty \frac{(at^\rho)^n}{\Gamma(\alpha n+1)}\} = \frac{1}{s}\sum_{n=0}^\infty \frac{\Gamma(\rho n+1)}{\Gamma(\alpha n+1)}\left(\frac{a}{s^\rho}\right)^n \\ \alpha>0 \quad , \quad \rho>0 \quad \mathrm{Re}(s-a)>0 \tag{59.5.4}$$

it follows that

$$\int_0^\infty E_\alpha(ax^\rho)\nu(t,\alpha+x)dx = \sum_{n=0}^\infty \frac{\Gamma(\rho n+1)a^n}{\Gamma(\alpha n+1)}\mu(t.\rho n+1,\alpha) \tag{60.5.4}$$

For $\alpha = \rho = 2$, the Mittag-Leffler function, $E_2(at^2)$, reduces to cosh (at) and therefore we have

$$\int_0^\infty \cosh(ax)\nu(t,\alpha+x)dx = \sum_{n=0}^\infty a^n\mu(t,2n+1,\alpha) = \\ \frac{1}{2a}\{e^{-\alpha a}\nu(e^a t,\alpha)-e^{\alpha a}\nu(e^{-a}t,\alpha)\} \tag{61.5.4}$$

Comparing (61.5.4) with (29.2.2) gives

$$\sum_{n=0}^\infty [\mu(t,n+1,\alpha)-2\mu(t,2n+1,\alpha)] = e^{-\alpha}\nu(\frac{t}{e},\alpha) \tag{62.5.4}$$

In the case of the generalized Mittag-Leffler function, the Laplace transform is given in (A-51)

$$L\{E_{\alpha,\beta}(at^\rho)\} = L\{\sum_{n=0}^\infty \frac{(at^\rho)^n}{\Gamma(\alpha n+\beta)}\} = \frac{1}{s}\sum_{n=0}^\infty \frac{\Gamma(\rho n+1)}{\Gamma(\alpha n+\beta)}\left(\frac{a}{s^\rho}\right)^n \tag{63.5.4}$$

In particular

$$L\{\sqrt{t}E_{1,3/2}(t)\} = L\{e^t erf(\sqrt{t})\} = \frac{\sqrt{s}}{s-1} - \frac{1}{\sqrt{s}} \tag{64.5.4}$$

$$L\{t^\beta E_{2,\beta}(-t^2)\} = L\{\frac{t^{\beta-1}}{\Gamma(\beta)} {}_1F_2(1;\frac{\beta}{2},\frac{\beta+1}{2};\frac{t^2}{4}\} = \frac{1}{s^{\beta-2}(s^2+1)} \tag{65.5.4}$$

and therefore

$$\int_0^\infty e^x erf(\sqrt{x})\nu(t,\alpha+x)dx = \nu(te) * \mu(t,\frac{1}{2},\alpha-1) \tag{66.5.4}$$

$$\begin{aligned}&\int_0^\infty x^{\beta-1} {}_1F_2(1;\frac{\beta}{2},\frac{\beta+1}{2};\frac{x^2}{4})\nu(t,\alpha+x)dx = \\ &\Gamma(\beta)\mu(t,\beta-1,\alpha-1) * Tc(t,\frac{1}{\pi})\end{aligned} \tag{67.5.4}$$

5.5. Infinite Integrals Associated with the Inverse Laplace Transforms of $1/[\ln(s^2+a^2)]^n$ with n = 1 and 2

In the book devoted to the life and work of Ramanujan [55], Hardy postulated that the integrals

$$\frac{1}{(1-x)^\alpha} - \int_0^\infty \frac{x^u\Gamma(u+\alpha)}{\Gamma(\alpha)\Gamma(u+1)}du = \int_0^\infty \frac{(1+xu)^{-\alpha}}{u[\pi^2+(\ln u)^2]}du \tag{1.5.5}$$

can be evaluated in terms of elementary functions if α is a positive integer. These and other related integrals were derived by Apelblat and Kravitsky [5] using the inverse Laplace transforms of the type $1/[\ln(s^2+a^2)]^n$, n = 1,2. Evidently, for a = 0 they are directly related to the integral representation of the Volterra function $\nu(t)$.

If the inversion integral expressing $f(t)$ in terms of the Laplace transform $F(s)$

$$f(t) = L^{-1}\{F(s)\} = \frac{1}{2\pi i}\int_{c-i\infty}^{c+i\infty} e^{st}F(s)ds \tag{2.5.5}$$

is written, as a result of complex integration, in the form

$$f(t)=\int_0^\infty g(t,u)du+h(t)\ f(t)=\int_0^\infty g(t,u)du+h(t) \tag{3.5.5}$$

and the Laplace transforms $L\{g(t,u)\} = G(s)$ and $L\{h(t)\} = H(s)$ are known then we have

$$\int_0^\infty G(s,u)du = F(s)-H(s) \tag{4.5.5}$$

Thus, using (3.5.5) and the Laplace transforms $G(s)$ and $H(s)$ it is possible to evaluate integrals in (4.5.5). The functional form of $g(t,u)$ and $h(t)$ depends on the chosen contour of the complex integration, and therefore this permits to obtain a number of different integrals with the use of the same $F(s)$.

In the first example we will start with the evaluation of

$$f(t)=L^{-1}\{F(s)\}=\frac{1}{2\pi i}\int_{c-i\infty}^{c+i\infty}\frac{e^{st}}{\ln(s^2+a^2)}ds \tag{5.5.5}$$

The integrand in (5.5.5) has the branch points s_1 and s_2 from $s^2 + a^2 = 0$ and the removable singularities p_1 and p_2 from $s^2 + a^2 = 1$. The location of $s_{1,2}$ and $p_{1,2}$ depends on the value of a. Four different Bromwich contours need to be chosen, for $a > 1$ with $s_{1,2} = \pm ia$ and $p_{1,2}=\pm i\sqrt{a^2-1}$; for $a = 1$ with $s_{1,2} = \pm i$ and $p_{1,2} = 0$; for $0 < a < 1$ with $s_{1,2} = \pm ia$ and $p_{1,2}=\pm\sqrt{1-a^2}$ and for $a = 0$ with $s_{1,2} = 0$ and $p_{1,2} = 1$. The final results of complex integration are

$$f(t)=2\int_0^\infty\frac{\sin(tu)}{\pi^2+[\ln(u^2-a^2)]^2}du+\frac{\sin(t\sqrt{a^2-1})}{\sqrt{a^2-1}} \tag{6.5.5}$$

$$a>1$$

$$f(t)=2\int_0^\infty\frac{\sin(tu)}{\pi^2+[\ln(u^2-1)]^2}du+t \tag{7.5.5}$$

$$a=1$$

$$f(t)=2\int_0^\infty\frac{\sin(tu)}{\pi^2+[\ln(u^2-1)]^2}du+\frac{\sinh(t\sqrt{1-a^2})}{\sqrt{1-a^2}} \tag{8.5.5}$$

$$0<a<1$$

$$f(t)=\frac{1}{2}\left\{\int_0^\infty \frac{e^{-tu}}{\pi^2+[\ln(u)]^2}du+e^t\right\}=\frac{1}{2}\nu'(t) \qquad (9.5.5)$$

$$a=0$$

As can be seen, all four expressions for $f(t)$ have the form of (3.5.5). Using the Laplace transforms of sine and hyperbolic sine functions and applying (4.5.5), for example to (8.5.5), we have

$$2\int_0^\infty \frac{u}{(s^2+u^2)\{\pi^2+[\ln(u^2-a^2)]^2\}}du+\frac{1}{s^2+a^2-1}=\frac{1}{\ln(s^2+a^2)} \qquad (10.5.5)$$

Changing the variable of integration and denoting $\alpha = s^2 + a^2$, the integral in (10.5.5) takes equivalent form

$$\int_0^\infty \frac{1}{(\alpha+u)[\pi^2+(\ln u)^2]}du=\int_{-\infty}^\infty \frac{e^t}{(\pi^2+t^2)(\alpha+e^t)}dt=\frac{1}{\ln\alpha}-\frac{1}{\alpha-1} \qquad (11.5.5)$$

$$\alpha\neq 1 \quad , \quad |\arg\alpha|<\pi$$

and for $\alpha = 1$ is

$$\int_0^\infty \frac{1}{(1+u)[\pi^2+(\ln u)^2]}du=\int_{-\infty}^\infty \frac{1}{(\pi^2+t^2)(1+e^{-t})}dt=\frac{1}{2} \qquad (12.5.5)$$

By a repeated differentiation of the integrand in (11.5.5) with regard to the parameter α we have

$$A_n(\alpha)=\int_0^\infty \frac{1}{(\alpha+u)^{n+1}[\pi^2+(\ln u)^2]}du=\int_{-\infty}^\infty \frac{e^t}{(\pi^2+t^2)(\alpha+e^t)^{n+1}}dt=$$

$$\frac{(-1)^n}{n!}\frac{d^n}{d\alpha^n}\left(\frac{1}{\ln\alpha}\right)-\frac{1}{(\alpha-1)^{n+1}} \qquad (13.5.5)$$

$$n=0,1,2,3,\ldots$$

For example, for n = 1 and n = 2 it follows from (13.5.5) that

$$\int_0^\infty \frac{1}{(\alpha+u)^2[\pi^2+(\ln u)^2]}du=\frac{1}{\alpha(\ln\alpha)^2}-\frac{1}{(\alpha-1)^2} \qquad (14.5.5)$$

$$\int_0^{\infty} \frac{1}{(1+u)^2[\pi^2+(\ln u)^2]} du = \frac{1}{12} \tag{15.5.5}$$

$$\int_0^{\infty} \frac{1}{(\alpha+u)^3[\pi^2+(\ln u)^2]} du = \frac{2+\ln\alpha}{2\alpha^2(\ln\alpha)^3} - \frac{1}{(\alpha-1)^3} \tag{16.5.5}$$

$$\int_0^{\infty} \frac{1}{(1+u)^3[\pi^2+(\ln u)^2]} du = \frac{1}{24} \tag{17.5.5}$$

Combining (11.5.5) with (14.5.5) and (12.5.5) with (15.5.5) yields

$$\int_0^{\infty} \frac{u}{(\alpha+u)^2[\pi^2+(\ln u)^2]} du = \int_{-\infty}^{\infty} \frac{e^{2t}}{(\pi^2+t^2)(\alpha+e^t)^2} dt = \frac{-1+\ln\alpha}{(\ln\alpha)^2} - \frac{1}{(\alpha-1)^2} \tag{18.5.5}$$

and for α = 1

$$\int_0^{\infty} \frac{u}{(1+u)^2[\pi^2+(\ln u)^2]} du = \frac{5}{12} \tag{19.5.5}$$

Differentiation of the integrand in (18.5.5) with regard to the parameter α gives

$$\int_0^{\infty} \frac{u}{(\alpha+u)^3[\pi^2+(\ln u)^2]} du = \frac{-2+\ln\alpha}{2\alpha(\ln\alpha)^3} - \frac{1}{(\alpha-1)^3} \qquad \alpha \neq 1 \tag{20.5.5}$$

The partial decomposition as applied to (11.5.5) gives

$$\int_0^{\infty} \frac{1}{(\alpha+u)(\beta+u)[\pi^2+(\ln u)^2]} du = \int_{-\infty}^{\infty} \frac{e^t}{(\pi^2+t^2)(\alpha+e^t)(\beta+e^t)} dt = \frac{1}{(\alpha-\beta)}\left(\frac{1}{\ln\beta} - \frac{1}{\ln\alpha}\right) - \frac{1}{(\alpha-1)(\beta-1)} \tag{21.5.5}$$

which for β = 0 and β = 1 are

$$\int_0^{\infty} \frac{1}{u(\alpha+u)[\pi^2+(\ln u)^2]} du = \int_{-\infty}^{\infty} \frac{1}{(\pi^2+t^2)(\alpha+e^t)} dt = \frac{1}{(\alpha-1)} - \frac{1}{\alpha \ln \alpha} \tag{22.5.5}$$

$$\int_0^{\infty} \frac{1}{(1+u)(\alpha+u)[\pi^2+(\ln u)^2]} du = \frac{\alpha+1}{2(\alpha-1)^2} - \frac{1}{(\alpha-1)\ln \alpha} \tag{23.5.5}$$

Substitution $\alpha = i\alpha$ and $\beta = -i\alpha$ into (21.5.5) yields

$$\int_0^{\infty} \frac{1}{(\alpha^2+u^2)[\pi^2+(\ln u)^2]} du = \frac{\pi}{2\alpha[\frac{\pi^2}{4}+(\ln \alpha)^2]} - \frac{1}{\alpha^2+1} \qquad \alpha > 0 \tag{24.5.5}$$

Repeated differentiation of (22.5.5) with regard to the parameter α gives

$$B_n(\alpha) = \int_0^{\infty} \frac{1}{u(\alpha+u)^{n+1}[\pi^2+(\ln u)^2]} du = \int_{-\infty}^{\infty} \frac{1}{(\pi^2+t^2)(\alpha+e^t)^{n+1}} dt = \frac{1}{(\alpha-1)^{n+1}} - \frac{(-1)^n}{n!} \frac{d^n}{d\alpha^n}\left(\frac{1}{\alpha \ln \alpha}\right) \qquad n = 0,1,2,3\ldots \tag{25.5.5}$$

The integrals in (13.5.5) and (25.5.5) satisfy the recursion formula

$$B_n(\alpha) = A_{n+1}(\alpha) + \alpha B_{n+1}(\alpha) \qquad n = 0,1,2,3,\ldots \tag{26.5.5}$$

In the second example, the inverse Laplace transform is

$$f(t) = L^{-1}\{F(s)\} = \frac{1}{2\pi i} \int_{c-i\infty}^{c+i\infty} \frac{e^{st}}{[\ln(s^2+a^2)]^2} ds \tag{27.5.5}$$

and the result of the complex integration for $0 < a < 1$, is

$$f(t) = 4\int_a^\infty \frac{\sin(tu)\ln(u^2 - a^2)}{\{\pi^2 + [\ln(u^2 - a^2)]^2\}^2} du + \frac{(1-2a^2)\sinh(t\sqrt{1-a^2})}{2(1-a^2)^{3/2}} + \frac{t\cosh(t\sqrt{1-a^2})}{2(1-a^2)} \quad (28.5.5)$$

$$0 < a < 1$$

Denoting α = s^2 + a^2, using (4.5.5), the Laplace transforms of *f*(*t*) and with the help of simple mathematical operations the following logarithmic integrals are available

$$C_n(\alpha) = \int_0^\infty \frac{\ln u}{(\alpha+u)^{n+1}[\pi^2 + (\ln u)^2]^2} du = \frac{(-1)^n}{2}\frac{d^n}{d\alpha^n}\left(\frac{1}{(\ln\alpha)^2} - \frac{\alpha}{(\alpha-1)^2}\right) \quad (29.5.5)$$

$$n = 0,1,2,3,\ldots$$

and

$$D_n(\alpha) = \int_0^\infty \frac{\ln u}{u(\alpha+u)^{n+1}[\pi^2 + (\ln u)^2]^2} du = \frac{(-1)^n}{2}\frac{d^n}{d\alpha^n}\left(\frac{1}{(\alpha-1)^2} - \frac{1}{\alpha(\ln\alpha)^2}\right) \quad (30.5.5)$$

$$n = 0,1,2,3,\ldots$$

The integrals in (29.5.5) and in (30.5.5) are interrelated in a similar way as in (26.5.5)

$$C_n(\alpha) = C_{n+1}(\alpha) + \alpha D_{n+1}(\alpha) \quad (31.5.5)$$

$$n = 0,1,2,3,\ldots$$

For example for n = 1 from (29.5.5) and (30.5.5) we have

$$\int_0^\infty \frac{\ln u}{(\alpha+u)^2[\pi^2 + (\ln u)^2]^2} du = \int_{-\infty}^\infty \frac{te^t}{(\pi^2 + t^2)^2(\alpha + e^t)^2} dt = \frac{1}{\alpha(\ln\alpha)^3} - \frac{\alpha+1}{2(\alpha-1)^3} \quad (32.5.5)$$

$$\int_0^\infty \frac{\ln u}{u(\alpha+u)^2[\pi^2+(\ln u)^2]^2}du = \int_{-\infty}^\infty \frac{t}{(\pi^2+t^2)^2(\alpha+e^t)^2}dt = \frac{1}{2}\left(\frac{2}{(\alpha-1)^3}-\frac{2+\ln\alpha}{\alpha^2(\ln\alpha)^3}\right) \quad (33.5.5)$$

$$\int_0^\infty \frac{u\ln u}{(\alpha+u)^2[\pi^2+(\ln u)^2]^2}du = \int_{-\infty}^\infty \frac{te^{2t}}{(\pi^2+t^2)^2(\alpha+e^t)^2}dt = \frac{-2+\ln\alpha}{2(\ln\alpha)^3}+\frac{\alpha}{(\alpha-1)^3} \quad (34.5.5)$$

As mentioned above, for a = 0, the integral in (9.5.5) is associated with the derivative of the Volterra function ν(*t*) and this fact permits to evaluate additional group of integrals. If the Laplace transforms $L\{g(t)\} = G(s)$ and $L\{\nu(t)g(t)\} = R(s)$ are known then we have

$$\int_0^\infty \frac{G(s+u)}{u[\pi^2+(\ln u)^2]}du = G(s-1)-R(s) \quad (35.5.5)$$

For example, in the case of $f(t) = \sinh(at)$ the Laplace transform is $G(s) = a/(s^2 - a^2)$ and

$$G(s+u) = \frac{a}{(s+u)^2-a^2}$$

$$L\{\sinh(at)f(t)\} = \frac{1}{2}[F(s-a)-F(s+a)] \quad (36.5.5)$$

$$R(s) = L\{\sinh(at)\nu(t)\} = \frac{1}{2}\{\frac{1}{(s-a)\ln(s-a)}-\frac{1}{(s+a)\ln(s+a)}\}$$

Using (35.5.5), (36.5.5) and introducing $s = (\alpha+\beta)/2$ and $a = (\alpha-\beta)/2$ we have

$$\int_0^\infty \frac{1}{u(\alpha+u)(\beta+u)[\pi^2+(\ln u)^2]}du = \frac{1}{\alpha\beta-\alpha-\beta+1}-\frac{1}{\alpha-\beta}\left(\frac{1}{\beta\ln\beta}-\frac{1}{\alpha\ln\alpha}\right) \quad (37.5.5)$$

As expected, when α = β, (37.5.5) reduces to (25.5.5) with n = 1.

In the case of the trigonometric functions, $g(t) = \sin(at)$, $G(s) = a/(s^2 + a^2)$ and $g(t) = \cos(at)$, $G(s) = s/(s^2 + a^2)$, the Laplace transforms $R(s)$ can be evaluated from

$$L\{\sin(at)f(t)\} = \frac{1}{2i}[F(s-ia) - F(s+ia)]$$
$$L\{\cos(at)f(t)\} = \frac{1}{2i}[F(s-ia) + F(s+ia)] \tag{38.5.5}$$

and therefore

$$L\{\sin(at)\nu(t)\} = \frac{s\tan^{-1}(\frac{a}{s}) + a\ln(\sqrt{s^2+a^2})}{(s^2+a^2)\{[\tan^{-1}(\frac{a}{s})]^2 + [\ln(\sqrt{s^2+a^2}]^2\}} \tag{39.5.5}$$

$$L\{\cos(at)\nu(t)\} = \frac{-a\tan^{-1}(\frac{a}{s}) + s\ln(\sqrt{s^2+a^2})}{(s^2+a^2)\{[\tan^{-1}(\frac{a}{s})]^2 + [\ln(\sqrt{s^2+a^2}]^2\}} \tag{40.5.5}$$

Denoting $s = \alpha$ and $a = \beta$, from (35.5.5),(39.5.5) and (40.5.5) we have

$$\int_0^\infty \frac{1}{[(\alpha+u)^2+\beta^2][\pi^2+(\ln u)^2]}du =$$
$$\frac{\tan^{-1}(\frac{\beta}{\alpha})}{\beta\{[\tan^{-1}(\frac{\beta}{\alpha})]^2 + [\ln(\sqrt{\alpha^2+\beta^2})]^2\}} - \frac{1}{(\alpha-1)^2+\beta^2} \tag{41.5.5}$$

and

$$\int_0^\infty \frac{1}{u[(\alpha+u)^2+\beta^2][\pi^2+(\ln u)^2]}du =$$
$$\frac{1}{(\alpha-1)^2+\beta^2} - \frac{\alpha\tan^{-1}(\frac{\beta}{\alpha}) + \beta\ln(\sqrt{\alpha^2+\beta^2}}{\beta(\alpha^2+\beta^2)\{[\tan^{-1}(\frac{\beta}{\alpha})]^2 + [\ln(\sqrt{\alpha^2+\beta^2})]^2\}} \tag{42.5.5}$$

5.6. The Integral Representations of the Volterra Functions as the Asymptotic Limit of the Bessel Function $J_\lambda(\lambda x)$.

The *impulse function* δ(*x*), the *Dirac delta function* is widely applied in solving physical and engineering problems with a sudden excitation of systems. Since the delta function is defined by giving a rule for integrating its product with a continuous function *f*(*x*) and not as an ordinary function, the impulse function is approximated by the sequence of functions converging to it, the *delta sequences* [60,96]. One of the delta sequences was derived by Lamborn [72] who showed that the *shifted Dirac function* δ(*x*-1) can be represented by the asymptotic limit of the Bessel function of the first kind $J_\lambda(\lambda x)$

$$\delta(x-1)=\lim_{\lambda\to\infty}[\lambda J_\lambda(\lambda x)] \tag{1.5.6}$$

If the Lamborn sequence is multiplied by a function *f*(*tx*) and integrated from zero to infinity with respect to variable *x* then we have [9-11]

$$f(t)=\int_0^\infty \delta(x-1)f(tx)dx=\lim_{\lambda\to\infty}[\lambda\int_0^\infty J_\lambda(\lambda x)f(tx)dx] \tag{2.5.6}$$

The representation of *f*(*t*) in (2.5.6) is derived using the *sifting property* of the delta function [96] and by assuming that the right-hand side integrals exist and the order of the limit and the integration can be reversed (for more details and applications of (2.5.6) see Appendix C).

If the reciprocal gamma function f(*t*) = 1/Γ(*t* + α +1) is introduced into (2.5.6) we have

$$\frac{1}{\Gamma(\xi+\alpha+1)}=\lim_{\lambda\to\infty}[\lambda\int_0^\infty \frac{J_\lambda(\lambda x)}{\Gamma(\xi x+\alpha+1)}dx] \tag{3.5.6}$$

Multiplication of both sides of (3.5.6) by $t^{\xi+\alpha}$ and integration from zero to infinity with respect to variable ξ gives the Volterra function ν(*t*,α)

$$\begin{aligned}\nu(t,\alpha)&=\int_0^\infty \frac{t^{\xi+\alpha}}{\Gamma(\xi+\alpha+1)}d\xi=\\ &\lim_{\lambda\to\infty}\{\lambda[\int_0^\infty J_\lambda(\lambda x)(\int_0^\infty \frac{t^{\xi+\alpha}}{\Gamma(\xi x+\alpha+1)}d\xi)]dx\}\end{aligned} \tag{4.5.6}$$

The change in of the integration variable in (4.5.6) yields the final result

$$\nu(t,\alpha) = t^{\alpha} \lim_{\lambda\to\infty}[\lambda \int_0^{\infty} \frac{t^{-\alpha/x}}{x} J_{\lambda}(\lambda x)\nu(t^{1/x},\alpha)dx] \tag{5.5.6}$$

$\mathrm{Re}\,\alpha > -1$

and in the same way it is possible to obtain

$$\mu(t,\beta,\alpha) = t^{\alpha} \lim_{\lambda\to\infty}[\lambda \int_0^{\infty} \frac{t^{-\alpha/x}}{x^{\beta+1}} J_{\lambda}(\lambda x)\mu(t^{1/x},\beta,\alpha)dx] \tag{6.5.6}$$

$\mathrm{Re}\,\alpha > -1 \quad , \quad \mathrm{Re}\,\beta > -1$

Starting from

$$\frac{1}{\Gamma(\xi+t+1)} = \lim_{\lambda\to\infty}[\lambda \int_0^{\infty} \frac{J_{\lambda}(\lambda x)}{\Gamma(\xi+tx+1)} dx] \tag{7.5.6}$$

we have after few manipulations

$$\nu(a,t) = \int_0^{\infty} \frac{a^{\xi+t}}{\Gamma(\xi+t+1)} d\xi = \lim_{\lambda\to\infty}[\lambda \int_0^{\infty} a^{t(1-x)} J_{\lambda}(\lambda x)\nu(a,tx)dx] \tag{8.5.6}$$

$t > -1$

From

$$\frac{1}{\xi[\pi^2+(\ln\xi)^2]} = \lim_{\lambda\to\infty}[\lambda \int_0^{\infty} \frac{J_{\lambda}(\lambda x)}{\xi x[\pi^2+(\ln\xi x)^2]} dx] \tag{9.5.6}$$

and the Laplace transform

$$L\{\frac{1}{\xi[\pi^2+(\ln\xi)^2]};t\} = e^t - \nu(t) \tag{10.5.6}$$

it is possible to obtain

$$\nu(t) = e^t - \lim_{\lambda\to\infty}\{\lambda \int_0^{\infty} \frac{J_{\lambda}(\lambda tx)}{x}[e^{1/x} - \nu(\frac{1}{x})]dx\} \tag{11.5.6}$$

Comparing (11.5.6) with the Ramanujan representation of $\nu(t)$ in (9.1.1) we have

$$\int_0^\infty \frac{e^{-tu}}{u[\pi^2+(\ln u)^2]}du = \lim_{\lambda\to\infty}\{\lambda\int_0^\infty \frac{J_\lambda(\lambda tx)}{x}[e^{1/x}-\nu(\frac{1}{x})]dx\} \tag{12.5.6}$$

The results in (5.5.6) and (6.5.6) can be generalized by considering the Volterra functions of composed variables, $\nu[\phi(t),\alpha]$ or $\mu[\phi(t),\beta,\alpha]$. Starting with (3.5.6) and using (4.5.6) the function $\nu[\phi(t),\alpha]$ is represented by

$$\nu[\phi(t),\alpha) = \int_0^\infty \frac{\phi(t)^{\xi+\alpha}}{\Gamma(\xi+\alpha+1)}d\xi = \\ \lim_{\lambda\to\infty}\{\lambda[\int_0^\infty J_\lambda(\lambda x)(\int_0^\infty \frac{\phi(t)^{\xi+\alpha}}{\Gamma(\xi x+\alpha+1)}d\xi)]dx\} \tag{13.5.6}$$

but

$$\int_0^\infty \frac{\phi(t)^{\xi+\alpha}}{\Gamma(\xi x+\alpha+1)}d\xi = \frac{\phi(t)^{\alpha-\alpha/x}}{x}\int_0^\infty \frac{[\phi(t)]^{(y+\alpha)/x}}{\Gamma(y+\alpha+1)}dy = \\ \frac{\phi(t)^{\alpha-\alpha/x}}{x}\nu[\phi(t)^{1/x},\alpha] \tag{14.5.6}$$

and therefore

$$\nu[\phi(t),\alpha] = \phi(t)^\alpha \lim_{\lambda\to\infty}\{\lambda\int_0^\infty \frac{J_\lambda(\lambda x)}{x}\phi(t)^{-\alpha/x}\nu[\phi(t)^{1/x},\alpha]dx\} \tag{15.5.6}$$

For example, for $\phi(t) = e^{-t}$ we have

$$\nu[e^{-t},\alpha] = e^{-\alpha t}\lim_{\lambda\to\infty}\{\lambda\int_0^\infty \frac{J_\lambda(\lambda x)}{x}e^{\alpha t/x}\nu(e^{-t/x},\alpha)dx\} \tag{16.5.6}$$

The limit of double integral of a function $f(\xi;\tau)$ where τ is a parameter can be written in two equivalent forms

$$\lim_{\lambda\to\infty}\{\lambda\int_0^\infty\int_0^\infty e^{-\xi x}J_\lambda(\lambda x)f(\xi;\tau)dxd\xi\}=$$

$$\lim_{\lambda\to\infty}\{\lambda\int_0^\infty J_\lambda(\lambda x)[\int_0^\infty e^{-\xi x}f(\xi;\tau)d\xi]dx\}= \tag{17.5.6}$$

$$\lim_{\lambda\to\infty}\{\lambda\int_0^\infty f(\xi;\tau)[\int_0^\infty e^{-\xi x}J_\lambda(\lambda x)dx]d\xi\}$$

The inner integrals in (17.5.6) are the Laplace transforms of $f(\xi;\tau)$ and $J_\lambda(\lambda x)$ and therefore

$$\lim_{\lambda\to\infty}\{\lambda\int_0^\infty J_\lambda(\lambda x)F[f(\xi;\tau);\xi]dx\}=$$

$$\lim_{\lambda\to\infty}\{\lambda\int_0^\infty f(\xi;\tau)F[J_\lambda(\lambda x);x]dx\} \tag{18.5.6}$$

Considering that

$$\int_0^\infty \delta(x-1)F(x;\tau)dx=$$

$$\lim_{\lambda\to\infty}\{\lambda\int_0^\infty J_\lambda(\lambda x)F(x;\tau)dx\}=F(1;\tau) \tag{19.5.6}$$

and that the Laplace transform of $J_\lambda(\lambda x)$ is

$$L\{J_\lambda(\lambda x);\xi\}=\frac{\lambda^\lambda}{\sqrt{\xi^2+\lambda^2}[\xi+\sqrt{\xi^2+\lambda^2}]^\lambda} \tag{20.5.6}$$

we have

$$\lim_{\lambda\to\infty}\{\lambda^{\lambda+1}\int_0^\infty\frac{f(\xi;\tau)}{\sqrt{\xi^2+\lambda^2}[\xi+\sqrt{\xi^2+\lambda^2}]^\lambda}d\xi\}=F(1;\tau) \tag{21.5.6}$$

Using (17.4.3), (34.4.1), (66.4.3) and (27.4.4), the following examples of (21.5.6) are presented here

$$\lim_{\lambda\to\infty}\{\lambda^{\lambda+1}\int_0^\infty \frac{1}{\Gamma(\xi+\alpha+1)\sqrt{\xi^2+\lambda^2}[\xi+\sqrt{\xi^2+\lambda^2}]^\lambda}d\xi\} = e^\alpha \nu(\frac{1}{e},\alpha) \quad (22.5.6)$$

$$\lim_{\lambda\to\infty}\{\lambda^{\lambda+1}\int_0^\infty \frac{\nu(a,\xi)}{\sqrt{\xi^2+\lambda^2}[\xi+\sqrt{\xi^2+\lambda^2}]^\lambda}d\xi\} = \nu(a)-\nu(\frac{a}{e}) \quad (23.5.6)$$

$$\lim_{\lambda\to\infty}\{\lambda^{\lambda+1}\int_0^\infty \frac{\nu(a\sqrt{\xi},\alpha)}{\sqrt{\xi^2+\lambda^2}[\xi+\sqrt{\xi^2+\lambda^2}]^\lambda}d\xi\} = \sqrt{\pi}\,a\nu(\frac{a^2}{4},\frac{\alpha-1}{2}) \quad (24.5.6)$$

$$\lim_{\lambda\to\infty}\{\lambda^{\lambda+1}\int_0^\infty \frac{\nu[a(1-e^{-\xi/b},\alpha)]}{\sqrt{\xi^2+\lambda^2}[\xi+\sqrt{\xi^2+\lambda^2}]^\lambda}d\xi\} = \frac{\Gamma(b+1)}{a^b}\nu(a,\alpha+b) \quad (25.5.6)$$

$$a>0 \ , \ b>0 \ , \ \mathrm{Re}\,\alpha>-1$$

In the case of the Fourier cosine and sine transformations, the expressions similar to that in (21.5.6) are

$$\lim_{\lambda\to\infty}\{\lambda\int_0^\infty \frac{\cos[\lambda\sin^{-1}(\frac{\xi}{\lambda})]f(\xi;\tau)}{\sqrt{\lambda^2-\xi^2}}d\xi\} = F_c(1;\tau) \quad (26.5.6)$$

and

$$\lim_{\lambda\to\infty}\{\lambda\int_0^\infty \frac{\sin[\lambda\sin^{-1}(\frac{\xi}{\lambda})]f(\xi;\tau)}{\sqrt{\lambda^2-\xi^2}}d\xi\} = F_s(1;\tau) \quad (27.5.6)$$

Introducing $\nu(a,t)$ into (26.5.6) and (27.5.6) and using (12.5.4) and (14.5.4) we have

$$\lim_{\lambda\to\infty}\{\lambda\int_0^\infty \frac{\cos[\lambda\sin^{-1}(\frac{\xi}{\lambda})]\nu(a,\xi)}{\sqrt{\lambda^2-\xi^2}}d\xi\} = Ts(a,\frac{1}{\pi}) \quad (28.5.6)$$

$$\lim_{\lambda\to\infty}\{\lambda\int_0^\infty \frac{\sin[\lambda\sin^{-1}(\frac{\xi}{\lambda})]\nu(a,\xi)}{\sqrt{\lambda^2-\xi^2}}d\xi\} = \nu(a)-Tc(a,\frac{1}{\pi}) \quad (29.5.6)$$

APPENDIX A.
THE LAPLACE TRANSFORMATION AND THE MITTAG-LEFFLER TYPE FUNCTIONS

A brief description of properties and relevant literature associated with the following functions

$$E_{\alpha}(z)=\sum_{n=0}^{\infty}\frac{z^n}{\Gamma(\alpha n+1)} \qquad \text{(A-1)}$$

$$E_{\alpha,\beta}(z)=\sum_{n=0}^{\infty}\frac{z^n}{\Gamma(\alpha n+\beta)} \qquad \text{(A-2)}$$

is presented in Chapter 18.1 of *Higher Transcendental Functions*, volume 3, pp. 206-212 [43]. The definition of $E_{\alpha,\beta}(z)$ as used there and in (A-1), differs from that introduced by Humbert and Agarwal [2,65] by the factor $z^{(\beta-1)/\alpha}$. These authors also considered similar functions

$$_{m}E_{\alpha,\beta}(z)=\sum_{n=0}^{\infty}\frac{z^n}{(n!)^m\,\Gamma(\alpha n+\beta)} \qquad \text{(A-3)}$$

$$m=1,2,3,\ldots$$

We attempt to give here only relations which are connected with the Laplace transformation and reducible cases of the Mittag-Leffler and related functions. The presented material is for the most part new and extends that given by Humbert and Agarwal.

In many cases the Mittag-Leffler function $E_{\alpha}(t^{\rho})$ reduces to elementary or special functions. If a known series expansion is compared with (A-1), particular values of α and ρ can be identified. It is easy to recognize that

$$E_1(t)=e^t \qquad \text{(A-4)}$$

$$E_2(t^2) = \cosh t \tag{A-5}$$

$$E_2(-t^2) = \cos t \tag{A-6}$$

However, reducible cases can be evaluated systematically by application of the Laplace transformation. Using (A-1), the Laplace transform of $E_\alpha(t^\rho)$ is

$$L\{E_\alpha(at^\rho)\} = \frac{1}{s}\sum_{n=0}^{\infty}\frac{\Gamma(\rho n+1)}{\Gamma(\alpha n+1)}(\frac{a}{s^\rho})^n \tag{A-7}$$

But if $\rho = \alpha$, then the series becomes

$$L\{E_\alpha(at^\alpha)\} = \frac{s^{\alpha-1}}{s^\alpha - a} \tag{A-8}$$

For example for $\alpha = 2$ and $a = \pm 1$

$$L\{E_2(t^2)\} = \frac{s}{s^2-1} \tag{A-9}$$

$$L\{E_2(-t^2)\} = \frac{s}{s^2+1} \tag{A-10}$$

and the inverse transforms are

$$E_2(t^2)\} = L^{-1}\{\frac{s}{s^2-1}\} = \cosh t \tag{A-11}$$

$$E_2(-t^2)\} = L^{-1}\{\frac{s}{s^2+1}\} = \cos t \tag{A-12}$$

which are evidently identical with (A-5) and (A-6).

In a similar way, it is possible to continue with $\alpha = 3$ and $\alpha = 4$

$$E_3(t^3)\} = L^{-1}\{\frac{s^2}{s^3-1}\} = \frac{1}{3}[e^t + 2e^{-t/2}\cos(\frac{\sqrt{3}}{2}t)] \tag{A-13}$$

$$E_3(-t^3)\} = L^{-1}\{\frac{s^2}{s^3+1}\} = \frac{1}{3}[e^{-t} + 2e^{t/2}\cos(\frac{\sqrt{3}}{2}t)] \tag{A-14}$$

$$E_4(t^4)\} = L^{-1}\{\frac{s^3}{s^4-1}\} = \frac{1}{2}[\cosh t + \cos t] \qquad \text{(A-15)}$$

$$E_4(-t^4)\} = L^{-1}\{\frac{s^3}{s^4+1}\} = \cos(\frac{t}{\sqrt{2}})\cosh(\frac{t}{\sqrt{2}}) \qquad \text{(A-16)}$$

or in a general form

$$E_n(\pm t^n)\} = L^{-1}\{\frac{s^{n-1}}{s^n \mp 1}\} \quad , \quad n = 1,2,3,\ldots \qquad \text{(A-17)}$$

where $E_n(\pm t^n)$ are expressed in terms of elementary functions and these functions are associated with *the hyperbolic and trigonometric functions of order n* [43].

The Mittag-Leffler functions are reduced to other special functions for non integer values of α. For example if $\alpha = 1/2$, then

$$E_{1/2}(\pm\sqrt{t})\} = L^{-1}\{\frac{1}{\sqrt{s}(\sqrt{s}\mp 1)}\} = e^t[1 \pm erf(\sqrt{t})] \qquad \text{(A-18)}$$

Representation of $E_{1/4}(\pm t^{1/4})$ is more difficult to establish because the inverse transforms are not tabulated. However, in this case, (A-8) can be written in the form

$$E_{1/4}(\pm t^{1/4})\} = L^{-1}\{\frac{1}{s^{3/4}(s^{1/4}\mp 1)}\} = \frac{1}{\sqrt{s}(\sqrt{s}-1)} \pm \frac{1}{s^{1/4}(s-1)} \pm \frac{1}{s^{3/4}(s-1)} \qquad \text{(A-19)}$$

The inverse transform of the first term is given in (A-18) and other terms can be inverted using the convolution theorem

$$L^{-1}\{\frac{1}{s^{1/4}(s-1)}\} = \int_0^t \frac{e^{(t-u)}}{\Gamma(\frac{1}{4})u^{3/4}}du = \frac{e^t\gamma(\frac{1}{4},t)}{\Gamma(\frac{1}{4})} \qquad \text{(A-20)}$$

$$L^{-1}\{\frac{1}{s^{3/4}(s-1)}\} = \int_0^t \frac{e^{(t-u)}}{\Gamma(\frac{3}{4})u^{1/4}}du = \frac{e^t\gamma(\frac{3}{4},t)}{\Gamma(\frac{3}{4})} \qquad \text{(A-21)}$$

Therefore

$$E_{1/4}(\pm t^{1/4})\} = e^{t}\{1 + erf(\sqrt{t}) \pm \frac{\gamma(\frac{1}{4},t)}{\Gamma(\frac{1}{4})} \pm \frac{\gamma(\frac{3}{4},t)}{\Gamma(\frac{3}{4})}\} \tag{A-22}$$

where $\gamma(a,t)$ is the incomplete gamma function.

The Laplace transform of the Mittag-Leffler function in (A-8) is a starting point for a number of functional relations satisfied by $E_{\alpha}(t^{\alpha})$. From

$$L\{t\} = \frac{1}{s^{2}} = \frac{s^{\alpha-1}(s^{\alpha}-1)}{(s^{\alpha}-1)s^{\alpha+1}} = \frac{s^{\alpha-1}}{(s^{\alpha}-1)}\{\frac{1}{s} - \frac{1}{s^{\alpha+1}}\} \tag{A-23}$$

and using the convolution theorem we have

$$t = \int_{0}^{t} E_{\alpha}(u^{\alpha})\{1 - \frac{(t-u)^{\alpha}}{\Gamma(\alpha+1)}\}du \tag{A-24}$$

which can be written as

$$t = E_{\alpha}(t^{\alpha}) * \{1 - \frac{t^{\alpha}}{\Gamma(\alpha+1)}\} \tag{A-25}$$

For a positive integer n

$$L\{\frac{t^{n}}{n!}\} = \frac{1}{s^{n+1}} = \frac{s^{\alpha-1}}{(s^{\alpha}-1)}\{\frac{1}{s^{n}} - \frac{1}{s^{\alpha+n}}\} \tag{A-26}$$

it follows that

$$\frac{t^{n}}{n!} = E_{\alpha}(t^{\alpha}) * \{\frac{t^{n-1}}{(n-1)!} - \frac{t^{\alpha+n-1}}{\Gamma(\alpha+n)}\} \tag{A-27}$$

The infinite sum of expressions of this type gives

$$e^{t} - 1 = E_{\alpha}(t^{\alpha}) * \{e^{t} - t^{\alpha-1}[E_{1,\alpha}(t) - \frac{1}{\Gamma(\alpha)}]\} \tag{A-28}$$

The analogous to (A-27) relation, but for non-integer β, is

$$\frac{t^{\beta}}{\Gamma(\beta+1)} = E_{\alpha}(t^{\alpha}) * \{\frac{t^{\beta-1}}{\Gamma(\beta)}[1 - \frac{t^{\alpha}B(\alpha,\beta)}{\Gamma(\alpha)}]\} \tag{A-29}$$

It follows from

$$L\{E_{\alpha}(t^{\alpha}) - 1\} = \{\frac{s^{\alpha-1}}{s^{\alpha}-1} - \frac{1}{s}\} = \frac{s^{\alpha-1}}{(s^{\alpha}-1)} \cdot \frac{1}{s^{\alpha}} \tag{A-30}$$

that

$$E_{\alpha}(t^{\alpha}) - 1 = \int_0^t E_{\alpha}(u^{\alpha}) \frac{(t-u)^{\alpha-1}}{\Gamma(\alpha)} du \tag{A-31}$$

Starting with

$$\frac{s^{\alpha-1}}{(s^{\alpha}-1)} = \frac{s^{2\alpha-1}}{(s^{2\alpha}-1)} + \frac{s^{2\alpha-1}}{(s^{2\alpha}-1)} \cdot \frac{1}{s^{\alpha}} \tag{A-32}$$

the following relation can be established

$$E_{\alpha}(t^{\alpha}) = E_{2\alpha}(t^{2\alpha}) + \int_0^t E_{2\alpha}(u^{2\alpha}) \frac{(t-u)^{\alpha-1}}{\Gamma(\alpha)} du \tag{A-33}$$

In particular, for α = 1/4, and by introducing (A-18) and (A-22) into (A-33) we have

$$\int_0^t e^{-(t-u)} \frac{[1+erf(\sqrt{t})]}{(t-u)^{3/4}} du = \gamma(\frac{1}{4},t) + \frac{\Gamma(\frac{1}{4})}{\Gamma(\frac{3}{4})} \gamma(\frac{3}{4},t) \tag{A-34}$$

Considering that

$$L\{tE_{\alpha}(t^{\alpha})\} = -\frac{d}{ds}\{\frac{s^{\alpha-1}}{(s^{\alpha}-1)}\} = (\frac{\alpha-1}{s^{\alpha}} - 1) \cdot \frac{s^{\alpha-1}}{(s^{\alpha}-1)} \cdot \frac{s^{\alpha-1}}{(s^{\alpha}-1)} \tag{A-35}$$

the inverse of (A-35) gives

$$tE_\alpha(t^\alpha) = \{\frac{(\alpha-1)t^{\alpha-1}}{\Gamma(\alpha)} - 1\} * E_\alpha(t^\alpha) * E_\alpha(t^\alpha) \tag{A-36}$$

If the Laplace transform of $E_\alpha(t^\alpha)$ is written in the form

$$F(s) = \frac{s^{\alpha-1}}{s^\alpha - 1} = \frac{1}{\sqrt{s}} \cdot \frac{(\sqrt{s})^{2\alpha-1}}{[(\sqrt{s})^{2\alpha} - 1]} \tag{A-37}$$

then by using the operational rule

$$L^{-1}\{\frac{1}{\sqrt{s}} F(\sqrt{s})\} = \frac{1}{\sqrt{\pi t}} \int_0^\infty e^{-u^2/4t} f(u)du \tag{A-38}$$

we have

$$E_\alpha(t^\alpha) = \frac{1}{\sqrt{\pi t}} \int_0^\infty e^{-u^2/4t} E_{2\alpha}(u^{2\alpha})du \tag{A-39}$$

Similarly from the operational rule

$$L^{-1}\{\frac{1}{s^2} F(\frac{1}{s})\} = \int_0^\infty \sqrt{\frac{t}{u}} J_1(2\sqrt{tu}) f(u)du \tag{A-40}$$

and

$$L\{1 - E_\alpha(t^\alpha)\} = \frac{1}{s} - \frac{s^{\alpha-1}}{s^\alpha - 1} = \frac{1}{s^2} \frac{(1/s)^{\alpha-1}}{[(1/s)^\alpha - 1]} \tag{A-41}$$

it follows that

$$\frac{1 - E_\alpha(t^\alpha)}{\sqrt{t}} = \int_0^\infty J_1(2\sqrt{tu}) E_\alpha(u^\alpha) \frac{du}{\sqrt{u}} \tag{A-42}$$

Finally, a number of Laplace transforms of the Mittag-Leffler functions based on (A-7) are presented

$$L\{E_1(\sqrt{t})\} = \frac{1}{s} e^{1/8s} D_{-2}(-\frac{1}{\sqrt{2s}}) \tag{A-43}$$

$$L\{E_1(t^{\rho})\} = \frac{1}{s} \int_0^{\infty} e^{(u/s)^{\rho} - u} du \tag{A-44}$$

$$0 < \rho < 1$$

$$L\{E_2(t)\} = \frac{1}{s}\{1 + \sqrt{\frac{\pi}{4s}} e^{1/4s} erf(\frac{1}{2\sqrt{s}})\} \tag{A-45}$$

$$L\{E_4(-t^2)\} = \frac{1}{s}\{1 - \sqrt{\frac{\pi}{2s}}[\sin(\frac{1}{4s}) C(\frac{1}{2\sqrt{s}}) - \cos(\frac{1}{4s}) S(\frac{1}{2\sqrt{s}})]\} \tag{A-46}$$

$$L\{E_{\alpha+1}(t^{\alpha})\} = \frac{1}{s} \int_0^1 (\frac{u}{s})^{\alpha} e^{(u/s)^{\alpha}(1-u)} du \tag{A-47}$$

Now, consider the $E_{\alpha,\beta}(t)$ function, which is defined in (A-2) and satisfies the following relations

$$E_{\alpha,1}(t) = E_{\alpha}(t) \tag{A-48}$$

$$t^{\alpha} E_{\alpha,\alpha+1}(t^{\alpha}) = E_{\alpha}(t^{\alpha}) - 1 \tag{A-49}$$

$$E_{\alpha,\beta}(t) = \frac{1}{\Gamma(\beta)} + t E_{\alpha,\alpha+\beta}(t) \tag{A-50}$$

The Laplace transform of the Mittag-Leffler type function is

$$L\{E_{\alpha,\beta}(at^{\rho})\} = \frac{1}{s} \sum_{n=0}^{\infty} \frac{\Gamma(\rho n + 1)}{\Gamma(\alpha n + \beta)} (\frac{a}{s^{\rho}})^n \tag{A-51}$$

$$\alpha > 0, \quad \rho > 0, \quad \text{Re}(s - a) > 0$$

The series can be written in the closed form for

$$L\{E_{2,2}(t)\} = \sqrt{\frac{\pi}{s}} e^{1/4s} erf(\frac{1}{2\sqrt{s}}) \tag{A-52}$$

$$L\{E_{4,2}(-t^2)\} = \sqrt{\frac{2\pi}{s}}\{\sin(\frac{1}{4s})S(\frac{1}{2\sqrt{s}}) + \cos(\frac{1}{4s})C(\frac{1}{2\sqrt{s}})\} \qquad \text{(A-53)}$$

$$L\{E_{\alpha,n+2}(t^\alpha)\} = \frac{1}{n!s}\int_0^1 \frac{(1-u)^n}{[1-(\frac{u}{s})^\alpha]}du \qquad \text{(A-54)}$$

$$n = 1,2,3,\ldots$$

and

$$L\{t^{\beta-1}E_{\alpha,\beta}(\pm t^\alpha)\} = \frac{s^{\alpha-\beta}}{s^\alpha \mp 1} \qquad \text{(A-55)}$$

If $E_\alpha(t^\alpha)$ is reduced to elementary or special function, then $E_{\alpha,\beta}(t^\alpha)$ is expressed by the convolution integral of that function

$$t^{\beta-1}E_{\alpha,\beta}(\pm t^\alpha) = E_\alpha(\pm t^\alpha) * \frac{t^{\beta-2}}{\Gamma(\beta-1)} \qquad \text{(A-56)}$$

$$\beta > 1$$

These integrals are easily evaluated when α and β are positive integers

$$t^{m-1}E_{n,m}(\pm t^n) = \int_0^t E_n(\pm t^n)\frac{(t-u)^{m-2}}{(m-1)!}du \qquad \text{(A-57)}$$

$$m = 2,3,4,\ldots, n = 1,2,3,\ldots$$

For example using (A-4) - (A-6) we immediately have

$$tE_{1,2}(t) = \int_0^t e^u du = e^t - 1 \qquad \text{(A-58)}$$

$$tE_{1,2}(-t) = \int_0^t e^{-u} du = 1 - e^{-t} \qquad \text{(A-59)}$$

$$t^2E_{1,3}(t) = \int_0^t e^u(t-u)du = e^t - t - 1 \qquad \text{(A-60)}$$

$$t^2 E_{1,3}(-t) = \int_0^t e^{-u}(t-u)du = t - 1 + e^t \tag{A-61}$$

$$tE_{2,2}(t^2) = \int_0^t \cosh u du = \sinh t \tag{A-62}$$

$$tE_{2,2}(-t^2) = \int_0^t \cos u du = \sin t \tag{A-63}$$

and in more elaborate cases

$$t^2 E_{3,3}(t^3) = \frac{1}{3}[e^t - 2e^{-t/2}\cos(\frac{\sqrt{3}}{2}t - \frac{\pi}{3})] \tag{A-64}$$

$$t^2 E_{3,3}(-t^3) = \frac{1}{3}[e^{-t} - 2e^{t/2}\cos(\frac{\sqrt{3}}{2}t + \frac{\pi}{3})] \tag{A-65}$$

$$t E_{4,2}(t^4) = \frac{1}{2}(\sinh t + \sin t) \tag{A-66}$$

$$t E_{4,2}(-t^4) = \frac{1}{\sqrt{2}}[\cos(\frac{t}{\sqrt{2}})\sinh(\frac{t}{\sqrt{2}}) + \sin(\frac{t}{\sqrt{2}})\cosh(\frac{t}{\sqrt{2}})] \tag{A-67}$$

$$t^2 E_{4,3}(t^4) = \frac{1}{2}[\cosh t - \cos t] \tag{A-68}$$

$$t^2 E_{4,3}(-t^4) = \sin(\frac{t}{\sqrt{2}})\sinh(\frac{t}{\sqrt{2}}) \tag{A-69}$$

$$t^3 E_{4,4}(t^4) = \sqrt{2}[\sinh(\frac{t}{\sqrt{2}}) - \sin(\frac{t}{\sqrt{2}})] \tag{A-70}$$

$$t^3 E_{4,4}(-t^4) = \frac{1}{\sqrt{2}}[\cos(\frac{t}{\sqrt{2}})\sinh(\frac{t}{\sqrt{2}}) - \sin(\frac{t}{\sqrt{2}})\cosh(\frac{t}{\sqrt{2}})] \tag{A-71}$$

Evidently, these results can also be obtained by the direct inversion of (A-55) which is useful when an evaluation of the convolution integrals is less convenient. Sometimes, the inverse transform can be easily determined or even found in tables of Laplace transforms as happens in the following cases

$$t^{1/2} E_{1,3/2}(t) = L^{-1}\{\frac{1}{\sqrt{s}(s-1)}\} = e^{t} erf(\sqrt{t}) \tag{A-72}$$

$$t^{3/2} E_{1,5/2}(t) = L^{-1}\{\frac{1}{s^{3/2}(s-1)}\} = e^{t} erf(\sqrt{t}) - 2\sqrt{\frac{t}{\pi}} \tag{A-73}$$

$$t^{\beta-1} E_{2,\beta}(-t^2) = L^{-1}\{\frac{1}{s^{\beta-2}(s^2+1)}\} = \frac{t^{\beta-1}}{\Gamma(\beta)} {}_1F_2(1;\frac{\beta}{2},\frac{\beta+1}{2};-\frac{t^2}{4}) \tag{A-74}$$

$$t^{\beta-1} E_{3,\beta}(-t^3) = L^{-1}\{\frac{s^{3-\beta}}{s^3+1}\} = \frac{t^{\beta-1}}{\Gamma(\beta)} {}_1F_3(1;\frac{\beta}{3},\frac{\beta+1}{3},\frac{\beta+1}{3};-\frac{t^3}{27}) \tag{A-75}$$

Functional relations associated with $E_{\alpha,\beta}(t^{\alpha})$ can be established in a similar way as those of the Mittag-Leffler function $E_{\alpha}(t^{\alpha})$. From

$$\frac{1}{s^{\beta+1}} = \frac{s^{\alpha-\beta}}{s^{\alpha}-1}\left\{\frac{1}{s} - \frac{1}{s^{\alpha+1}}\right\} \tag{A-76}$$

it follows that

$$\frac{t^{\beta}}{\Gamma(\beta+1)} = \int_0^t u^{\beta-1} E_{\alpha,\beta}(u^{\alpha})\left\{1 - \frac{(t-u)^{\alpha}}{\Gamma(\alpha+1)}\right\} du \tag{A-77}$$

Considering that

$$\frac{s^{\alpha-\beta}}{s^{\alpha}-1} = \frac{s^{2\alpha-\beta}}{s^{2\alpha}-1} + \frac{s^{2\alpha-\beta}}{(s^{2\alpha}-1)} \cdot \frac{1}{s^{\alpha}} \tag{A-78}$$

we have

$$E_{\alpha,\beta}(t^{\alpha}) = E_{2\alpha,\beta}(t^{2\alpha}) + \int_0^t (\frac{u}{t})^{\beta-1} E_{2\alpha,\beta}(u^{2\alpha}) \frac{(t-u)^{\alpha-1}}{\Gamma(\alpha)} du \tag{A-79}$$

Application of the operational rule (A-38) to

$$\frac{s^{\alpha-(\beta+1)/2}}{s^{\alpha}-1} = \frac{(\sqrt{s})^{2\alpha-\beta}}{\sqrt{s}[(\sqrt{s})^{2\alpha}-1]} \tag{A-80}$$

gives

$$t^{(\beta-1)/2}E_{\alpha,(\beta+1)/2}(t^{\alpha})=\frac{1}{\sqrt{\pi t}}\int_0^{\infty}e^{-u^2/4t}u^{\beta-1}E_{2\alpha,\beta}(u^{2\alpha})du \tag{A-81}$$

and from the operational rule

$$L^{-1}\{\frac{1}{s^{2\lambda+1}}F(\frac{1}{s})\}=\int_0^{\infty}(\frac{t}{u})^{\lambda}J_{2\lambda}(2\sqrt{tu})f(u)du \tag{A-82}$$

with

$$\frac{1}{s^{\beta}}-\frac{s^{\alpha-\beta}}{s^{\alpha}-1}=\frac{(1/s)^{\alpha-\beta}}{s^{2\beta}[(1/s)^{\alpha}-1]} \tag{A-83}$$

we have

$$\frac{1}{\sqrt{t}}\left[\frac{1}{\Gamma(\beta)}-E_{\alpha,\beta}(t^{\alpha})\right]=\int_0^{\infty}J_{2\beta-1}(2\sqrt{tu})E_{\alpha,\beta}(u^{\alpha})\frac{du}{\sqrt{u}} \tag{A-84}$$

which reduces to (A-41) for $\beta = 1$.

If $f(t) = t^{\beta-1}E_{\alpha,\beta}(t^{\alpha})$, $\alpha > 0$ and $\beta \geq n+1$, $n = 1, 2, 3, \ldots$ then it follows from (A-2) that $f(0) = f'(0) = \ldots = f^{(n-1)}(0) = 0$ and

$$f^{(n)}(t)=L^{-1}\{\frac{s^{\alpha-(\beta-n)}}{s^{\alpha}-1}\} \tag{A-85}$$

where the inverse transform of (A-85) is

$$\frac{d^n}{dt^n}\left(t^{\beta-1}E_{\alpha,\beta}(t^{\alpha})\right)=t^{\beta-n-1}E_{\alpha,\beta-n}(t^{\alpha}) \tag{A-86}$$

For n = 1, it can be written in the form

$$t\frac{d}{dt}E_{\alpha,\beta}(t^{\alpha})=E_{\alpha,\beta-1}(t^{\alpha})-(\beta-1)E_{\alpha,\beta}(t^{\alpha}) \qquad \alpha>0\ ,\quad \beta\geq 2 \tag{A-87}$$

If $\alpha = n$ and $\beta = n+1$, then (A-86) becomes

$$\frac{d^n}{dt^n}\left(t^n E_{n,n+1}(t^n)\right) = E_{n,1}(t^n) \qquad n = 1,2,3,\ldots \tag{A-88}$$

But from (A-48) and (A-49)

$$t^n E_{n,n+1}(t^n) = E_n(t^n) - 1 \tag{A-89}$$

and therefore

$$\frac{d^n}{dt^n} E_n(t^n) = E_n(t^n) \qquad n = 1,2,3,\ldots \tag{A-90}$$

which has already been mentioned in [43].

The Wright function

$$\phi(\alpha,\beta;t) = \sum_{n=0}^{\infty} \frac{t^n}{n!\Gamma(\alpha n+\beta)} \tag{A-91}$$

is a particular case of the generalized Mittag-Leffler function which was introduced by Humbert and Agarwal [65]. From (A-3) and (A-91) it follows that $\phi(\alpha,\beta;t) = {}_1E_{\alpha,\beta}(t)$ and

$$L\{\phi(\alpha,\beta;t)\} = \frac{1}{s} E_{\alpha,\beta}(\frac{1}{s}) \tag{A-92}$$

$$L\{t^{\beta-1}\phi(\alpha,\beta;t^\alpha)\} = \frac{1}{s^\beta} e^{1/s^\alpha} \tag{A-93}$$

For $\alpha = 1$, (A-93) leads to the modified Bessel functions, because

$$L^{-1}\{\frac{1}{s^\beta} e^{1/s}\} = t^{(\beta-1)/2} I_{(\beta-1)/2}(2\sqrt{t}) \qquad \operatorname{Re}\beta > 0 \tag{A-94}$$

and therefore

$${}_1E_{1,\beta}(t) = \phi(1,\beta;t)\} = t^{(1-\beta)/2} I_{(\beta-1)/2}(2\sqrt{t}) \tag{A-95}$$

which is equivalent to (56.4.5).

In general

$$L\{{}_mE_{\alpha,\beta}(t)\} = \frac{1}{s}\left({}_{m-1}E_{\alpha,\beta}(\frac{1}{s})\right) \qquad \text{(A-96)}$$

$$m = 1,2,3,\ldots$$

$$L\{{}_mE_{\alpha,1}(t^{\alpha})\} = \frac{1}{s}\left({}_{m-1}E_{1,1}(\frac{1}{s^{\alpha}})\right) \qquad \text{(A-97)}$$

and introducing the generalized hypergeometric function

$$L\{{}_mE_{1,\beta}(t)\} = \frac{1}{\Gamma(\beta)s}\,{}_0F_{m-1}(;\beta,1,1,\ldots 1;\frac{1}{s}) \qquad \text{(A-98)}$$

or in the form

$$L\{{}_mE_{1,\beta}(t)\} = \frac{1}{\Gamma(\beta)s}\,{}_1F_m(1;\beta,1,1,\ldots 1;\frac{1}{s}) \qquad \text{(A-99)}$$

Since the confluent hypergeometric function is given by

$${}_1F_1(1;\beta;\frac{1}{s}) = (\beta-1)(\frac{1}{s})^{1-\beta}e^{1/s}\gamma(\beta-1,\frac{1}{s}) \qquad \text{(A-100)}$$

then from (A-99) we have

$$L\{{}_1E_{1,\beta}(t)\} = \frac{1}{\Gamma(\beta-1)}(\frac{1}{s})^{2-\beta}e^{1/s}\gamma(\beta-1,\frac{1}{s}) \qquad \text{(A-101)}$$

and in particular

$$L\{{}_1E_{1,1/2}(t)\} = \frac{1}{\sqrt{\pi}\,s}[1+\sqrt{\frac{\pi}{s}}\,e^{1/s}erf(\frac{1}{\sqrt{s}})] \qquad \text{(A-102)}$$

$$L\{{}_1E_{1,3/2}(t)\} = \frac{1}{\sqrt{s}}e^{1/s}erf(\frac{1}{\sqrt{s}}) \qquad \text{(A-103)}$$

which can be extended to β being any half an odd integer because of the recurrence relation satisfied by the incomplete gamma function is [1]

$$\gamma(\beta,\frac{1}{s}) = (\beta-1)\gamma(\beta-1,\frac{1}{s}) - (\frac{1}{s})^{\beta-1}e^{-1/s} \qquad \text{(A-104)}$$

Introducing

$$_0F_1(;\beta;\frac{1}{s}) = \Gamma(\beta)(\frac{1}{s})^{(1-\beta)/2} I_{\beta-1}(\frac{2}{\sqrt{s}}) \tag{A-105}$$

into (A-98) we have

$$L\{_2E_{1,\beta}(t)\} = (\frac{1}{s})^{(3-\beta)/2} I_{\beta-1}(\frac{2}{\sqrt{s}}) \tag{A-106}$$

If β is half an odd integer, then the Laplace transforms in (A-106) are expressed by elementary functions, for example

$$L\{_2E_{1,1/2}(t)\} = \frac{1}{\sqrt{\pi s}}\cosh(\frac{2}{\sqrt{s}}) \tag{A-107}$$

$$L\{_2E_{1,3/2}(t)\} = \frac{1}{\sqrt{\pi s}}\sinh(\frac{2}{\sqrt{s}}) \tag{A-108}$$

and for β being a positive integer by the Bessel functions.

The series representing the Laplace transforms of the generalized Mittag-Leffler functions

$$L\{_mE_{\alpha,\beta}(at^{\rho})\} = \frac{1}{s}\sum_{n=0}^{\infty}\frac{\Gamma(\rho n+1)}{(n!)^m\Gamma(\alpha n+\beta)}(\frac{a}{s^{\rho}})^n \tag{A-109}$$

$$L\{t^{\beta-1}{}_mE_{\alpha,\beta}(at^{\alpha})\} = \frac{1}{s^{\beta}}\,{}_{m-1}E_{1,1}(\frac{a}{s^{\alpha}}) = \frac{1}{s^{\beta}}\,{}_0F_{m-1}(;1,1,\ldots,1;\frac{a}{s^{\alpha}}) \tag{A-110}$$

can be evaluated in the closed form for some values of parameters α, β, ρ and m.

$$L\{_2E_{1,1}(t^2)\} = \frac{1}{s}e^{2/s^2} I_0(\frac{2}{s^2}) \tag{A-111}$$

$$L\{_3E_{1,1}(-t^2)\} = \frac{1}{s}[J_0(\frac{2}{s^2})]^2 \tag{A-112}$$

Direct use of (A-3) gives

$$_2E_{2,1}(t)\} = [ber(2t^{1/4})]^2 + [bei(2t^{1/4})]^2 \tag{A-113}$$

$$_2E_{2,1}(-t)\} = J_0(2t^{1/4})I_0(2t^{1/4}) \tag{A-114}$$

APPENDIX B.
SUMMATION FORMULAS

B-1. CAUCHY'S RESIDUE THEOREM AND SUMMATION FORMULAS

Given a Laplace transform $\Phi(s)$, the Bromwich-Wagner integral (18.3.7) provides the original function $f(t)$. This implies appropriate choices for integral contours, evaluation of residues and familiarity with the *Laurent expansions*. At this point it seems desirable to recall some useful calculation techniques.

An analytic function $f(z)$ admits in a neighborhood of a pole z = a the representation (*Laurent expansion*)

$$f(z) = \frac{a_m}{(z-a)^m} + \frac{a_{m-1}}{(z-a)^{m-1}} + \ldots + \frac{a_1}{z-a} + \phi(z) \tag{B-1}$$

where $\phi(z)$ denotes an analytic function regular at z = a. Then $f(z)$ is said to have *a pole of order n at $z = a$* and the coefficient a_1 is called the *residue* of $f(z)$ at z = a.

$$a_1 = \frac{1}{2\pi i}\int_C f(z)dz \tag{B-2}$$

The integration contour C includes the pole z = a but not any other singular point of $f(z)$.

Cauchy's residue theorem states: If Γ denotes a *simple* (i.e. deprived of multiple points), *closed, rectifiable curve*, then

$$\int_\Gamma f(z)dz = 2\pi i \sum_k \rho_k \tag{B-3}$$

and $\sum_k \rho_k$ denotes the sum of all residues situated in the interior of Γ.

Evaluation of residues is often made easier through recourse to appropriate formulas. Thus, if $f(z)$ admits a simple pole z = a we may write $f(z) = h(z)/g(z)$ with $h(a) \neq 0$, $g(a) = 0$ and $g(z)$ is *holomorphic function* in some neighborhood of z = a. Let $\xi = z - a$, then the residue ρ_a is

$$\rho_a = \lim_{\xi \to 0}\{\frac{h(a+\xi)}{[g(a+\xi)-g(a)]/\xi}\} = \frac{h(a)}{g'(a)} \qquad \text{(B-4)}$$

while if $f(z) = h(z)/z{\cdot}g(z)$ with $h(a) \neq 0$, $g(a) \neq 0$, the residue at pole $z = 0$ is $\rho_a = h(0)/g(0)$. Furthermore, if $g(z)$ admits simple poles at $z = a_j$, j = 1,2,3, ..., m, while $h(a_j) \neq 0$, the corresponding residues of $f(z)$ are

$$\rho_j = \lim_{z \to a_j}[\frac{(z-a_j)h(z)}{zg(z)}] \qquad \text{(B-5)}$$
$$j = 1,2,3,...m$$

Now, the bracketed expression takes the undetermined form 0/0 and L'Hospital rule gives

$$\rho_j = \lim_{z \to a_j}[\frac{(z-a_j)h'(z)+h(z)}{zg'(z)+g(z)}] = \frac{h(a_j)}{a_j g'(a_j)} \qquad \text{(B-6)}$$

Examples

1. Let a be a real number, a ≠ 0 and n a positive integer. The function $f(z) = e^{nz}/(z^2+a^2)$ admits two simple poles $z_1 = ia$, $z_2 = -ia$ and the residues are

$$\rho_1 = \lim_{z \to ia}(z-ia)f(z) = \frac{e^{ina}}{2ia}$$
$$\rho_2 = \lim_{z \to -ia}(z+ia)f(z) = -\frac{e^{-ina}}{2ia} \qquad \text{(B-7)}$$

If C denotes a circle $|z|$ = R > a, then we have

$$\frac{1}{2\pi i}\int_C \frac{e^{nz}}{z^2+a^2}dz = \frac{e^{ina}-e^{-ina}}{2ia} = \frac{\sin(na)}{a} \qquad \text{(B-8)}$$

2. Consider $f(z) = e^z/z\cdot\cos z$. The origin $z = 0$ and the points $\pm 1/2, \pm 3/2, \ldots, \pm (2n+1)/2, \ldots$ are simple poles. If C is a circle centered at the origin and R satisfying the condition: $(m+1/2)\pi < R < (m+3/2)\pi$ where m is a positive integer we obtain the summation formula

$$1+\frac{2}{\pi}\sum_{k=-m}^{k=m}\frac{(-1)^k e^{(2k+1)/2}}{2k+1}=\frac{1}{2\pi i}\int_C \frac{e^z}{z\cos\pi z}dz \quad \text{(B-9)}$$

3. The function $f(z) = \coth z = \cosh z/\sinh z$, admits simple poles, namely the points 0, $\pm\pi i, \pm 2\pi i, \ldots, \pm n\pi i$. The residue is the same for each pole, $\rho = 1$.

The case of a pole of order m at $z = a$ when $m > 1$ will be considered now.

As $(z\text{-a})^m f(z)$ must be holomorphic in a neighborhood of $z = a$ one may set

$$\begin{aligned} p(z) &= (z-a)^m f(z) = \\ & p(a)+(z-a)p'(a)+\ldots+\frac{(z-a)^{m-1}p^{(m-1)}(a)}{(m-1)!}+\ldots \end{aligned} \quad \text{(B-10)}$$

so that the residue of $f(z)$ is

$$\rho_a = \frac{1}{(m-1)!}\lim_{z\to a}\frac{d^{m-1}}{dz^{m-1}}[(z-a)^m f(z)] \quad \text{(B-11)}$$

The concept of the residue at point $z = \infty$ proves very useful in several applications of Cauchy's theorem on residues. Let

$$\begin{aligned} & f(z) = \phi(z^{-1}) = \phi(\zeta) \\ & \zeta = z^{-1} \end{aligned} \quad \text{(B-12)}$$

Now, the behaviour of $\phi(\zeta)$ in a neighborhood of $\zeta = 0$ will determine the behaviour of $f(z)$ in a neighborhood of $z = \infty$. If $\phi(\zeta)$ admits an expansion

$$\begin{aligned} & \phi(\zeta) = a_k\zeta^k + a_{k+1}\zeta^{k+1}+\ldots \\ & a_k \neq 0 \end{aligned} \quad \text{(B-13)}$$

$z = \infty$ is said to be a zero of order k of $f(z)$. And if $\zeta = 0$ is a pole of order m of $\phi(\zeta)$, $f(z)$ is said to admit a pole of order m at $z = \infty$. Moreover, $z = \infty$ is said to be an essential singular point of $f(z)$ whenever $\zeta = 0$ is an essential singular point of $\phi(\zeta)$. In such way, it seems natural to introduce the residue at $z = \infty$ whenever $f(z)$ admits a pole at $z = \infty$ and the following theorems are easily proved.

If $z = \infty$ is an isolated singularity of $f(z)$, the residue at infinity is given by the integral $\frac{1}{2\pi i}\int_C f(z)dz$ where C denotes a sufficiently large circle described in the clockwise sense.

If $f(z)$ admits a finite number of singularities, the sum of its residues (the residue at infinity included) is equal zero.

After these preliminary considerations related to evaluation of residues, the first group of summation formulas obtained through the Cauchy residues theorem will be discussed. [32,75,105].

Theorem

Given:

a) a bounded domain (D) in the plane of complex variable z.

b) a function $f(z)$, holomorphic in (D).

c) a meromorphic function $g(z)$ admitting N simple poles at the points a, a+1, a+2, ..., a+N-1, situated in (D) and is deprived of other poles in (D).

Then the summation formula is given by

$$\frac{1}{2\pi i}\int_C f(z)g(z)dz = \sum_{k=0}^{N-1} c_k f(a+k) \tag{B-14}$$

where C is a contour situated in the interior of (D) and c_k denotes the residue at the pole a+k. This result can be generalized in several ways. Whenever (D) remains bounded and the singular points of $f(z)$ and $g(z)$ are poles, their number must be finite. Otherwise these poles will admit an accumulation point at finite distance corresponding to an essential singular point. In other words, the summation formula consists of a finite number of terms:

$$\frac{1}{2\pi i}\int_C f(z)g(z)dz = S + \sum_k c_k f(a+k) \tag{B-15}$$

where S denotes the sum of residues corresponding to the product $f(z)g(z)$, wherever $f(z)$ admits singular points in the interior of C, which must not coincide with the poles of $g(z)$.

B-2. The Principle of the Argument

A very interesting case arises when $f(z)$ is a *logarithmic derivative*

$$f(z) = \frac{d}{dz}[\ln \phi(z)] = \frac{\phi'(z)}{\phi(z)} \tag{B-16}$$

More specifically consider $\int_C \frac{\phi'(z)}{\phi(z)} dz$ where the integration contour C being a closed curve deprived of multiple points, while $\phi(z)$ is meromorphic in the interior of C but deprived of poles and zeros on C. Thus, the only singularities of the logarithmic derivative $\phi'(z)/\phi(z)$ situated in the closed domain are the zeros and the poles of $\phi(z)$. If $z = a$ is a zero of order k of $f(z)$ one may write $\phi(z) = (z-a)^k \psi(z)$ where $\psi(z)$ is a holomorphic function with $|\psi(z)| > 0$ in a neighborhood of $z = a$ and then

$$\frac{\phi'(z)}{\phi(z)} = \frac{k}{z-a} + \frac{\psi'(z)}{\psi(z)} \tag{B-17}$$

and the residue of the logarithmic derivative $\phi'(z)/\phi(z)$ at its simple pole $z = a$ is k.

In the same way, if $z = b$ is a pole of order m of $\phi(z)$, the logarithmic derivative $\phi'(z)/\phi(z)$ admits at this point a simple pole with residue -m. If $a_1, ..., a_m$ denote the zeros, $b_1, ..., b_n$ denote the poles of $\phi(z)$ in the domain interior to the contour C then

$$\frac{1}{2\pi i} \int_C \frac{\phi'(z)}{\phi(z)} dz = \sum_{j=1}^{m} k_j - \sum_{j=1}^{n} l_j \tag{B-18}$$

where k_j is the order of the zero a_j and l_j is the order of the pole b_j.

Theorem

Let C be a simple (i.e. deprived of multiple points), closed, rectifiable curve, $\phi(z)$ a meromorphic function deprived of zeros and poles on C then

$$\frac{1}{2\pi i} \int_C \frac{\phi'(z)}{\phi(z)} dz = Z - P \tag{B-19}$$

where Z is the total number of zeros and P is the total number of poles lying in the domain interior to C.

Now, it will be convenient to connect this result with a geometric interpretation called *principle of the argument* [57-59]. Since the left-hand side of (B-19) is

$$\frac{1}{2\pi i}\int_C d[\ln\phi(z)]dz = \frac{1}{2\pi i}[\ln\phi(z)]_C = \frac{1}{2\pi i}[\ln(\phi(z)(+i\arg\phi(z)]_C =$$
$$\frac{1}{2\pi}[\arg\phi(z)]_C \qquad \text{(B-20)}$$

the variation of $\left|\ln\phi(z)\right|$ being null after the image of the complex variable z has traversed the closed integration contour C. Hence, it is possible to conclude that the argument of $\phi(z)$ increases by a multiple of 2π when z encircles C

$$\Delta_C[\arg\phi(z)] = 2\pi(Z-P) \qquad \text{(B-21)}$$

which is valid for the logarithmic derivative of a meromorphic function. The rigorous proof of this result is based on the *Heine-Borel lemma* [105].

The equation (B-21) leads to a new summation formula, if $\phi(z)$ satisfies already mentioned conditions and $g(z)$ is a holomorphic function in the interior of the closed integration contour C and as well as on C, then we have

$$\frac{1}{2\pi i}\int_C \frac{\phi'(z)}{\phi(z)}g(z)dz = \sum_{j=1}^{m} k_j g(a_j) - \sum_{j=1}^{n} l_j g(b_j) \qquad \text{(B-22)}$$

where k_j, l_j, a_j and b_j have the same meaning as previously.

B-3. Further Examples of Summation Formulas

1. Poisson's Summation Formula

Let a function $f(z)$ be holomorphic and periodic in a strip (B), a < Im z < b. It is easy to verify that the coefficients of its Fourier series can be inferred from the Laurent expansion of a function $F(z)$ associated with $f(z)$ as follows. Suppose first that the period of $f(z)$ is *one*, i.e. $f(z+1) = f(z)$. The change of variable $\zeta = e^{2\pi iz}$ transforms the strip (B) in an annulus (i.e. a ring-shaped region bounded by two concentric circles) with $|\zeta| = e^{-2\pi a} = R_1$ and $|\zeta| = e^{-2\pi b} = R_2$. Now, the correspondence between the strip and the annulus is not *one to one*: a fixed value of ζ corresponds to infinity values of z. In other words, $\zeta = 0$ is a *branch point*. A convenient method overcoming this difficulty is a modification of the annulus domain through building of a *barrier* (or "*cut*") which erases the "many-valued character" of an expression. In this case, such a barrier may be a half line issuing from the origin $\zeta = 0$. Let

$$F(\zeta) = f[\frac{1}{2\pi i}(\ln\frac{\zeta}{\zeta_0} + \ln\zeta_0)] \qquad \text{(B-23)}$$

where $\zeta_0 \neq 0$, denotes some fixed point in the annulus. Thus, $[\ln(\zeta/\zeta_0) + \ln \zeta_0]$, becomes a holomorphic function in the annulus $0 \leq \arg \zeta_0 < 2\pi$. Such a barrier, (half line, arg z = constant = θ_0 = arg z_0, issuing from the center of the annulus) insures that both the Laurent expansion and the Fourier series

$$F(\zeta) = \sum_{m=-\infty}^{\infty} A_m \zeta^m$$
$$f(z) = \sum_{m=-\infty}^{\infty} A_m e^{2\pi imz} \qquad \text{(B-24)}$$

are valid. Hence, the following theorem comes: if $f(z)$ is holomorphic and periodic with the period one in an unbounded strip a < Im z < b, its Fourier series is uniformly convergent in any compact subset of this strip. The choice of the period one is not essential but is convenient. The *Poisson summation formula* is a corollary of this theorem. Let $\phi(z)$ denotes a function holomorphic throughout an unbounded strip -a < Im z < a, a > 0 and periodic $\phi(z+1) = \phi(z)$. If the series

$$\Phi(z) = \sum_{m=-\infty}^{\infty} \phi(z+m) \qquad \text{(B-25)}$$

converges uniformly in every closed set including the segment $0 \leq x = \text{Re}\, z \leq 1$, its sum $\Phi(z)$ is holomorphic by virtue of the Weierstrass theorem on uniformly convergent series of holomorphic functions. Consequently its Fourier expansion

$$\Phi(z) = \sum_{m=-\infty}^{\infty} A_m e^{-2\pi imx}$$
$$A_m = \int_0^1 \Phi(x) e^{-2\pi imx} dx \qquad \text{(B-26)}$$

is uniformly convergent in the segment $0 \leq x \leq 1$. This permits term-by-term integration and therefore

$$A_m = \int_0^1 \Phi(x)e^{-2\pi imx}dx = \int_0^1 [\sum_{m=-\infty}^{\infty} \phi(x+m)e^{-2\pi imx}]dx =$$

$$\sum_{m=-\infty}^{\infty} [\int_0^1 \phi(x+m)e^{-2\pi imx}dx] = \sum_{m=-\infty}^{\infty} [\int_m^{m+1} \phi(x)e^{-2\pi imx}dx] = \tag{B-27}$$

$$\int_{-\infty}^{\infty} \phi(x)e^{-2\pi imx}dx$$

Introducing (B-27) into (B-26) and setting $x = 0$ one obtains the *Poisson summation formula*

$$\sum_{m=-\infty}^{\infty} \phi(m) = \sum_{m=-\infty}^{\infty} [\int_{-\infty}^{\infty} \phi(x)e^{-2\pi imx}dx] \tag{B-28}$$

One should note that this summation formula deals with a periodic function defined in a unbounded domain, while the formula (B-22) dealt with summations of finite number of terms, involving boundary conditions for bounded domains.

Poisson's formula leads also to an interesting result [105], that the *thetâ function*

$$\vartheta(t) = \sum_{m=-\infty}^{\infty} e^{-\pi m^2 t} \tag{B-29}$$

satisfies the functional equation

$$f(t) = t^{-1/2} f(t^{-1}) \tag{B-30}$$

2. Meromorphic Functions and Summation Formulas

We start with applications of the proceeding theorems.

The function $\pi \cot \pi z = \dfrac{d}{dz} \ln \sin \pi z$ is a logarithmic derivative. Euler's formulas for $\sin u = (e^{iu} - e^{-iu})/2i$ and $\cos u = (e^{iu} + e^{-iu})/2$ enable to write

$$\pi \cot \pi z = -\pi i \frac{e^{i\pi z} + e^{-i\pi z}}{e^{i\pi z} - e^{-i\pi z}} = -\pi i \frac{e^{2i\pi z} + 1}{e^{2i\pi z} - 1} \tag{B-31}$$

Now, by setting $w = e^{i2\pi z}$ one obtains two expansions

$$\pi \cot \pi z = -\pi i \frac{w+1}{w-1} = -\pi i(1 + 2w + 2w^2 + \ldots)$$
$$\pi \cot \pi z = -\pi i \frac{1+w^{-1}}{1-w^{-1}} = \pi i(1 + \frac{2}{w} + \frac{2}{w^2} + \ldots) \tag{B-32}$$

The first expansion is valid for $|w| < 1$, i.e. Im $z > 0$ and the second expansion for $|w| > 1$ and Im $z < 0$. Finally, equations (B-32) take the form

$$\pi \cot \pi z = -\pi i\,(1 + 2\sum_{m=1}^{\infty} e^{i2\pi mz})$$
$$\operatorname{Im} z > 0$$
$$\pi \cot \pi z = \pi i\,(1 + 2\sum_{m=1}^{\infty} e^{-i2\pi mz}) \tag{B-33}$$
$$\operatorname{Im} z < 0$$

$\pi \cot \pi z$ is a meromorphic function, its poles are simple and located at points $z_n = (2n+1)/2$, n = 0, ±1, ±2,... . The corresponding Fourier expansions converge uniformly on every compact set. It is possible to obtain the first formula in (B-33) through (B-22) by setting $g(z) = 1$, $\phi(z) = \sin \pi z$ and the contour C is a rectangle with vertices $(N+1/2) \pm i\lambda$ and $-(N+1/2) \pm i\lambda$, N is positive integer, $\lambda > 0$ and then let $N \to \infty$.

When a meromorphic function admits an infinite number of poles the only possible accumulation point of these poles is $z \to \infty$. This is the case of the circular functions tan z, cot z, sec z, csec z and the hyperbolic functions tanh z, coth z, sech z, csch z. In both cases of circular and hyperbolic functions one deals with simple combinations of exponential functions. The main difficulty encountered at this point is the generalization (B-15) to unbounded domains. In solving this problem Bernoull's numbers and polynomials play an important role as well as some other sequences of numbers and polynomials. The next two sections will deal with these subjects.

B-4. Bernoulli Numbers

The sequence $\{B_n\}$ of *Bernoulli numbers* is a sequence of coefficients appearing in the Taylor series of $(z/2)\cdot\coth(z/2)$, explicitly

$$\frac{z}{2}\coth\frac{z}{2} = 1 + \sum_{n=1}^{\infty} (-1)^{n-1} B_n \frac{z^{2n}}{(2n)!} \tag{B-34}$$

Before going further two preliminary considerations are essential:

a) (z/2)·coth(z/2) is an even function of z and $\lim_{z\to 0}[\frac{z}{2}\coth\frac{z}{2}] = 1$

b) $\frac{z}{2}\coth\frac{z}{2} = \frac{z}{2}(\frac{e^{z/2}+e^{-z/2}}{e^{z/2}-e^{-z/2}}) = \frac{z}{2}(\frac{e^z+1}{e^z-1}) = \frac{z}{2} + \frac{z}{e^z-1}$

Let $f(z) = z/(e^z - 1)$ then (B-34) becomes

$$f(z) = 1 - \frac{z}{2} + \sum_{n=1}^{\infty}(-1)^{n-1}B_n\frac{z^{2n}}{(2n)!} \tag{B-35}$$

The right hand-side of (B-35) is the Taylor series of the meromorphic function $f(z)$ which is regular at the origin provided that we agree to define

$$f(0) = \lim_{z\to 0}\frac{z}{e^z-1} = 1 \tag{B-36}$$

Moreover

$$f'(0) = \lim_{h\to 0}[\frac{f(h)-f(0)}{h}] = -\frac{1}{2} \tag{B-37}$$

$f(z)$ admits an infinity of simple poles situated at the points $a_n = 2n\pi i$, n = ±1, ±2, ... and the radius of convergence of the power series (B-35) is 2π. The series offers a possibility of computing the terms of the sequence $\{B_n\}$ for small values of n since the series expansion of $[f(z)]^{-1}$ is known

$$[f(z)]^{-1} = \frac{e^z-1}{z} = 1 + \frac{z}{2!} + \frac{z^2}{3!} + \ldots + \frac{z^n}{(n+1)!} + \ldots \tag{B-38}$$

The multiplication of (B-35) by (B-38) gives unity and by equating the coefficients of the expansion it is possible to obtain the values of the first terms of the sequence: $B_1 = 1/6$, $B_2 = 1/30$, $B_3 = 1/42$, $B_4 = 1/30$, $B_5 = 5/66$. A recurrence formula is known for the Bernoulli numbers which enables the determination of B_n for large values of n.

Legendre (1752-1833) obtained integral representations for Bernoulli numbers starting with the integrals $\int_C \frac{e^{\pm iaz}}{e^{2\pi z}-1}dz$, $a > 0$, when the integration contour is a rectangle whose vertices are the points 0, R, R + i, i , indented at z = 0 and z = i (points corresponding to simple poles of the integrand). Then putting R→ ∞, we obtain

$$\int_0^\infty \frac{\sin(ax)}{e^{2\pi x}-1}dx = \frac{1}{4}\coth\frac{a}{2} - \frac{1}{2a} \tag{B-39}$$

Introducing $a = 2p$ and $x = s/2$ into (B-39) the integral becomes

$$\int_0^\infty \frac{\sin(ps)}{e^{\pi s}-1}ds = \frac{1}{2}\coth p - \frac{1}{2p} = -\frac{1}{2p} + \frac{1}{2p}[\frac{2p}{2}\coth\frac{2p}{2}] \tag{B-40}$$

and using (B-34) it takes the form

$$\begin{aligned}\int_0^\infty \frac{\sin(ps)}{e^{\pi s}-1}ds &= -\frac{1}{2p} + \frac{1}{2p}[1+\sum_{n=1}^\infty (-1)^{n-1}B_n\frac{(2p)^{2n}}{(2n)!}] = \\ &B_1\frac{(2p)}{2!} - B_2\frac{(2p)^3}{4!} + B_3\frac{(2p)^5}{6!} - \ldots\end{aligned} \tag{B-41}$$

Now

$$\frac{\partial^n}{\partial s^n}\left(\int_0^\infty \frac{\sin(ps)}{e^{\pi s}-1}ds\right) = \int_0^\infty \frac{s^n \sin(ps+\frac{\pi n}{2})}{e^{\pi s}-1}ds \tag{B-42}$$

Thus, the successive differentiations of the integral and the series in (B-41) with putting $p = 0$ affords a method for finding integral representations of Bernoulli numbers. Introducing $s = 2t$ into (B-42) one has the *Legendre formula*

$$B_n = 4n\int_0^\infty \frac{t^{n-1}}{e^{2\pi t}-1}dt \tag{B-43}$$

Legendre's formal method can be justified by taking into account that in a neighborhood of $p = 0$ the integral (B-42) is uniformly convergent [120].

At this point it appears adequate to illustrate the part played by Bernoulli numbers in expansions of meromorphic functions [105].

a) Introduction of $z = 2i\zeta$ into (B-34) yields

$$\zeta\cot\zeta = 1 - \sum_{n=1}^\infty B_n\frac{2^n\zeta^{2n}}{(2n)!} \qquad |\zeta| < \pi \tag{B-44}$$

b) This formula leads to the expansion of tan z.

From

$$\cot 2z = (\tan 2z)^{-1} = [\frac{2\tan z}{1-(\tan z)^2}]^{-1} = \frac{1}{2}(\cot z - \tan z)$$
$$\tan z = \cot z - 2\cot 2z = z^{-1}(z\cot z - 2z\cot 2z) \tag{B-45}$$

we have

$$\tan z = \sum_{n=1}^{\infty} B_n \frac{(2^{4n}-2^{2n})z^{2n-1}}{(2n)!}$$
$$|z| < \frac{\pi}{2} \tag{B-46}$$

c) Since dln cos z /dz = -tan z , the integration along a segment issuing from the origin gives

$$\ln\cos z = -\sum_{n=1}^{\infty} B_n \frac{(2^{4n}-2^{2n})z^{2n}}{(2n)!\,2n}$$
$$|z| < \frac{\pi}{2} \tag{B-47}$$

d) Similarly from dln (sin z/z)/dz = cot z - $1/z$ = (z cot z - 1)/z after integration term by term of (B-44) we has

$$\ln\frac{\sin z}{z} = -\sum_{n=1}^{\infty} B_n \frac{(2z)^{2n}}{(2n)!\,2n} \tag{B-48}$$
$$|z| < \pi$$

Bernoulli numbers afford simple expressions for the sums of series with the general term $u_n = n^{-2m}$, m = 1,2,3, ...and this is perhaps their most striking property.

Along with the expansion (B-34) for (z/2)·coth(z/2) there is the well known expansion

$$\coth\frac{z}{2} = \frac{2}{z} + \sum_{n=1}^{\infty} \frac{4z}{z^2+4n^2\pi^2} \tag{B-49}$$

Consequently

$$\frac{z}{2}\coth\frac{z}{2} = 1 + \sum_{n=1}^{\infty}(-1)^{n-1}B_n\frac{z^{2n}}{(2n)!} = 1 + \sum_{n=1}^{\infty}\frac{2z^2}{z^2+4n^2\pi^2} \qquad \text{(B-50)}$$
$$|z| < 2\pi$$

but

$$\frac{2z^2}{z^2+4n^2\pi^2} = \frac{z^2}{2\pi^2n^2}[1 - \frac{z^2}{4\pi^2n^2} + \frac{z^4}{16\pi^4n^4} - \ldots] \qquad \text{(B-51)}$$
$$|z| < 2\pi$$

and for $|z| < 2\pi$ the right-hand expression in (B-50) becomes an absolutely convergent double series. The order of summation can be changed and therefore

$$\frac{z}{2}\coth\frac{z}{2} = 1 + \frac{z^2}{2\pi^2}\sum_{n=1}^{\infty}\frac{1}{n^2} - \frac{z^4}{2^3\pi^4}\sum_{n=1}^{\infty}\frac{1}{n^4} + \frac{z^6}{2^5\pi^6}\sum_{n=1}^{\infty}\frac{1}{n^6} - \ldots \qquad \text{(B-52)}$$

Hence

$$\sum_{n=1}^{\infty}\frac{1}{n^2} = \pi^2 B_1$$
$$\sum_{n=1}^{\infty}\frac{1}{n^4} = \frac{\pi^4}{3}B_2 \qquad \text{(B-53)}$$
$$\sum_{n=1}^{\infty}\frac{1}{n^6} = \frac{2\pi^6}{45}B_3$$

the general form of (B-53) is therefore

$$\sum_{n=1}^{\infty}\frac{1}{n^{2m}} = \frac{2^{2m-1}\pi^{2m}}{(2m)!}B_m \qquad \text{(B-54)}$$

These formulas ratify the necessity of a factor $(-1)^{n-1}$ in (B-34) in order to restore Bernoulli numbers as a positive sequence. The Bernoulli sequence is not the only conspicuous sequence of numbers met in classical analysis. Amongst them it seems appropriate to quote:

Euler's numbers E_n are defined by

$$\operatorname{sech} z = (\cosh z)^{-1} = 1 + \sum_{n=1}^{\infty} E_n \frac{z^{2n}}{(2n)\,!} \qquad \text{(B-55)}$$

$$|z| < \frac{\pi}{2}$$

Gregory's numbers i_n are defined by

$$\frac{z}{\ln(1-z)} = \sum_{n=1}^{\infty} i_n z^n \qquad \text{(B-56)}$$

$$|z| < 1$$

Bourguet's numbers β_n are defined by

$$\frac{1}{\Gamma(z+1)} = \sum_{n=1}^{\infty} \beta_n z^n \qquad \text{(B-57)}$$

$$|z| < \infty$$

Very often such sequences are those of coefficients that appear in the Taylor and Mac Laurin expansions. It is unusual that the sequences $f^{(n)}(z_0)/n!, f^{(n)}(0)/n!$ correspond to explicit functions of n and they are obtained through term by term integration or differentiation. Such is the case of $f(z) = \sin^{-1}z$ or $f(z) = \tan^{-1}z$.

A method, sometimes efficient, consists in finding the expansion $f(z) = \sum_n a_n z^n$ when the expansion $g(z) = [f(z)]^{-1} = \sum_n b_n z^n$, $b_0 \neq 0$ is known. Since $f(z)g(z) = 1$, it follows that

$$\begin{aligned} &a_0 b_0 = 1 \\ &a_0 b_1 + a_1 b_3 = 0 \\ &\cdots\cdots\cdots\cdots\cdots\cdots \\ &a_0 b_n + a_1 b_{n-1} + \ldots + a_n b_0 = 0 \end{aligned} \qquad \text{(B-58)}$$

and therefore

$$a_0 = \frac{1}{b_0}$$

$$a_1 = -\frac{b_1}{b_0^2}$$

$$a_2 = \frac{b_1^2 - b_0 b_2}{b_0^3} \qquad \text{(B-59)}$$

....

$$a_n = \frac{(-1)^n}{b_0^{n+1}} \begin{vmatrix} b_1 & b_0 & 0 & ...0 \\ b_2 & b_1 & b_0 & ...0 \\ ... & ... & ... & ...0 \\ b_n & b_{n-1} & b_{n-2} & ...b_1 \end{vmatrix}$$

B-5. Bernoulli Polynomials

Let

$$F(t,x) = t\frac{e^{xt} - 1}{e^t - 1} \qquad \text{(B-60)}$$

$$|z| < 2\pi$$

The power series expansion of $F(t,x)$ with t considered as the variable and x as a parameter, is obtained as a product of two Taylor series:

$$F(t,x) = [e^{xt} - 1][t(e^t - 1)^{-1}] =$$

$$(xt + \frac{x^2t^2}{2!} + \frac{x^3t^3}{3!} + \ldots)(1 - \frac{t}{2} + B_1\frac{t^2}{2!} - B_2\frac{t^4}{4!} + \ldots) = \qquad \text{(B-61)}$$

$$\sum_{n=1}^{\infty} B_n(x)\frac{t^n}{n!}$$

where $B_n(x)$ denotes nth degree polynomial

$$\begin{aligned}
&B_1 = x \\
&B_2 = x^2 - x \\
&B_3 = x^3 - \frac{3}{2}x^2 + \frac{1}{2}x \\
&\ldots\ldots\ldots\ldots\ldots\ldots\ldots\ldots \\
&B_n(x) = x^n - \frac{n}{2}x^{n-1} + \binom{n}{2}B_1 x^{n-2} - \binom{n}{4}B_2 x^{n-4} + \ldots
\end{aligned} \tag{B-62}$$

the last term being of the second degree if n is even and the first degree if n is odd. This sequence of polynomials $\{B_n(x)\}$ is the sequence of *Bernoulli polynomials.* In his posthumous book *Ars Conjectandi* (1713), Jacob Bernoulli (1654-1703) defined these polynomials by the formula $B_n(x) = \sum_{s=1}^{x-1} s^n$, $x, s, n \in N$.

The definition of the generating functions (B-60) and (B-61) implies

$$\begin{aligned}
F(t, x+1) - F(t, x) &= t\left[\frac{e^{(x+1)t}}{e^t - 1} - \frac{e^{xt}}{e^t - 1}\right] = te^{xt} \\
F(t, x+1) - F(t, x) &= \sum_{n=1}^{\infty} \frac{B_n(x+1) - B_n(x)}{n!}
\end{aligned} \tag{B-63}$$

Consequently we have

$$te^{xt} = \sum_{n=1}^{\infty} \frac{x^{n-1}}{(n-1)!} = \sum_{n-1}^{\infty} \frac{B_n(x+1) - B_n(x)}{n!} \tag{B-64}$$

and the functional relation

$$B_n(x+1) - B_n(x) = nx^{n-1} \tag{B-65}$$

Successive differentiation of (B-65) gives

$$\begin{aligned}
&B_n'(x+1) - B_n'(x) = n(n-1)x^{n-2} \\
&B_n^{(m)}(x+1) - B_n^{(m)}(x) = n(n-1)\ldots(n-m)x^{n-m-1} \\
&B_n^{(n-k)}(x+1) - B_n^{(n-k)}(x) = n(n-1)\ldots kx^{k-1}
\end{aligned} \tag{B-66}$$

Since $B_n(0) = 0$, n = 1,2,3,... it follows that

$$\begin{aligned}
&B_n(1) = 0 \\
&B_n^{(n-m)}(0) = B_n^{(n-m)}(1) \\
&B_n^{(n-1)}(0) = B_n^{(n-1)}(1) + n! \\
&n = 2,3,\ldots \quad , \quad m > 1
\end{aligned} \tag{B-67}$$

Using (B-67) and putting $x = 1$ in (B-62) one gets the recurrent formula

$$0 = 1 - \frac{n}{2} + \binom{n}{2} B_1 - \binom{n}{4} B_2 + \binom{n}{6} B_3 - \ldots \tag{B-68}$$

which allows the computation of Bernoulli numbers for large values of n.

B-6. The Darboux Formula

Let

a) $f(z)$ to be a function holomorphic in a neighborhood of a point z = a.

b) λ to be a real number, $0 \le \lambda \le 1$ and $\phi(\lambda)$ a polynomial of nth degree.

The sum $F(z,\lambda)$

$$F(z,\lambda) = \sum_{k=1}^{n} (-1)^k (z-a)^k \phi^{(n-k)}(\lambda) f^{(k)}[a + \lambda(z-a)] \tag{B-69}$$

is a function of z, holomorphic in some neighborhood of z = a and λ is considered as a parameter . The derivative of this sum, $\partial F / \partial \lambda$, reduces to two terms

$$\begin{aligned}
&\frac{\partial F(z,\lambda)}{\partial \lambda} = -(z-a)\phi^{(n)}(\lambda) f'[a + \lambda(z-a)] + \\
&(-1)^n (z-a)^{n+1} \phi(\lambda) f^{(n+1)}[a + \lambda(z-a)]
\end{aligned} \tag{B-70}$$

Setting $\phi^{(n)}(\lambda) = \phi^{(n)}(0)$ and integrating both terms with respect to λ between $\lambda = 0$ and $\lambda = 1$, we obtain

$$\sum_{k=1}^{n}(-1)^k(z-a)^k[\phi^{(n-k)}(1)f^{(k)}(z)-\phi^{(n-k)}(0)f^{(k)}(a)]=$$
$$-\phi^{(n)}(0)\int_0^z f'(\lambda)d\lambda+(-1)^n(z-a)^{n+1}\int_0^1\phi(\lambda)f^{(n+1)}[a+\lambda(z-a)]d\lambda \qquad \text{(B-71)}$$

which can be rewritten as

$$\phi^{(n)}(0)[f(z)-f(a)]=$$
$$\sum_{k=1}^{n}(-1)^{k-1}(z-a)^k[\phi^{(n-k)}(1)f^{(k)}(z)-\phi^{(n-k)}(0)f^{(k)}(a)]+ \qquad \text{(B-72)}$$
$$(-1)^n(z-a)^{n+1}\int_0^1\phi(\lambda)f^{(n+1)}[a+\lambda(z-a)]d\lambda$$

This expression is called the *Darboux formula* [120]. If $\phi(\lambda) = (\lambda - 1)^n$ is introduced into (B-72) then we have

$$f(z)=f(a)+\sum_{k=1}^{n}\frac{(z-a)^k}{k!}f^{(k)}(a)+$$
$$(-1)^n\frac{(z-a)^{n+1}}{n!}\int_0^1(\lambda-1)^n f^{(n+1)}[a+\lambda(z-a)]d\lambda \qquad \text{(B-73)}$$

This is the Taylor expansion in which the remainder appears as a definite integral.

The *Euler-MacLaurin summation formula* is also a particular case of the Darboux formula where $\phi(\lambda) = B_n(\lambda)$ is the nth Bernoulli polynomial.

B-7. Plana's Summation Formula

The function $g(z) = \pi \cot \pi z$ is a meromorphic function, its singularities are simple poles situated at $z = 0, \pm 1, \pm 2, \ldots$ and an essential singularity at $z = \infty$. Moreover $g(z)$ is a logarithmic derivative, $\pi\cot\pi z=\dfrac{d}{dz}\ln\sin\pi z$, thus allowing the use of the argument method (B-15)

$$\sum_{k=n}^{k=n+m} f(k)=\frac{1}{2\pi i}\int_C(\pi\cot\pi z)f(z)dz-S_C \qquad \text{(B-74)}$$

The poles z = n, n+1, ... , n+m being situated in the interior of the integration contour C and S_C denotes the sum of residues of $\pi \cot \pi z$ at singular points of $f(z)$ situated in the interior of C. $S_C = 0$ if $f(z)$ is holomorphic in the considered region.

Plana's summation formula results if the integration contour C is the rectangle (R) having for vertices the points n - $i\lambda$, n + m - $i\lambda$, n + $i\lambda$, n + m + $i\lambda$, $\lambda > 0$, n, m are integers and m > n. The point z traversing (R) according to the sense of positive rotation. (R) is indented at z = n and z = m.

In a neighborhood of the pole z = n the product $f(z)g(z)$ admits the Laurent expansion

$$f(z)g(z) = (\pi \cot \pi z) f(z) = \frac{f(n)}{z-n} + h(z) \tag{B-75}$$

where $h(z)$ is holomorphic in a neighborhood of the pole z = n of $g(z)$. If γ denotes the half circle indenting the rectangle (R) at z = n, then

$$\int_{\gamma} (\pi \cot \pi z) f(z) dz = \int_{\pi/2}^{-\pi/2} \frac{f(n)}{\rho e^{i\theta}} d(\rho e^{i\theta}) + \alpha_n(\rho) = -\pi i f(n) + \alpha_n(\rho) \tag{B-76}$$

A similar expression is available at the pole z = n + m.

Since $\lim_{\rho \to 0} \alpha_n(\rho) = 0$ and $\lim_{\rho \to 0} \alpha_{n+m}(\rho) = 0$, (B-74) reduces to

$$\sum_{k=n}^{k=n+m} f(k) = \frac{1}{2}[f(n) + f(n+m)] + \frac{1}{2\pi i} \int_{(R)} (\pi \cot \pi z) f(z) dz \tag{B-77}$$

The Plana summation formula is a consequence of this result.

Let $\Omega(z)$ be defined as

$$\begin{gathered} \Omega(z) = i + \cot \pi z = \frac{2i}{1 - e^{2\pi i z}} \\ \operatorname{Im} z > 0 \\ \Omega(z) = -i + \cot \pi z = \frac{2i}{e^{2\pi i z} - 1} \\ \operatorname{Im} z < 0 \end{gathered} \tag{B-78}$$

and R_1 and R_2 are contours, then

$$\int_{R_1} f(z)dz - \int_{R_2} f(z)dz = 0$$

$$\int_{R_1} f(z)dz + \int_{R_2} f(z)dz = 2\int_{n}^{n+m} f(x)dx \qquad \text{(B-79)}$$

$$x = \operatorname{Re} z$$

and (B-77) may be rewritten as

$$\sum_{k=n}^{k=n+m} f(k) = \frac{1}{2}[f(n) + f(n+m)] + \int_{n}^{n+m} f(x)dx + \frac{1}{2i}\int_{R} \Omega(z) f(z)dz \qquad \text{(B-80)}$$

where R is the integration contour, a rectangle with vertices, n - $i\lambda$, n + m - $i\lambda$, n + m + $i\lambda$, n + $i\lambda$, traversed in the sense of positive rotations. If $f(z)$ is holomorphic in the strip n $\leq x =$ Re $z \leq$ n+m and if $\lim_{y\to\infty} e^{-2\pi|y|} f(x+iy) = 0$, $y = \operatorname{Im} z$, the convergence being uniform on n $\leq x$ $\leq$ n+m. It can be shown that the integral in (B-80) is Q(n+m) - Q(n), where

$$Q(x) = \frac{1}{i}\int_{0}^{\infty} \frac{f(x+iy) - f(x-iy)}{e^{2\pi y} - 1} dy \qquad \text{(B-81)}$$

In a particular case when: m $\rightarrow \infty$, f(n+m) $\rightarrow \infty$ and Q(n+m) $\rightarrow$ 0 we have *Plana's summation formula*

$$\sum_{k=n}^{\infty} f(k) = \frac{1}{2} f(n) + \int_{n}^{\infty} f(x)dx - Q(n) \qquad \text{(B-82)}$$

Applications of (B-82) to the expansion of Euler's constant in series of Bernoulli numbers and that the Euler-MacLaurin formula is a consequence of the Plana summation formula [105].

APPENDIX C. APPLICATION OF THE SHIFTED DIRAC DELTA FUNCTION δ (X -1) TO THE EVALUATION OF THE ASYMPTOTIC RELATIONS, LIMITS OF SERIES, INTEGRALS AND INTEGRAL REPRESENTATIONS OF ELEMENTARY AND SPECIAL FUNCTIONS

C-1. INTRODUCTION

The impulse functions were introduced to mathematics by Cauchy, Poisson, Hermite, Helmholtz, Kirchhoff, Kelvin and Heaviside and they are widely recognized as a powerful tool in many fields of physics, engineering and applied mathematics. The notation $\delta(x)$ is associated with the name of Dirac who was first to use the impulse function in quantum mechanics. Properties and a short history of the $\delta(x)$-function are given in the books of van der Pol and Bremmer [96] and Hoskins [60]. The Dirac delta function is not an ordinary function because it is defined in the following way

$$\delta(x-x_0)=\begin{cases}0 & \text{for all } x\neq x_0\\ \infty & \text{for } x=x_0\end{cases} \tag{C-1}$$

The most important property of the shifted Dirac delta function, $\delta(x\text{-}x_0)$, is called the *sifting property* or the *sampling property*. If $f(x)$ is continuously differentiable in the neighborhood of point x_0, $-a < x_0 < a$, then the sifting property of $\delta(x\text{-}x_0)$ (or $\delta(x)$ for $x_0 = 0$) is expressed by

$$\int_{-a}^{a}\delta(x-x_0)f(x)dx=f(x_0) \tag{C-2}$$

The property itself is independent of the actual values of the limits of integration ± a depending only on the behaviour of the integrand near the point x_0. Thus, the delta function is defined by giving a rule for integrating its product with a continuous function $f(x)$. In the similar way are defined derivatives of the delta function

$$\int_{-a}^{a} \delta^{(n)}(x-x_0) f(x) dx = (-1)^n f^{(n)}(x_0) \qquad \text{(C-3)}$$

$$n = 0, 1, 2, \ldots$$

The impulse function is approximated by the sequence of functions converging to it, which are called the *delta sequences.* Four delta sequences for $\delta(x\text{-}1)$ are presented here

$$\delta(x-1) = \lim_{\lambda \to \infty} \left(\frac{\lambda}{\pi[1 + \lambda^2 (x-1)^2]} \right) = \lim_{\varepsilon \to 0} \left(\frac{\varepsilon}{\pi[\varepsilon^2 + (x-1)^2]} \right) \qquad \text{(C-4)}$$

$$\delta(x-1) = \lim_{\lambda \to \infty} \left\{ \frac{\lambda}{\pi} \left(\frac{\sin[\lambda(x-1)]}{x-1} \right)^2 \right\} \qquad \text{(C-5)}$$

$$\delta(x-1) = \lim_{\lambda \to \infty} \left(\frac{\lambda e^{-|\lambda(x-1)|^n}}{2\Gamma(1 + \frac{1}{n})} \right) \qquad \text{(C-6)}$$

$$n = 1, 2, 3, \ldots$$

$$\delta(x-1) = \lim_{\lambda \to \infty} [\lambda J_\lambda(\lambda x)] \qquad \text{(C-7)}$$

The first two delta sequences were considered in works of Cauchy, Poisson and Hermite. The particular case of (C-6) for n = 2 is called the "heat source" of Kelvin. The fourth sequence contains the Bessel function of the first kind $J_\lambda(z)$ and was introduced by Lamborn [72]. Representations of $\delta(x\text{-}1)$ are also possible by applying the *orthogonality identities*, for example, by using the Bessel functions

$$\delta(x-1) = \int_0^\infty x u J_\lambda(u) J_\lambda(xu) du \qquad \text{(C-8)}$$

and

$$\delta(x-1) = \int_0^\infty x u Y_\lambda(u) H_\lambda(xu) du \qquad \text{(C-9)}$$

where $Y_\lambda(z)$ and $H_\lambda(z)$ are the Bessel functions of the second kind and the *Struve* functions respectively.

Using mainly the delta sequence of Lamborn (C-7), Apelblat [9-11] derived a large number of integral representations of the mathematical constants, the elementary and special functions, the first derivatives, the integral transforms of these representations, the asymptotic and functional relations and the limits of series. The Bessel and Struve finite and infinite integrals are available from the orthogonality identities (C-8) and (C-9). These relations and procedures as applied to elementary and special functions are illustrated here while the case of the Volterra functions was already considered in Section 5.6.

C-2. The Basic Relations

If the sifting property of the shifted Dirac delta function $\delta(x\text{-}1)$ is combined with the delta sequence of Lamborn (C-7) we have

$$f(t)=\int_0^\infty f(tx)\delta(x-1)dx=\lim_{\lambda\to\infty}[\lambda\int_0^\infty f(tx)J_\lambda(\lambda x)dx] \tag{C-10}$$

i.e. the integral representation of function $f(t)$ is given as the asymptotic limit of the infinite integral of the product of $f(tx)$ and the Bessel function $J_\lambda(\lambda x)$. If the right-hand integral in (C-10) can be evaluated in the closed form and the result of integration denoted by $\Phi(t,\lambda)$ then (C-10) can be written as

$$f(t)=\lim_{\lambda\to\infty}[\lambda\Phi(t,\lambda)] \tag{C-11}$$

This limit can be considered as the generalization of the l'Hospital's rule.

Similarly, in the case of the first derivative of $f(t)$ we have

$$tf'(t)=\frac{1}{2}\lim_{\lambda\to\infty}\{\lambda^2\int_0^\infty[J_{\lambda+1}(\nu x)-J_{\lambda-1}(\nu x)]f(tx)dx\} \tag{C-12}$$

If $f(tx)$ is replaced by $g(x) = xf(tx)$ then

$$f(t)+tf'(t)=\frac{1}{2}\lim_{\nu\to\infty}\left\{\lambda^2\int_0^\infty[J_{\lambda+1}(\lambda x)-J_{\lambda-1}(\lambda x)]xf(tx)dx\right\} \tag{C-13}$$

In relations (C-10) - (C13) the same function $f(t)$ appears on both sides of equations. Introduction of integral transforms enlarges the usefulness of these relations.

If integral transformation of a function $f(\xi;\tau)$ where τ is a parameter is defined by

$$T\{f(\xi;\tau)\} = \int_0^\infty K(s,\xi)f(\xi;\tau)d\xi = T(s,\tau) \tag{C-14}$$

where $K(s,\xi)$ is the kernel of integral transformation, then using (C-10) (see also (18.5.6)) we have

$$\begin{aligned} &\lim_{\lambda\to\infty}[\lambda\int_0^\infty f(\xi;\tau)T\{J_\lambda(\lambda x);\xi\}d\xi] = \\ &\lim_{\lambda\to\infty}[\lambda\int_0^\infty J_\lambda(\lambda x)T\{f(\xi;\tau);x\}dx] = T(1,\tau) \end{aligned} \tag{C-15}$$

Since for the Laplace transform, the Fourier cosine and sine transforms and the Stieltjes transform of $J_\lambda(\lambda x)$ are known

$$L\{J_\lambda(\lambda x);\xi\} = \frac{\lambda^\lambda}{\sqrt{\lambda^2+\xi^2}[\xi+\sqrt{\lambda^2+\xi^2}]^\lambda} \tag{C-16}$$

$$\mathrm{Re}\,\lambda > -1$$

$$F_c\{J_\lambda(\lambda x);\xi\} = \frac{\cos[\lambda\sin^{-1}(\frac{\xi}{\lambda})]}{\sqrt{\xi^2-\lambda^2}} \tag{C-17}$$

$$0 \le \xi < \lambda$$

$$F_s\{J_\lambda(\lambda x);\xi\} = \frac{\sin[\lambda\sin^{-1}(\frac{\xi}{\lambda})]}{\sqrt{\xi^2-\lambda^2}} \tag{C-18}$$

$$0 \le \xi < \lambda$$

$$S\{J_\lambda(\lambda x);\xi\} = \frac{\pi}{\sin(\pi\lambda)}[\mathrm{J}_\lambda(\lambda\xi) - J_\lambda(\lambda\xi)] = \int_0^\infty e^{-\lambda\zeta-\lambda\xi\sinh\zeta}d\zeta \tag{C-19}$$

$$\mathrm{Re}\,\xi > 0$$

it follows from (C-15) that

$$\lim_{\lambda\to\infty}\left\{\lambda^{\lambda+1}\int_0^\infty \frac{f(\xi;\tau)}{\sqrt{\lambda^2+\xi^2}[\xi+\sqrt{\lambda^2+\xi^2}]^\lambda}d\xi\right\}=$$
$$\lim_{\lambda\to\infty}[\lambda\int_0^\infty J_\lambda(\lambda x)F\{x;\tau\}dx]=F(1;\tau) \tag{C-20}$$

$$\lim_{\lambda\to\infty}\{\lambda\int_0^\infty \frac{\cos[\lambda\sin^{-1}(\frac{\xi}{\lambda})]f(\xi;\tau)}{\sqrt{\lambda^2-\xi^2}}d\xi\}=$$
$$\lim_{\lambda\to\infty}[\lambda\int_0^\infty J_\lambda(\lambda x)F_c(x;\tau)dx]=F_c(1;\tau) \tag{C-21}$$

$$\lim_{\lambda\to\infty}\{\lambda\int_0^\infty \frac{\sin[\lambda\sin^{-1}(\frac{\xi}{\lambda})]f(\xi;\tau)}{\sqrt{\lambda^2-\xi^2}}d\xi\}=$$
$$\lim_{\lambda\to\infty}[\lambda\int_0^\infty J_\lambda(\lambda x)F_s(x;\tau)dx]=F_s(1;\tau) \tag{C-22}$$

$$\lim_{\lambda\to\infty}\{\lambda\int_0^\infty f(\xi;\tau)[\int_0^\infty e^{-\nu\zeta-\lambda\xi\sinh\zeta}d\zeta]d\xi\}=$$
$$\lim_{\nu\to\infty}[\lambda\int_0^\infty J_\lambda(\lambda x)S(x;\tau)\,dx]=S(1;\tau) \tag{C-23}$$

If the function $f(\xi;\tau)$ is substituted by the kernels of integral transformations, $f(\xi;\tau)=K(\xi,\tau)$ and taking into account that

$$\int_0^\infty \delta(x-1)J_\lambda(\lambda x)dx=\lim_{\lambda\to\infty}\{\lambda\int_0^\infty K(\xi;\tau)J_\lambda(\lambda\xi)d\xi\}=K(1;\tau) \tag{C-24}$$

we have from (C-16) - (C-19)

$$\lim_{\lambda\to\infty}\left\{\frac{\lambda^\lambda}{\sqrt{\lambda^2+\tau^2}[\tau+\sqrt{\lambda^2+\tau^2}]^\lambda}\right\}=e^{-\tau} \tag{C-25}$$

$$\lim_{\lambda\to\infty}\left\{\frac{\lambda\cos[\lambda\sin^{-1}(\frac{\tau}{\lambda})]}{\sqrt{\tau^2-\lambda^2}}\right\}=\cos\tau \tag{C-26}$$

$$\lim_{\lambda\to\infty}\left\{\frac{\lambda\sin[\lambda\sin^{-1}(\frac{\tau}{\lambda})]}{\sqrt{\tau^2-\lambda^2}}\right\}=\sin\tau \tag{C-27}$$

$$\lim_{\lambda\to\infty}\left\{\lambda\int_0^\infty e^{-\lambda\zeta-\lambda\tau\sinh\zeta}\,d\zeta\right\}=\frac{1}{\tau+1} \tag{C-28}$$

The Laplace transformation, the Fourier cosine and sine transformations and the Hankel *J*, *Y*, *K* and *H*-transformations satisfy the following similarity rule

$$T\{f(at);s\}=\frac{1}{a}T(\frac{s}{a}) \tag{C-29}$$

$$a>0$$

and therefore

$$T(s)=\lim_{\lambda\to\infty}\{\lambda[\int_0^\infty J_\lambda(\lambda x)\frac{1}{x}T(\frac{s}{x})dx]\} \tag{C-30}$$

On the other hand it follows from (C-10) that

$$T(s)=\lim_{\lambda\to\infty}\{\lambda[\int_0^\infty J_\lambda(\lambda\xi)T(s\xi)d\xi]\} \tag{C-31}$$

and comparing (C-30) with (C-31) we have two equivalent forms of the limit

$$\lim_{\lambda\to\infty}\{\lambda[\int_0^\infty J_\lambda(\lambda x)\frac{1}{x}T(\frac{s}{x})dx]\}=\lim_{\lambda\to\infty}\{\lambda[\int_0^\infty J_\lambda(\lambda\xi)T(s\xi)d\xi]\}=T(s) \tag{C-32}$$

In the generalized Stieltjes transformation

$$S\{f(x);\rho\} = \int_0^\infty \frac{f(x)}{(s+x)^\rho}dx = S(s;\rho) \tag{C-33}$$

the similarity rule is

$$S\{f(ax);\rho\} = a^{\rho-1}S(sa;\rho) \qquad a > 0 \tag{C-34}$$

and therefore

$$\lim_{\lambda\to\infty}\{\lambda[\int_0^\infty x^{\rho-1}J_\lambda(\lambda x)S(sx;\rho)dx]\} = \lim_{\lambda\to\infty}\{\lambda[\int_0^\infty J_\lambda(\lambda\xi)S(s\xi;\rho)d\xi]\} = S(s;\rho) \tag{C-35}$$

For ρ = 1, both integrands are identical in (C-35).

If the sifting property of the delta function is applied to other delta sequences, then the equivalent of (C-10) for the Cauchy function is

$$f(t) = \frac{1}{\pi}\lim_{\lambda\to 0}\{\varepsilon\int_0^\infty \frac{f(tx)}{[\varepsilon^2+(x-1)^2]}dx\} \tag{C-36}$$

In the case of the delta sequence (C-6) used by Poisson, Kirchhoff and Kelvin, for n = 1 and n = 2 we have

$$f(t) = \frac{1}{2}\lim_{\lambda\to\infty}\{\lambda\int_0^\infty e^{-\lambda|x-1|}f(tx)dx\} \tag{C-37}$$

$$f(t) = \frac{1}{\sqrt{\pi}}\lim_{\lambda\to\infty}\{\lambda\int_0^\infty e^{-\lambda^2(x-1)^2}f(tx)dx\} \tag{C-38}$$

The function representation in (C-37) can be expressed in a more proper form by writing the integral as a sum of two integrals

$$\int_0^\infty e^{-\lambda|x-1|}f(tx)dx = \int_0^1 e^{-\lambda|x-1|}f(tx)dx + \int_1^\infty e^{-\lambda(x-1)}f(tx)dx \tag{C-39}$$

but

$$\int_0^1 e^{-\lambda|x-1|} f(tx)dx = \int_0^1 e^{-\lambda(1-x)} f(tx)dx \qquad \text{(C-40)}$$

and

$$\int_1^\infty e^{-\lambda(x-1)} f(tx)dx = \int_0^\infty e^{-\lambda\xi} f[t(1+\xi)]d\xi \qquad \text{(C-41)}$$

However, if the right-hand side integral in (C-41) is treated as the Laplace transform we have

$$\int_0^\infty e^{-\lambda\xi} f[t(1+\xi)]d\xi = \frac{e^\lambda}{t}[F(\frac{\lambda}{t}) - t\int_0^1 e^{-\lambda x} f(tx)dx] \qquad \text{(C-42)}$$

and finally

$$f(t) = \frac{1}{2t}\lim_{\lambda\to\infty}\{\lambda e^\lambda F(\frac{\lambda}{t}) + 2t\lambda\int_0^1 \sinh[\lambda(x-1)]f(tx)dx\} \qquad \text{(C-43)}$$

If integral transformations are combined with (C-38), the analogous relation to that in (C-15) is

$$\lim_{\lambda\to\infty}[\lambda\int_0^\infty f(\xi;\tau)T\{e^{-\lambda^2(x-1)^2};\xi\}d\xi] =$$
$$\lim_{\lambda\to\infty}[\lambda\int_0^\infty e^{-\lambda^2(x-1)^2} T\{f(\xi;\tau);x\}dx] = \sqrt{\pi}T(1,\tau) \qquad \text{(C-44)}$$

For example, for the Laplace transformation [52, 3.462.1]

$$L\{e^{-\lambda^2(x-1)^2};\xi\} = \frac{\sqrt{\pi}}{2\lambda}e^{-\lambda^2+[(\xi-2\lambda^2)/2\lambda]^2} erfc(\frac{\xi-2\lambda^2}{2\lambda}) \qquad \text{(C-45)}$$

it follows from (C-44) that

$$\lim_{\lambda\to\infty}\{e^{-\lambda^2}\int_0^\infty f(\xi;\tau)e^{-[(\xi-2\lambda^2)/2\lambda]^2} erfc(\frac{\xi-2\lambda^2}{2\lambda})d\xi\} = 2F(1;\tau) \qquad \text{(C-46)}$$

Thus, elementary and special functions can be represented as the limit of infinite integral containing the complementary error function.

If the sifting property of the delta Dirac function is applied to a function $g(t) = t^{1/2}f(t)$ and combined with the orthogonality identity (C-8) , the result can be written in the form

$$t^{3/2} f(t) = t g(t) = \int_0^\infty u^{1/2} J_\lambda(u) G(\frac{u}{t};\lambda) du \tag{C-47}$$

where $G(s;\lambda) = G_\lambda\{g(t), s\} = G_\lambda\{t^{1/2}f(t), s\}$ is the Hankel transform of order λ, defined by

$$G_\lambda\{g(\xi);s\} = \int_0^\infty (s\xi)^{1/2} J_\lambda(s\xi) g(\xi) d\xi = G(s;\lambda)$$

$$\text{Re}\,\lambda > -\frac{1}{2} \tag{C-48}$$

The relation (C-47) can be treated as a kind of the inversion formula. For $\lambda = \pm 1/2$, the Hankel transforms are reduced to the Fourier transforms.

Similarly, using (C-9) we have

$$t^{3/2} f(t) = t g(t) = \int_0^\infty u^{1/2} Y_\lambda(u) H(\frac{u}{t};\lambda) du \tag{C-49}$$

where

$$H_\lambda\{g(\xi);s\} = \int_0^\infty (s\xi)^{1/2} H_\lambda(s\xi) g(\xi) d\xi = H(s;\lambda)$$

$$\text{Re}\,\lambda > -\frac{1}{2} \tag{C-50}$$

is the H-transform of order λ of the function $g(t) = t^{1/2}f(t)$. In this integral transform the Struve function serves as the kernel.

If the Hankel and Struve transforms of $g(t)$ are available, the Bessel integrals of the type (C-47) and (C-49) can be evaluated. Evidently, the Hankel transforms also permit evaluation of limits from

$$t f(t) = \lim_{\lambda\to\infty} \{\lambda^{1/2} G(\frac{\lambda}{t};\lambda)\} \tag{C-51}$$

C-3. Illustrative Examples

The basic relations presented in the section C-2 will be now applied to variety of elementary and special functions to show the usefulness of derived above results [9-11].

The asymptotic expressions based on (C-11) are available if the Bessel integrals of $f(tx)$

$$\Phi(t;\lambda) = \int_0^\infty J_\lambda(\lambda x) f(tx) dx \tag{C-52}$$

are known.

Asymptotic formulas

For example from [119, 1.301 and 1.502]

$$\int_0^\infty J_\lambda(\lambda x) \ln(tx) dx = \frac{1}{\lambda}[\ln(\frac{2t}{\lambda}) + \psi(\frac{\lambda+1}{2})] \tag{C-53}$$

it follows that

$$\lim_{\lambda\to\infty} \{\ln(\frac{2}{\lambda}) + \psi(\frac{\lambda+1}{2})\} = 0 \tag{C-54}$$

or denoting $x = (\lambda+1)/2$, the asymptotic formula for the psi-function takes the form

$$\psi(x) \sim \ln\left(x - \frac{1}{2}\right) \quad \text{as} \quad x \to \infty \tag{C-55}$$

Few cases of the modified Bessel functions will be considered. For example, from [1, 11.4.28]

$$\int_0^\infty J_\lambda(\lambda x) e^{-t^2x^2} dx = \frac{\sqrt{\pi}}{2t} e^{-\lambda^2/8t^2} I_{\lambda/2}(\frac{\lambda^2}{8t^2}) \tag{C-56}$$

we have that for the modified Bessel function of the first kind and of the order $\lambda/2$ the asymptotic expression is

$$\lim_{\lambda\to\infty} \{\lambda e^{-\lambda^2/8t^2} I_{\lambda/2}(\frac{\lambda^2}{8t^2})\} = \frac{2t}{\sqrt{\pi}} e^{-t^2} \tag{C-57}$$

and combining (C-12) with (C-56) leads to

$$\lim_{\lambda\to\infty}\{\lambda^2 e^{-\lambda^2/8t^2}[I_{(\lambda-1)/2}(\frac{\lambda^2}{8t^2})-I_{(\lambda+1)/2}(\frac{\lambda^2}{8t^2})]\}=\frac{8t^3}{\sqrt{\pi}}e^{-t^2} \tag{C-58}$$

Using [52, 6.571.1]

$$\int_0^\infty J_\lambda(\lambda x)\frac{[tx+\sqrt{a^2+t^2x^2}\,]^\mu}{\sqrt{a^2+t^2x^2}}dx=\frac{a^\mu}{t}I_{(\lambda-\mu)/2}(\frac{a\lambda}{2t})K_{(\lambda+\mu)/2}(\frac{a\lambda}{2t})$$
$$a>0\ ,\ t>0\ ,\ \mathrm{Re}\,\lambda>-1\ ,\ \mathrm{Re}\,\mu<\frac{3}{4} \tag{C-59}$$

we have

$$\lim_{\lambda\to\infty}\{\lambda I_{(\lambda-\mu)/2}(\frac{a\lambda}{2t})K_{(\lambda+\mu)/2}(\frac{a\lambda}{2t})\}=\frac{a^{-\mu}t[t+\sqrt{a^2+t^2}\,]^\mu}{\sqrt{a^2+t^2}} \tag{C-60}$$

Denoting $\tau = a/t$, $\lambda = 2\sigma$ and $\mu = 2\rho$ (C-60) becomes

$$\lim_{\sigma\to\infty}\{\sigma I_{\sigma-\rho}(\sigma\tau)K_{\sigma+\rho}(\sigma\tau)\}=\frac{[1+\sqrt{1+\tau^2}\,]^{2\rho}}{2\tau^{2\rho}\sqrt{1+\tau^2}} \tag{C-61}$$

In particular, for $\rho = 0$, the asymptotic formula for the product of the modified Bessel functions with large orders is

$$\sigma I_\sigma(\sigma\tau)K_\sigma(\sigma\tau)\sim\frac{1}{2\sqrt{1+\tau^2}}\quad\text{as}\quad\sigma\to\infty \tag{C-62}$$

which is the first term in the asymptotic expansion [1, 9.7.5]

$$\sigma I_\sigma(\sigma\tau)K_\sigma(\sigma\tau)\sim\frac{1}{2\sqrt{1+\tau^2}}\left\{1-\frac{1}{2}\frac{(4\sigma^2-1)}{(2\sigma\tau)^2}+\frac{3}{8}\frac{(4\sigma^2-1)(4\sigma^2-9)}{(2\sigma\tau)^4}-\ldots\right\}$$
$$\sigma\to\infty \tag{C-63}$$

These results can be extended to the products of the Bessel functions but with more complex arguments. Using the Hankel transform [85, 1.5.19]

$$G_\lambda\{\frac{e^{-\alpha(\beta^2+t^2)^{1/2}}}{[t(\beta^2+t^2)]^{1/2}};s\} = s^{1/2} I_{\lambda/2}[\theta_1(s)]K_{\lambda/2}[\theta_2(s)]$$

$$\theta_1(s) = \frac{\beta}{2}[\sqrt{\alpha^2+s^2}-\alpha] \quad , \quad \theta_2(s) = \frac{\beta}{2}[\sqrt{\alpha^2+s^2}+\alpha] \qquad \text{(C-64)}$$

$$\text{Re}\,\alpha > 0 \quad , \quad \text{Re}\,\lambda > -1$$

we have

$$\lim_{\lambda\to\infty}\{\lambda I_{\lambda/2}[\theta_1(\frac{\lambda}{t})]K_{\lambda/2}[\theta_2(\frac{\lambda}{t})]\} = \frac{te^{-\alpha(\beta^2+t^2)}}{(\beta^2+t^2)^{1/2}} \qquad \text{(C-65)}$$

For $\alpha = 0$, $\beta = 1$, $\lambda = 2\sigma$ and $\tau = 1/t$, the limit (C-65) reduces to (C-62). Similarly, from [85, 1.5.15]

$$G_\lambda\{t^{-3/2}e^{-\alpha t-\beta/t};s\} = 2s^{1/2}J_\lambda[\omega_1(s)]K_\lambda[\omega_2(s)]$$

$$\omega_1(s) = \{2\beta[\sqrt{\alpha^2+s^2}-\alpha]\}^{1/2} \quad , \quad \omega_2(s) = \{2\beta[\sqrt{\alpha^2+s^2}+\alpha]\}^{1/2} \qquad \text{(C-66)}$$

$$\text{Re}\,\alpha > 0 \quad , \quad \text{Re}\,\beta > 0$$

it follows that

$$\lim_{\lambda\to\infty}\{\lambda J_\lambda[\omega_1(\frac{\lambda}{t})]K_\lambda[\omega_2(\frac{\lambda}{t})]\} = \frac{e^{-\alpha t-\beta/t}}{2} \qquad \text{(C-67)}$$

and for $\alpha = 0$, $\beta = 1$ and $\tau = 1/t$, the limit (C-67) reduces to

$$\lambda J_\lambda(\sqrt{2\lambda\tau})K_\lambda(\sqrt{2\lambda\tau}) \sim \frac{e^{-\tau}}{2} \quad \text{as} \quad \lambda\to\infty \qquad \text{(C-68)}$$

Now, few examples of the hypergeometric and the confluent hypergeometric functions. Starting from [1, 13.2.2-3]

$$\int_0^\infty J_\lambda(\lambda x)x^{\mu-1}e^{-tx}dx = \frac{\Gamma(\lambda+\mu)\lambda^\lambda}{2^\lambda\Gamma(\lambda+1)t^{\lambda+\mu}}\,{}_2F_1(\frac{\lambda+\mu}{2},\frac{\lambda+\mu+1}{2};\lambda+1;-\frac{\lambda^2}{t^2}) \qquad \text{(C-69)}$$

$$\text{Re}(\lambda+\mu) > 0 \quad , \quad t > 0$$

we have from (C-10)

$$\lim_{\lambda\to\infty}\left\{\frac{\Gamma(\lambda+\mu)}{\Gamma(\lambda+1)}\left(\frac{\lambda}{2t}\right)^{\lambda+1} {}_2F_1\left(\frac{\lambda+\mu}{2},\frac{\lambda+\mu+1}{2};\lambda+1;-\frac{\lambda^2}{t^2}\right)\right\}=\frac{t^{\mu-1}e^{-t}}{2} \tag{C-70}$$

In particular, for $\mu = 0$, the hypergeometric function reduces to

$${}_2F_1\left(\frac{\lambda}{2},\frac{\lambda+1}{2};\lambda+1;-\frac{\lambda^2}{t^2}\right)=\left(\frac{2t}{t+\sqrt{t^2+\lambda^2}}\right)^{\lambda} \tag{C-71}$$

and therefore (C-70) reduces to the simple result

$$\lim_{\lambda\to\infty}\left(\frac{\lambda}{t+\sqrt{t^2+\lambda^2}}\right)^{\lambda}=e^{-t} \tag{C-72}$$

From [1, 11.4.28]

$$\int_0^\infty J_\lambda(\lambda x)x^{\mu-1}e^{-t^2x^2}dx=\frac{\Gamma\left(\dfrac{\lambda+\mu}{2}\right)\lambda^{\lambda}}{2^{\lambda+1}\Gamma(\lambda+1)t^{\lambda+\mu}}\,{}_1F_1\left(\frac{\lambda+\mu}{2};\lambda+1;-\frac{\lambda^2}{4t^2}\right) \tag{C-73}$$

it follows that

$$\lim_{\lambda\to\infty}\left\{\frac{\Gamma\left(\dfrac{\lambda+\mu}{2}\right)}{\Gamma(\lambda+1)}\left(\frac{\lambda}{2t}\right)^{\lambda+1}{}_1F_1\left(\frac{\lambda+\mu}{2};\lambda+1;-\frac{\lambda^2}{4t^2}\right)\right\}=t^{\mu-1}e^{-t^2} \tag{C-74}$$

Evidently, for $\mu = 1$, the confluent hypergeometric function can be reduced to the modified Bessel function as is expressed in (C-57).

From the Weber-Schafheitlin type integral [1, 11.4.34]

$$\int_0^\infty J_\lambda(\lambda x)J_\mu(tx)dx=$$

$$\frac{\Gamma\left(\dfrac{\lambda+\mu+1}{2}\right)t^{\mu}}{\Gamma(\mu+1)\Gamma\left(\dfrac{\lambda-\mu+1}{2}\right)\lambda^{\mu+1}}\,{}_2F_1\left(\frac{\lambda+\mu+1}{2},\frac{\mu-\lambda+1}{2};\mu+1;\frac{t^2}{\lambda^2}\right) \tag{C-75}$$

$$\mathrm{Re}(\mu)>-1 \quad , \quad 0<t<\lambda$$

we have

$$\lim_{\lambda\to\infty}\{\frac{\Gamma\left(\frac{\lambda+\mu+1}{2}\right)}{\Gamma\left(\frac{\lambda-\mu+1}{2}\right)\lambda^{\mu}}{}_2F_1(\frac{\lambda+\mu+1}{2},\frac{\mu-\lambda+1}{2};\mu+1;\frac{t^2}{\lambda^2})\}=$$

$$\Gamma(\mu+1)t^{-\mu}J_{\mu}(t) \tag{C-76}$$

$$\mathrm{Re}\,\mu>-1 \quad , \quad t>0$$

which for $\mu = 0$ reduces to

$$\lim_{\lambda\to\infty}\{{}_2F_1(\frac{\lambda+1}{2},\frac{1-\lambda}{2};1;\frac{t^2}{\lambda^2})\}=J_0(t) \tag{C-77}$$

If the representation of function based on the "heat source" of Kelvin (C-38) is combined with [52, 3.462.5]

$$\int_0^{\infty} xe^{-\lambda^2(x-1)^2}dx=\frac{e^{-\lambda^2}}{2\lambda^2}+\frac{\sqrt{\pi}}{2\lambda}[1+erf(\lambda)] \tag{C-78}$$

the asymptotic formula for the error function is

$$erf(\lambda)\,\square\,1-\frac{e^{-\lambda^2}}{\sqrt{\pi}\,\lambda} \quad \text{as} \quad \lambda\to\infty \tag{C-79}$$

The Hankel transform [85, 7.5.13]

$$G_{\lambda}\{t^{\mu-1}e^{-\alpha t^2};s\}=\frac{\Gamma\left(\frac{\lambda+\mu+1}{2}\right)}{\Gamma(\lambda+1)}\left(\frac{e^{-s^2/8\alpha}}{\alpha^{\mu/2}s^{1/2}}\right)M_{\mu/2,\lambda/2}\left(\frac{s^2}{4\alpha}\right) \tag{C-80}$$

$$\mathrm{Re}\,\alpha>0 \quad , \quad \mathrm{Re}(\lambda+\mu)>0$$

leads to the limit of the Whittaker function

$$\lim_{\lambda\to\infty}\{\frac{\Gamma\left(\frac{\lambda+\mu+1}{2}\right)}{\Gamma(\lambda+1)}e^{-\lambda^2/8\tau}M_{\mu/2,\lambda/2}\left(\frac{\lambda^2}{4\tau}\right)\}=\tau^{\mu}e^{-\tau^2} \tag{C-81}$$

where $\tau = \alpha t^2$. For $\mu = 0$, the Whittaker function is reduced to the modified Bessel function and (C-81) becomes (C-57).

Using the following Hankel transform [85, 1.5.3]

$$G_\lambda\{t^{n+1/2}e^{-\alpha t};s\} = (-1)^{n+1}s^{\lambda+1/2}\frac{d^{n+1}}{d\alpha^{n+1}}\left\{\frac{1}{\sqrt{\alpha^2+s^2}[\alpha+\sqrt{\alpha^2+s^2}]^\lambda}\right\} \tag{C-82}$$

$$\mathrm{Re}\,\alpha > 0 \quad , \quad \mathrm{Re}\,\lambda > 0 \quad , \quad n = 0,1,2,3...$$

the limit from (C-25) can be generalized

$$\lim_{\lambda\to\infty}\frac{d^{n+1}}{d\tau^{n+1}}\left\{\frac{\lambda^{\lambda+1}}{\sqrt{\tau^2+\lambda^2}[\tau+\sqrt{\tau^2+\lambda^2}]^\lambda}\right\} = (-1)^{n+1}e^{-\tau} \tag{C-83}$$

where $\tau = \alpha t$. For n = 0, the first derivative in (C-83) gives

$$\lim_{\lambda\to\infty}\left\{\frac{\lambda^{\lambda+1}[\tau+\lambda\sqrt{\tau^2+\lambda^2}]}{(\tau^2+\lambda^2)^{3/2}[\tau+\sqrt{\tau^2+\lambda^2}]^\lambda}\right\} = e^{-\tau} \tag{C-84}$$

Representations of Mathematical Constants and Special Functions

The limits expressing the mathematical constants and special functions are presented in this section. They are evaluated by using the Laplace transforms of different functions from (C-20)

$$L\{f(\xi;\tau);\lambda\} = F(\lambda;\tau)$$

$$\lim_{\lambda\to\infty}\left\{\lambda^{\lambda+1}\int_0^\infty\frac{f(\xi;\tau)}{\sqrt{\lambda^2+\xi^2}[\xi+\sqrt{\lambda^2+\xi^2}]^\lambda}d\xi\right\} = F(1;\tau) \tag{C-85}$$

The most interesting cases of (C-85) are these when special functions are represented by the limits of integrals with the integrands being elementary functions. If not presented here, the applied Laplace transforms of $f(\xi,\tau)$ functions are easily available in tables of integrals and integral transforms [1,7,8,29,31,44,52,56,80,85-88,102,119].

The Laplace transforms

$$L\left\{\frac{\sin t}{t}\right\} = \tan^{-1}\left(\frac{1}{s}\right) \tag{C-86}$$

$$L\{\ln t\} = -\frac{1}{s}(\gamma + \ln s) \tag{C-87}$$

together with (C-85) give representations of the π-number and the Euler constant γ

$$\frac{\pi}{4} = \lim_{\lambda \to \infty}\left\{\lambda^{\lambda+1}\int_0^{\infty}\frac{\sin\xi}{\sqrt{\lambda^2+\xi^2}[\xi+\sqrt{\lambda^2+\xi^2}]^{\lambda}}d\xi\right\} \tag{C-88}$$

$$\gamma = -\lim_{\lambda \to \infty}\left\{\lambda^{\lambda+1}\int_0^{\infty}\frac{\ln\xi}{\sqrt{\lambda^2+\xi^2}[\xi+\sqrt{\lambda^2+\xi^2}]^{\lambda}}d\xi\right\} \tag{C-89}$$

The Bernoulli numbers can be evaluated from

$$L\left\{\frac{t^{\mu-1}}{\sinh t}\right\} = 2^{1-\mu}\Gamma(\mu)\zeta(\mu,\frac{s+1}{2}) \tag{C-90}$$

$$\mu > 1$$

and

$$2^{1-\mu}\Gamma(\mu)\zeta(\mu,1) = 2^{1-\mu}\Gamma(\mu)\zeta(\mu) =$$

$$\lim_{\lambda \to \infty}\left\{\lambda^{\lambda+1}\int_0^{\infty}\frac{\xi^{\mu-1}}{\sqrt{\lambda^2+\xi^2}[\xi+\sqrt{\lambda^2+\xi^2}]^{\lambda}\sinh\xi}d\xi\right\} \tag{C-91}$$

but

$$\zeta(2n) = \frac{(2\pi)^{2n}}{2(2n)!}|B_{2n}| \tag{C-92}$$

$$n = 1, 2, 3, \ldots$$

and therefore

$$\frac{\pi^{2n}}{2n}|B_{2n}| = \lim_{\lambda \to \infty}\left\{\lambda^{\lambda+1}\int_0^{\infty}\frac{\xi^{2n-1}}{\sqrt{\lambda^2+\xi^2}[\xi+\sqrt{\lambda^2+\xi^2}]^{\lambda}\sinh\xi}d\xi\right\} \tag{C-93}$$

$$n = 1, 2, 3, \ldots$$

The representations of the gamma and related functions are given by

$$\Gamma(\mu+1)=\lim_{\lambda\to\infty}\left\{\lambda^{\lambda+1}\int_0^\infty \frac{\xi^{\lambda}}{\sqrt{\lambda^2+\xi^2}[\xi+\sqrt{\lambda^2+\xi^2}]^{\lambda}}d\xi\right\} \tag{C-94}$$

$\lambda>-1$

$$\Gamma(\mu,\alpha)=\lim_{\lambda\to\infty}\left\{\lambda^{\lambda+1}\int_\alpha^\infty \frac{\xi^{\lambda}}{\sqrt{\lambda^2+\xi^2}[\xi+\sqrt{\lambda^2+\xi^2}]^{\lambda}}d\xi\right\} \tag{C-95}$$

$\mu>0$

$$\psi(\mu)=-\gamma-\lim_{\lambda\to\infty}\left\{\lambda^{\lambda+1}\int_0^\infty \frac{\ln(e^{\xi/\mu}-1)}{\sqrt{\lambda^2+\xi^2}[\xi+\sqrt{\lambda^2+\xi^2}]^{\lambda}}d\xi\right\} \tag{C-96}$$

$$\psi^{(n)}(\mu)=\frac{(-1)^{n+1}}{\mu^{n+1}}\lim_{\lambda\to\infty}\left\{\lambda^{\lambda+1}\int_0^\infty \frac{\xi^{n}}{\sqrt{\lambda^2+\xi^2}[\xi+\sqrt{\lambda^2+\xi^2}]^{\lambda}\ln(1-e^{\xi/\mu})}d\xi\right\} \tag{C-97}$$

$n=1,2,3,\ldots$

$$B(\rho,\mu)=\frac{1}{\rho}\lim_{\lambda\to\infty}\left\{\lambda^{\lambda+1}\int_0^\infty \frac{(1-e^{\xi/\rho})^{\mu-1}}{\sqrt{\lambda^2+\xi^2}[\xi+\sqrt{\lambda^2+\xi^2}]^{\lambda}}d\xi\right\} \tag{C-98}$$

The Riemann zeta function is presented in (C-91), the generalized zeta function is given by

$$\zeta(\mu+1,\rho+1)=$$

$$\frac{1}{\Gamma(\mu+1)\rho^{\mu+1}}\lim_{\lambda\to\infty}\left\{\lambda^{\lambda+1}\int_0^\infty \frac{\xi^{\mu}}{\sqrt{\lambda^2+\xi^2}[\xi+\sqrt{\lambda^2+\xi^2}]^{\lambda}(e^{\xi/\rho}-1)}d\xi\right\} \tag{C-99}$$

$\mu>-1 \quad , \quad \rho>0$

The exponential integral and related functions are represented by

$$-e^{\alpha}Ei(-\alpha)=\lim_{\lambda\to\infty}\left\{\lambda^{\lambda+1}\int_0^\infty \frac{1}{\sqrt{\lambda^2+\xi^2}[\xi+\sqrt{\lambda^2+\xi^2}]^{\lambda}(\alpha+\xi)}d\xi\right\} \tag{C-100}$$

$$Ci(\alpha) = \lim_{\lambda\to\infty}\left\{\lambda^{\lambda+1}\int_0^\infty \frac{[\alpha\sin\alpha - \xi\cos\alpha]}{\sqrt{\lambda^2+\xi^2}[\xi+\sqrt{\lambda^2+\xi^2}]^\lambda(\alpha^2+\xi^2)}d\xi\right\} \tag{C-101}$$

$$Si(\alpha) = \frac{\pi}{2} - \lim_{\lambda\to\infty}\left\{\lambda^{\lambda+1}\int_0^\infty \frac{[\alpha\cos\alpha + \xi\sin\alpha]}{\sqrt{\lambda^2+\xi^2}[\xi+\sqrt{\lambda^2+\xi^2}]^\lambda(\alpha^2+\xi^2)}d\xi\right\} \tag{C-102}$$

The error function, the hypergeometric function, the Whittaker function and the parabolic cylinder function are expressed by

$$erf(\alpha) = \frac{1}{\pi}\lim_{\lambda\to\infty}\left\{\lambda^{\lambda+1}\int_0^\infty \frac{\sin(2\alpha\sqrt{\xi})}{\sqrt{\lambda^2+\xi^2}[\xi+\sqrt{\lambda^2+\xi^2}]^\lambda \xi}d\xi\right\} \tag{C-103}$$

$${}_2F_1(\mu,\alpha;\alpha+\beta+1;\rho) =$$

$$\frac{1}{\alpha B(\alpha,\beta+1)}\lim_{\lambda\to\infty}\left\{\lambda^{\lambda+1}\int_0^\infty \frac{(1-e^{-\xi/\alpha})^\beta}{\sqrt{\lambda^2+\xi^2}[\xi+\sqrt{\lambda^2+\xi^2}]^\lambda(1-\rho e^{-\xi/\alpha})^\mu}d\xi\right\} \tag{C-104}$$

$$\beta > -1 \quad , \quad |\rho| < 1$$

$$W_{\kappa,\mu}(\alpha) = \frac{e^{-\alpha/2}\alpha^{1/2-\mu}}{\Gamma(\mu-\kappa+\frac{1}{2})}\lim_{\lambda\to\infty}\left\{\lambda^{\lambda+1}\int_0^\infty \frac{\xi^{\mu-\kappa-\frac{1}{2}}(\alpha+\xi)^{\mu+\kappa-\frac{1}{2}}}{\sqrt{\lambda^2+\xi^2}[\xi+\sqrt{\lambda^2+\xi^2}]^\lambda}d\xi\right\} \tag{C-105}$$

$$\mathrm{Re}(\mu-\kappa+\frac{1}{2}) > 0$$

$$D_\mu(\alpha) = \frac{e^{-\alpha^2/4}}{\Gamma(\frac{1-\mu}{2})}\lim_{\lambda\to\infty}\left\{\lambda^{\lambda+1}\int_0^\infty \frac{\xi^{-(\mu+1)/2}(\alpha^2+2\xi)^{\mu/2}}{\sqrt{\lambda^2+\xi^2}[\xi+\sqrt{\lambda^2+\xi^2}]^\lambda}d\xi\right\} \tag{C-106}$$

$$\mu < 1$$

There is a number of representations associated with the Bessel functions and related functions. The modified Bessel functions are

$$I_{\mu}(\alpha)=\frac{e^{\alpha}}{\sqrt{\pi}\,\Gamma(\mu+\frac{1}{2})(2\alpha)^{\mu}}\lim_{\lambda\to\infty}\left\{\lambda^{\lambda+1}\int_{0}^{2\alpha}\frac{(2\alpha\xi-\xi^{2})^{\mu-1/2}}{\sqrt{\lambda^{2}+\xi^{2}}\,[\xi+\sqrt{\lambda^{2}+\xi^{2}}\,]^{\lambda}}d\xi\right\} \quad \text{(C-107)}$$

$$\mu>-\frac{1}{2}$$

$$K_{\mu}(\alpha)=\frac{\sqrt{\pi}\,e^{-\alpha}}{\Gamma(\mu+\frac{1}{2})(2\alpha)^{\mu}}\lim_{\lambda\to\infty}\left\{\lambda^{\lambda+1}\int_{0}^{\infty}\frac{(\xi^{2}+2\alpha\xi)^{\mu-1/2}}{\sqrt{\lambda^{2}+\xi^{2}}\,[\xi+\sqrt{\lambda^{2}+\xi^{2}}\,]^{\lambda}}d\xi\right\} \quad \text{(C-108)}$$

$$\mu>-\frac{1}{2}$$

The Struve functions $H_{\mu}(\alpha)$, $L_{\mu}(\alpha)$ the Anger function $\mathbf{J}_{\mu}(\alpha)$ and Weber function $E_{\mu}(\alpha)$ are represented by

$$H_{\mu}(\alpha)=$$

$$Y_{\mu}(\alpha)+\frac{1}{2^{\mu-1}\sqrt{\pi}\,\Gamma(\mu+\frac{1}{2})\alpha^{\mu}}\lim_{\lambda\to\infty}\left\{\lambda^{\lambda+1}\int_{0}^{\infty}\frac{(\alpha^{2}+\xi^{2})^{\mu-1/2}}{\sqrt{\lambda^{2}+\xi^{2}}\,[\xi+\sqrt{\lambda^{2}+\xi^{2}}\,]^{\lambda}}d\xi\right\} \quad \text{(C-109)}$$

$$\mu>-\frac{1}{2}\;,\quad \alpha>0$$

$$L_{\mu}(\alpha)=$$

$$I_{\mu}(\alpha)-\frac{1}{2^{\mu-1}\sqrt{\pi}\,\Gamma(\mu+\frac{1}{2})\alpha^{\mu}}\lim_{\lambda\to\infty}\left\{\lambda^{\lambda+1}\int_{0}^{\infty}\frac{(\alpha^{2}-\xi^{2})^{\mu-1/2}}{\sqrt{\lambda^{2}+\xi^{2}}\,[\xi+\sqrt{\lambda^{2}+\xi^{2}}\,]^{\lambda}}d\xi\right\} \quad \text{(C-110)}$$

$$\mu>-\frac{1}{2}\;,\quad \alpha>0$$

$$\mathrm{J}_{\mu}(\alpha)=$$

$$J_{\mu}(\alpha)-\frac{\alpha^{\mu}\sin(\pi\mu)}{\pi\mu}\lim_{\lambda\to\infty}\left\{\lambda^{\lambda+1}\int_{0}^{\infty}\frac{[(\xi+\sqrt{\alpha^{2}+\xi^{2}})^{-\mu}-\alpha^{-\mu}]}{\sqrt{\lambda^{2}+\xi^{2}}\,[\xi+\sqrt{\lambda^{2}+\xi^{2}}\,]^{\lambda}}d\xi\right\} \quad \text{(C-111)}$$

$$-\mathrm{E}_{\mu}(\alpha)=Y_{\mu}(\alpha)+$$
$$\frac{1}{\pi\alpha^{\mu}}\lim_{\lambda\to\infty}\left\{\lambda^{\lambda+1}\int_0^{\infty}\frac{[(\xi+\sqrt{\alpha^2+\xi^2})^{\mu}+\cos(\pi\mu)(\sqrt{\alpha^2+\xi^2}-\xi)^{\mu}]}{\sqrt{\lambda^2+\xi^2}[\xi+\sqrt{\lambda^2+\xi^2}]^{\lambda}\sqrt{\alpha^2+\xi^2}}d\xi\right\} \quad \text{(C-112)}$$

and the special cases of the Lommel functions

$$S_{0,\mu}(\alpha)=\frac{1}{\alpha\mu}\lim_{\lambda\to\infty}\left\{\lambda^{\lambda+1}\int_0^{\infty}\frac{\sinh[\mu\sinh^{-1}(\frac{\xi}{\alpha})]}{\sqrt{\lambda^2+\xi^2}[\xi+\sqrt{\lambda^2+\xi^2}]^{\lambda}}d\xi\right\} \quad \text{(C-113)}$$

$$S_{1,\mu}(\alpha)=\frac{1}{\alpha}\lim_{\lambda\to\infty}\left\{\lambda^{\lambda+1}\int_0^{\infty}\frac{\cosh[\mu\sinh^{-1}(\frac{\xi}{\alpha})]}{\sqrt{\lambda^2+\xi^2}[\xi+\sqrt{\lambda^2+\xi^2}]^{\lambda}}d\xi\right\} \quad \text{(C-114)}$$

The Fourier cosine or sine transformations can be applied to the Bessel functions of the first kind using (C-21) or (C-22), for example

$$J_{\mu}(\alpha)=\frac{1}{2^{\mu-1}\sqrt{\pi}\Gamma(\mu+\frac{1}{2})\alpha^{\mu}}\lim_{\lambda\to\infty}\{\lambda\int_0^{\alpha}\frac{\cos[\lambda\sin^{-1}(\frac{\xi}{\lambda})](\alpha^2-\xi^2)^{\mu-1/2}}{\sqrt{\lambda^2-\xi^2}}d\xi\} \quad \text{(C-115)}$$
$$\operatorname{Re}\mu>-\frac{1}{2}$$

$$Y_{-(\mu+1)}(\alpha)=$$
$$\frac{1}{2^{\mu-1}\sqrt{\pi}\Gamma(\mu+\frac{1}{2})\alpha^{\mu+1}}\lim_{\lambda\to\infty}\{\lambda\int_{\alpha}^{\infty}\frac{\sin[\lambda\sin^{-1}(\frac{\xi}{\lambda})]\xi(\xi^2-\alpha^2)^{\mu-1/2}}{\sqrt{\lambda^2-\xi^2}}d\xi\} \quad \text{(C-116)}$$
$$-\frac{1}{2}<\operatorname{Re}\mu<0$$

Some special functions can not be represented by the limits of integrals with the integrands being elementary functions and they should be replaced by other special functions. For example, this is the case of the complete elliptic integrals $K(k)$ and $E(k)$

$$K(k)=\frac{\pi}{2}\lim_{\lambda\to\infty}\left\{\lambda^{\lambda+1}\int_0^\infty\frac{[I_0(\frac{k\xi}{2})]^2}{\sqrt{\lambda^2+\xi^2}[\xi+\sqrt{\lambda^2+\xi^2}]^\lambda}d\xi\right\} \tag{C-117}$$

$$E(k)=\frac{\pi}{2}-\frac{\pi k}{2}\lim_{\lambda\to\infty}\left\{\lambda^{\lambda+1}\int_0^\infty\frac{I_0(k\xi)I_1(k\xi)}{\sqrt{\lambda^2+\xi^2}[\xi+\sqrt{\lambda^2+\xi^2}]^\lambda\xi}d\xi\right\} \tag{C-118}$$

Limits of Functional Series

If the sequence of functions $f_n(\rho;t)$ where ρ is a parameter converges to the sum $S(\rho;t)$

$$\sum_n^\infty f_n(\rho;t)=S(\rho;t) \tag{C-119}$$

and denoting

$$g_n(\rho;t)=\int_0^\infty J_\lambda(\lambda x)f_n(\rho;tx)dx \tag{C-120}$$

we have from (C-10) that

$$\lim_{\lambda\to\infty}\{\lambda\sum_n^\infty g_n(\rho;t)\}=S(\rho;t) \tag{C-121}$$

It is assumed the existence of integrals in (C-121) and that the order of summation, integration and the limit of the series can be reversed. This relation is illustrated by few examples starting with the series of elementary functions.

The Laplace transform of $f(x)=e^{-nx}/x$ is [102, 9.7.16]

$$\int_0^\infty J_\lambda(\lambda x)\frac{e^{-ntx}}{x}dx=\frac{1}{\lambda}\left[\frac{\lambda}{nt+\sqrt{(nt)^2+\lambda^2}}\right]^\lambda \tag{C-122}$$

and from (C-10) or (C-25) we have

$$\lim_{\lambda\to\infty}\left[\frac{\lambda}{nt+\sqrt{(nt)^2+\lambda^2}}\right]^\lambda=e^{-nt} \tag{C-123}$$

but

$$S(t)=\sum_{n=1}^{\infty}e^{-nt}=\frac{e^{-t}}{1-e^{-t}} \tag{C-124}$$

and therefore from (C-120)

$$\lim_{\lambda\to\infty}\{\lambda^{\lambda}\sum_{n=1}^{\infty}\frac{1}{[nt+\sqrt{(nt)^2+\lambda^2}]^{\lambda}}\}=\frac{e^{-t}}{1-e^{-t}} \tag{C-125}$$

Similarly, it is possible to obtain

$$\lim_{\lambda\to\infty}\{\lambda^{\lambda}\sum_{n=1}^{\infty}\frac{(-1)^n}{[nt+\sqrt{(nt)^2+\lambda^2}]^{\lambda}}=-\frac{e^{-t}}{1+e^{-t}} \tag{C-126}$$

Combining (C-124) and (C-125) we have

$$\lim_{\lambda\to\infty}\{\lambda^{\lambda}\sum_{n=1}^{\infty}\frac{1}{[2nt+\sqrt{(2nt)^2+\lambda^2}]^{\lambda}}\}=\frac{e^{-t}}{\sinh t} \tag{C-127}$$

and

$$\lim_{\lambda\to\infty}\{\lambda^{\lambda}\sum_{n=1}^{\infty}\frac{1}{[(2n-1)t+\sqrt{[(2n-1)t]^2+\lambda^2}]^{\lambda}}\}=\frac{1}{2\sinh t} \tag{C-128}$$

Considering that the modified Bessel function of the first kind, $I_{\lambda}(z)$, is represented in (C-57)

$$\lim_{\lambda\to\infty}\left\{\frac{\lambda e^{-\lambda^2/2nt^2}}{\sqrt{n}}I_{\lambda}\left(\frac{\lambda^2}{2nt^2}\right)\right\}=\frac{t}{\sqrt{\pi}}e^{-nt^2} \tag{C-129}$$

it follows immediately from (C-120) and (C-123) that

$$\lim_{\lambda\to\infty}\left\{\lambda\sum_{n=1}^{\infty}\frac{e^{-\lambda^2/2nt^2}}{\sqrt{n}}I_{\mu}\left(\frac{\lambda^2}{2nt^2}\right)\right\}=\frac{t}{\sqrt{\pi}}\frac{e^{-t^2}}{1-e^{-t^2}} \tag{C-130}$$

Since the infinite integrals of Bessel functions with trigonometric functions are known

$$\int_0^\infty J_\lambda(\lambda x)\sin(ax)dx = \frac{\sin[\lambda \sin^{-1}(\frac{a}{\lambda})]}{\sqrt{\lambda^2 - a^2}} \tag{C-131}$$
$$0 < a < \lambda$$

and

$$\int_0^\infty J_\lambda(\lambda x)\cos(ax)dx = \frac{\cos[\lambda \sin^{-1}(\frac{a}{\lambda})]}{\sqrt{\lambda^2 - a^2}} \tag{C-132}$$
$$0 < a < \lambda$$

it is possible to apply (C-121) to the Fourier expansions of functions.

Thus, from [86, 1.103 and 1.104]

$$\sum_{n=1}^\infty \frac{\rho^n}{n!}\sin(nx) = e^{\rho\cos x}\sin(\rho \sin x) \tag{C-133}$$

and

$$\sum_{n=1}^\infty \frac{\rho^n}{n!}\cos(nx) = e^{\rho\cos x}\cos(\rho \sin x) \tag{C-134}$$

one has

$$\lim_{\lambda\to\infty}\left\{\lambda \sum_{n=1}^\infty \frac{\rho^n}{n!}\frac{\sin[\lambda\sin^{-1}(nt/\lambda)]}{\sqrt{\lambda^2-(nt)^2}}\right\} = e^{\rho\cos t}\sin(\rho\sin t) \tag{C-135}$$

and

$$\lim_{\lambda\to\infty}\left\{\lambda \sum_{n=1}^\infty \frac{\rho^n}{n!}\frac{\cos[\lambda\sin^{-1}(nt/\lambda)]}{\sqrt{\lambda^2-(nt)^2}}\right\} = e^{\rho\cos t}\cos(\rho\sin t) \tag{C-136}$$

Similarly, from [86, 4.33 and 4.34]

$$\sum_{n=0}^\infty J_n(\rho)J_{n+1}(\rho)\sin[(2n+1)x] = \frac{1}{2}J_1(2\rho\sin x) \tag{C-137}$$

and

$$\sum_{n=0}^{\infty}(-1)^n J_n(\rho)J_{n+1}(\rho)\cos[(2n+1)x]=\frac{1}{2}J_1(2\rho\cos x) \tag{C-138}$$

using (C-121), (C-131) and (C-132) we have

$$\lim_{\lambda\to\infty}\left\{\lambda\sum_{n=0}^{\infty}J_n(\rho)J_{n+1}(\rho)\frac{\sin\{\lambda\sin^{-1}[(2n+1)t/\lambda]\}}{\sqrt{\lambda^2-(2n+1)^2t^2}}\right\}=\frac{1}{2}J_1(2\rho\sin t) \tag{C-139}$$

and

$$\lim_{\lambda\to\infty}\left\{\lambda\sum_{n=0}^{\infty}(-1)^n J_n(\rho)J_{n+1}(\rho)\frac{\cos\{\lambda\sin^{-1}[(2n+1)t/\lambda]\}}{\sqrt{\lambda^2-(2n+1)^2t^2}}\right\}=\frac{1}{2}J_1(2\rho\cos t) \tag{C-140}$$

Integrals of the Bessel Functions

As mentioned above, if the sifting property of the delta Dirac function is applied to a function $g(t) = t^{1/2}f(t)$ and combined with the orthogonality identity (C-8) or (C-9), the result can be written in the form

$$t^{3/2}f(t)=tg(t)=\int_0^{\infty}u^{1/2}J_{\lambda}(u)G(\frac{u}{t};\lambda)du \tag{C-141}$$

$$t^{3/2}f(t)=tg(t)=\int_0^{\infty}u^{1/2}Y_{\lambda}(u)H(\frac{u}{t};\lambda)du \tag{C-142}$$

where the Hankel transform and the H-transform of the function $g(t) = t^{1/2}f(t)$ are defined in (C-48) and (C-50) respectively. Available in the literature tables of Bessel transforms [1,44,52,80,85] permit easy generation of various but rather complex integrals.

From the Hankel function-transform pair [85, 4.23]

$$\begin{gathered} g(t)=\frac{t^{\lambda+1/2}}{(c^2+t^2)^{\mu}} \\ G(s;\lambda)=\frac{2^{1-\mu}c^{\lambda-\mu+1}s^{\mu-1/2}}{\Gamma(\mu)}K_{\lambda-\mu+1}(cs) \\ \operatorname{Re}\lambda>-1\quad,\quad \operatorname{Re}(\mu-\lambda)>0 \end{gathered} \tag{C-143}$$

it follows from (C-141) that

$$\int_0^\infty u^\mu J_\lambda(u) K_{\mu-\lambda+1}(\frac{cu}{t}) du = \frac{2^{\mu-1}\Gamma(\mu) c^{\mu-\lambda-1} t^{\mu+\lambda+1}}{(c^2+t^2)^\mu} \tag{C-144}$$

Denoting u = ax, b = ac/t and $\rho = \lambda^{-\mu+1}$, the integral in (C-144) becomes

$$\int_0^\infty x^{\lambda-\rho+1} J_\lambda(ax) K_\rho(bx) dx = \frac{2^{\lambda-\rho}\Gamma(\lambda-\rho+1) a^\lambda}{b^\rho (a^2+b^2)^{\lambda-\rho+1}}$$
$$\mathrm{Re}\,\lambda > -1 \quad , \quad \mathrm{Re}(\lambda - 2\rho) > -\frac{3}{2} \tag{C-145}$$

For $\lambda = \pm 1/2$, the integrals include the trigonometric functions

$$\int_0^\infty x^{1-\rho} \sin(ax) K_\rho(bx) dx = \frac{\Gamma(\frac{3}{2}-\rho)\sqrt{\pi}\, a}{(2b)^\rho (a^2+b^2)^{3/2-\rho}} \tag{C-146}$$

$$\int_0^\infty x^{-\rho} \cos(ax) K_\rho(bx) dx = \frac{\Gamma(\frac{1}{2}-\rho)\sqrt{\pi}}{2^{\rho+1} b^\rho (a^2+b^2)^{1/2-\rho}} \tag{C-147}$$

In the case of the *H*-transform-function pair

$$g(t) = t^{1/2} Y_\lambda(at^{1/2}) K_\lambda(at^{1/2})$$
$$H(s;\lambda) = \frac{1}{2s^{3/2}} e^{-a^2/2s}$$
$$\mathrm{Re}\,\lambda > -\frac{3}{2} \tag{C-148}$$

and denoting $\alpha = at^{1/2}$, we have from (C-142) that

$$\int_0^\infty \frac{e^{-\alpha^2/u}}{u} Y_\lambda(u) du = 2Y_\lambda(\alpha) K_\lambda(\alpha)$$
$$\mathrm{Re}\,\lambda > -\frac{3}{2} \tag{C-149}$$

Similarly as infinite integrals, the finite integrals of Bessel functions can be derived using the Hankel and *H*-transforms but with the limited strips of convergence.

From the Hankel function-transform pair

$$g(t)=t^{1/2}J_{\lambda/2-1/4}(\frac{at}{2})J_{\lambda/2+1/4}(\frac{at}{2})$$
$$G(s;\lambda)=\frac{2^{1/2}}{\pi[as(a-s)]^{1/2}}[u(s)-u(s-a)] \qquad \text{(C-150)}$$
$$\mathrm{Re}\,\lambda>-1$$

it follows from (C-141) that

$$\int_0^a \frac{J_\lambda(tx)}{(a-x)^{1/2}}dx=\pi(\frac{a}{2})^{1/2}J_{\lambda/2-1/4}(\frac{at}{2})J_{\lambda/2+1/4}(\frac{at}{2}) \qquad \text{(C-151)}$$
$$\mathrm{Re}\,\lambda>-1$$

For $\lambda = n \pm 1/2$, $n = 0,1,2,3,\ldots$ the integral in (C-151) becomes the integral of elementary functions, for example for $\lambda = \pm 1/2$ we have

$$\int_0^a \frac{\sin(tx)}{[x(a-x)]^{1/2}}dx=\pi\sin(\frac{at}{2})J_0(\frac{at}{2})] \qquad \text{(C-152)}$$

$$\int_0^a \frac{\cos(tx)}{[x(a-x)]^{1/2}}dx=\pi\cos(\frac{at}{2})J_0(\frac{at}{2}) \qquad \text{(C-153)}$$

Using the following *H*-transform –function pair

$$g(t)=t^{1/2}\{[J_{\lambda/2}(\frac{at}{2})]^2-[Y_{\lambda/2}(\frac{at}{2})]^2\}$$
$$H(s;\lambda)=\frac{4}{\pi}[s(s^2-a^2)]^{-1/2}u(s-a) \qquad \text{(C-154)}$$
$$-\frac{3}{2}<\mathrm{Re}\,\lambda<1$$

and denoting $\alpha = at$, (C-142) gives

$$\int_{\alpha}^{\infty}\frac{Y_{\lambda}(u)}{(u^2-\alpha^2)^{1/2}}du=\frac{\pi}{4}\{[J_{\lambda/2}(\frac{\alpha}{2})]^2-[Y_{\lambda/2}(\frac{\alpha}{2})]^2\}$$
$$-\frac{3}{2}<\operatorname{Re}\lambda<1 \tag{C-155}$$

In the final example, the following Hankel function transform pair is used

$$g(t)=t^{1/2-\lambda}J_{\lambda}(at)J_{\lambda}(bt)$$
$$G(s,\lambda)=$$
$$\frac{s^{1/2-\nu}\{[s^2-(a-b)^2][(a+b)^2-s^2]\}^{\lambda-1/2}}{2^{3\lambda-1}\sqrt{\pi}\Gamma(\lambda+\frac{1}{2})(ab)^{\lambda}}\{u[s-|a-b|]-u[s-(a+b)]\} \tag{C-156}$$
$$\operatorname{Re}\lambda>-\frac{1}{2}$$

which leads to

$$\int_{|a-b|}^{a+b}x^{1-\lambda}\{[x^2-(a-b)^2][(a+b)^2-x^2]\}^{\lambda-1/2}J_{\lambda}(tx)dx=$$
$$2^{3\lambda-1}\sqrt{\pi}\Gamma(\lambda+\frac{1}{2})(\frac{ab}{t})^{\lambda}J_{\lambda}(at)J_{\lambda}(bt) \tag{C-157}$$

or by denoting a = (α + β)/2 and b = (β – α)/2 the integral (C-157) takes the form

$$\int_{\alpha}^{\beta}x^{1-\lambda}\{[x^2-\alpha^2][\beta^2-x^2]\}^{\lambda-1/2}J_{\lambda}(tx)dx=$$
$$2^{3\lambda-1}\sqrt{\pi}\Gamma(\lambda+\frac{1}{2})(\beta^2-\alpha^2)^{\lambda}t^{-\lambda}J_{\lambda}\left(\frac{\alpha+\beta}{2}t\right)J_{\lambda}\left(\frac{\beta-\alpha}{2}t\right) \tag{C-158}$$
$$\operatorname{Re}\lambda>-\frac{1}{2}$$

For α = 0 and β = α, the integral (C-158) reduces to

$$\int_{0}^{\alpha}x^{\lambda}(\alpha^2-x^2)^{\lambda-1/2}J_{\lambda}(tx)dx=2^{3\lambda-1}\sqrt{\pi}\Gamma(\lambda+\frac{1}{2})\alpha^{2\lambda}t^{-\lambda}[J_{\lambda}\left(\frac{\alpha t}{2}\right)]^2$$
$$\operatorname{Re}\lambda>-\frac{1}{2} \tag{C-159}$$

APPENDIX D. TABLES OF VOLTERRA FUNCTIONS AND THE LIST OF SYMBOLS AND DEFINITIONS OF SPECIAL FUNCTIONS

Values of the Volterra functions for a limited number of parameters α and β were obtained by numerical evaluation of integrals as described in [3,6]. They always were round-out to five digits. The computed ranges of variable t, from $t = 0.0001$ to $t = 2.0$ for $\nu(t,\alpha)$ and from $t = 0.001$ to $t = 2.0$ for $\mu(t,\beta,\alpha)$, can be extended by applying the asymptotic expressions for $t \to 0$ and $t \to \infty$. They are presented in Chapter 2, Sections 2.4 and 2.5, equations (1.2.4), (7.2.4), (8.2.4) and (11.2.4) for the Volterra functions $\nu(t,\alpha)$ and (2.2.4) and (40.2.5) for the generalized Volterra functions $\mu(t,\beta,\alpha)$.

Table I. Values of the Volterra functions ν(*t*,α) for $t \leq 2.0$ and $-4.0 \leq \alpha \leq 1.0$

t/α	**1.00**	**0.95**	**0.90**	**0.85**	**0.80**
0.0001	0.00001	0.00002	0.00003	0.00004	0.00007
0.001	0.00014	0.00020	0.00028	0.00041	0.00059
0.005	0.00086	0.00115	0.00153	0.00204	0.00272
0.01	0.00195	0.00251	0.00324	0.00418	0.00537
0.02	0.00448	0.00560	0.00698	0.00870	0.01081
0.03	0.00738	0.00904	0.01105	0.01349	0.01645
0.04	0.01056	0.01276	0.01538	0.01852	0.02227
0.05	0.01400	0.01673	0.01996	0.02377	0.02828
0.06	0.01767	0.02093	0.02475	0.02923	0.03446
0.07	0.02156	0.02534	0.02975	0.03487	0.04082
0.08	0.02565	0.02996	0.03495	0.04070	0.04734
0.09	0.02994	0.03477	0.04033	0.04671	0.05403
0.10	0.03442	0.03977	0.04590	0.05290	0.06088
0.15	0.05949	0.06744	0.07635	0.08633	0.09747
0.20	0.08880	0.09934	0.11095	0.12378	0.13791
0.25	0.12221	0.13529	0.14958	0.16518	0.18216
0.30	0.15969	0.17531	0.19223	0.21052	0.23026
0.35	0.20126	0.21942	0.23893	0.25986	0.28227
0.40	0.24702	0.26770	0.28977	0.31330	0.33832
0.45	0.29706	0.32026	0.34487	0.37095	0.39853
0.50	0.35153	0.37723	0.40436	0.43296	0.46304
0.55	0.41056	0.43877	0.46840	0.49948	0.53203
0.60	0.47434	0.50505	0.53716	0.57070	0.60567
0.65	0.54310	0.57625	0.61084	0.64681	0.68416
0.70	0.61691	0.65259	0.68963	0.72801	0.76772
0.75	0.69612	0.73429	0.77377	0.81454	0.85656
0.80	0.78093	0.82157	0.86348	0.90662	0.95094
0.85	0.87158	0.91470	0.95902	1.0045	1.0511
0.90	0.96835	1.0139	1.0607	1.1085	1.1573
0.95	1.0715	1.1196	1.1687	1.2188	1.2699
1.00	1.1814	1.2319	1.2834	1.3359	1.3891
1.05	1.2983	1.3513	1.4052	1.4599	1.5154
1.10	1.4225	1.4780	1.5342	1.5912	1.6489
1.15	1.5545	1.6124	1.6710	1.7303	1.7901
1.20	1.6946	1.7549	1.8159	1.8774	1.9393
1.25	1.8431	1.9059	1.9692	2.0330	2.0970
1.30	2.0005	2.0658	2.1314	2.1974	2.2636
1.35	2.1673	2.2350	2.3030	2.3712	2.4395
1.40	2.3439	2.4141	2.4844	2.5548	2.6251
1.45	2.5308	2.6034	2.6760	2.7486	2.8210
1.50	2.7285	2.8035	2.8785	2.9533	3.0277
1.55	2.9376	3.0150	3.0923	3.1692	3.2457
1.60	3.1585	3.2384	3.3180	3.3971	3.4756
1.65	3.3920	3.4744	3.5562	3.6375	3.7180
1.70	3.6387	3.7234	3.8076	3.8940	3.9736
1.75	3.8991	3.9863	4.0728	4.1583	4.2429
1.80	4.1741	4.2637	4.3524	4.4401	4.5267
1.85	4.4643	4.5563	4.6473	4.7372	4.8257
1.90	4.7705	4.8650	4.9583	5.0502	5.1407
1.95	5.0936	5.1905	5.2860	5.3801	5.4725
2.00	5.4344	5.5337	5.6315	5.7276	5.8220

Table I. (Continued).

t/α	**0.75**	**0.70**	**0.65**	**0.60**	**0.55**
0.0001	0.00011	0.00018	0.00030	0.00047	0.00076
0.001	0.00084	0.00121	0.00174	0.00248	0.00355
0.005	0.00361	0.00479	0.00634	0.00838	0.01106
0.01	0.00690	0.00885	0.01132	0.01447	0.01845
0.02	0.01343	0.01664	0.02059	0.02544	0.03136
0.03	0.02002	0.02434	0.02953	0.03577	0.04325
0.04	0.02674	0.03205	0.03835	0.04582	0.05465
0.05	0.03359	0.03983	0.04716	0.05574	0.06578
0.06	0.04058	0.04770	0.05599	0.06561	0.07675
0.07	0.04771	0.05567	0.06487	0.07546	0.08763
0.08	0.05498	0.06376	0.07382	0.08533	0.09847
0.09	0.06240	0.07195	0.08285	0.09523	0.10930
0.10	0.06996	0.08027	0.09197	0.10520	0.12014
0.15	0.10990	0.12374	0.13911	0.15616	0.17502
0.20	0.15345	0.17049	0.18915	0.20955	0.23180
0.25	0.20006	0.22066	0.24236	0.26581	0.29110
0.30	0.25152	0.27439	0.29893	0.32521	0.35329
0.35	0.30624	0.33181	0.35903	0.38797	0.41866
0.40	0.36490	0.39306	0.42286	0.45431	0.48745
0.45	0.42764	0.45832	0.49058	0.52443	0.55989
0.50	0.49463	0.52774	0.56238	0.59854	0.63621
0.55	0.56604	0.60152	0.63845	0.67683	0.71663
0.60	0.64205	0.67984	0.71901	0.75953	0.80137
0.65	0.72287	0.76292	0.80426	0.84687	0.89067
0.70	0.80872	0.85097	0.89444	0.93906	0.98477
0.75	0.89982	0.94424	0.98978	1.0364	1.0839
0.80	0.99641	1.0430	1.0905	1.1390	1.1884
0.85	1.0988	1.1474	1.1970	1.2473	1.2984
0.90	1.2072	1.2579	1.3094	1.3616	1.4144
0.95	1.3219	1.3746	1.4280	1.4820	1.5365
1.00	1.4432	1.4979	1.5532	1.6090	1.6651
1.05	1.5715	1.6282	1.6853	1.7428	1.8005
1.10	1.7071	1.7657	1.8247	1.8839	1.9431
1.15	1.8503	1.9109	1.9716	2.0325	2.0933
1.20	2.0016	2.0640	2.1266	2.1891	2.2514
1.25	2.1613	2.2257	2.2899	2.3540	2.4178
1.30	2.3298	2.3960	2.4620	2.5277	2.5930
1.35	2.5077	2.5757	2.6434	2.7107	2.7773
1.40	2.6953	2.7651	2.8345	2.9033	2.9714
1.45	2.8931	2.9647	3.0358	3.1061	3.1756
1.50	3.1017	3.1751	3.2478	3.3197	3.3905
1.55	3.3216	3.3968	3.4712	3.5445	3.6167
1.60	3.5534	3.6304	3.7063	3.7811	3.8547
1.65	3.7977	3.8764	3.9539	4.0302	4.1050
1.70	4.0551	4.1355	4.2146	4.2923	4.3685
1.75	4.3263	4.4084	4.4890	4.5682	4.6456
1.80	4.6119	4.6957	4.7779	4.8585	4.9371
1.85	4.9128	4.9982	5.0820	5.1639	5.2438
1.90	5.2296	5.3168	5.4020	5.4853	5.5665
1.95	5.5632	5.6520	5.7388	5.8234	5.9058
2.00	5.9145	6.0050	6.0933	6.1793	6.2629

Table I. (Continued).

t/α	**0.50**	**0.45**	**0.40**	**0.35**	**0.30**
0.0001	0.00121	0.00193	0.00306	0.00485	0.00768
0.001	0.00505	0.00719	0.01019	0.01443	0.02038
0.005	0.01456	0.01913	0.02508	0.03281	0.04284
0.01	0.02348	0.02983	0.03782	0.04784	0.06038
0.02	0.03860	0.04742	0.05814	0.07113	0.08683
0.03	0.05220	0.06289	0.07562	0.09073	0.10865
0.04	0.06506	0.07731	0.09169	0.10852	0.12819
0.05	0.07748	0.09109	0.10690	0.12520	0.14635
0.06	0.08962	0.10447	0.12155	0.14115	0.16357
0.07	0.10159	0.01176	0.13580	0.15656	0.18014
0.08	0.11344	0.13046	0.14976	0.17158	0.19621
0.09	0.12522	0.14322	0.16351	0.18632	0.21190
0.10	0.13697	0.15589	0.17710	0.20083	0.22730
0.15	0.19584	0.21877	0.24397	0.27159	0.30179
0.20	0.25602	0.28230	0.31078	0.34154	0.37469
0.25	0.31832	0.34755	0.37885	0.41230	0.44794
0.30	0.38324	0.41511	0.44894	0.48475	0.52257
0.35	0.45114	0.48542	0.52154	0.55948	0.59924
0.40	0.52228	0.55881	0.59704	0.63693	0.67845
0.45	0.59695	0.63558	0.67576	0.71745	0.76059
0.50	0.67537	0.71599	0.75801	0.80138	0.84602
0.55	0.75780	0.80029	0.84405	0.88900	0.93505
0.60	0.84446	0.88875	0.93416	0.98060	1.0280
0.65	0.93561	0.98162	1.0286	1.0764	1.1251
0.70	1.0315	1.0791	1.1276	1.1768	1.2266
0.75	1.1324	1.1816	1.2315	1.2820	1.3330
0.80	1.2385	1.2893	1.3406	1.3923	1.4443
0.85	1.3502	1.4024	1.4551	1.5080	1.5611
0.90	1.4677	1.5214	1.5753	1.6294	1.6834
0.95	1.5913	1.6464	1.7016	1.7568	1.8118
1.00	1.7214	1.7779	1.8343	1.8905	1.9465
1.05	1.8583	1.9161	1.9737	2.0310	2.0878
1.10	2.0024	2.0614	2.1209	2.1785	2.2362
1.15	2.1539	2.2143	2.2741	2.3334	2.3919
1.20	2.3134	2.3749	2.4359	2.4961	2.5555
1.25	2.4811	2.5439	2.6059	2.6671	2.7272
1.30	2.6576	2.7216	2.7846	2.8467	2.9076
1.35	2.8433	2.9084	2.9725	3.0354	3.0971
1.40	3.0386	3.1049	3.1699	3.2338	3.2961
1.45	3.2441	3.3115	3.3775	3.4422	3.5053
1.50	3.4603	3.5287	3.5957	3.6612	3.7450
1.55	3.6876	3.7571	3.8251	3.8914	3.9559
1.60	3.9268	3.9973	4.0662	4.1333	4.1985
1.65	4.1783	4.2499	4.3197	4.3876	4.4534
1.70	4.4429	4.5155	4.5862	4.6548	4.7213
1.75	4.7212	4.7948	4.8664	4.9357	5.0028
1.80	5.0139	5.0885	5.1609	5.2310	5.2987
1.85	5.3217	5.3973	5.4705	5.5414	5.6097
1.90	5.6454	5.7220	5.7961	5.8677	5.9366
1.95	5.9858	6.0634	6.1383	6.2105	6.2800
2.00	6.3439	6.4224	6.4982	6.5711	6.6412

Table I. (Continued).

t/α	0.25	0.20	0.15	0.10	0.05
0.0001	0.01212	0.01908	0.02995	0.04688	0.07313
0.001	0.02872	0.04036	0.05657	0.07905	0.11014
0.005	0.05579	0.07247	0.09391	0.12350	0.15637
0.01	0.07604	0.09553	0.11972	0.14962	0.18648
0.02	0.10577	0.12853	0.15582	0.18842	0.22722
0.03	0.12981	0.15475	0.18403	0.21831	0.25830
0.04	0.15109	0.17769	0.20848	0.24401	0.28486
0.05	0.17069	0.19866	0.23067	0.26720	0.30874
0.06	0.18917	0.21830	0.25135	0.28872	0.33083
0.07	0.20684	0.23700	0.27095	0.30904	0.35163
0.08	0.22391	0.25498	0.28973	0.32846	0.37148
0.09	0.24051	0.27242	0.30789	0.34719	0.39059
0.10	0.25675	0.28942	0.32555	0.36538	0.40912
0.15	0.33471	0.37049	0.40927	0.45116	0.49625
0.20	0.41032	0.44850	0.48927	0.53269	0.57876
0.25	0.48582	0.52596	0.56836	0.61300	0.65985
0.30	0.56239	0.60421	0.64798	0.69366	0.74114
0.35	0.64079	0.68408	0.72905	0.77561	0.82364
0.40	0.72155	0.76616	0.81219	0.85983	0.90805
0.45	0.80511	0.85092	0.89791	0.94596	0.99495
0.50	0.89185	0.93876	0.98662	1.0353	1.0846
0.55	0.98209	1.0300	1.0787	1.1279	1.1776
0.60	1.0761	1.1250	1.1744	1.2242	1.2743
0.65	1.1743	1.2241	1.2742	1.3245	1.3748
0.70	1.2769	1.3275	1.3783	1.4290	1.4796
0.75	1.3842	1.4356	1.4870	1.5382	1.5890
0.80	1.4965	1.5486	1.6006	1.6522	1.7033
0.85	1.6141	1.6669	1.7195	1.7715	1.8229
0.90	1.7373	1.7908	1.8439	1.8963	1.9478
0.95	1.8665	1.9207	1.9742	2.0270	2.0787
1.00	2.0019	2.0567	2.1108	2.1639	2.2158
1.05	2.1440	2.1994	2.2539	2.3073	2.3595
1.10	2.2931	2.3491	2.4040	2.4578	2.5101
1.15	2.4495	2.5061	2.5615	2.6155	2.6680
1.20	2.6138	2.6709	2.7267	2.7810	2.8337
1.25	2.7862	2.8438	2.9000	2.9546	3.0074
1.30	2.9672	3.0254	3.0819	3.1368	3.1898
1.35	3.1573	3.2160	3.2729	3.3280	3.3812
1.40	3.3570	3.4161	3.4734	3.5288	3.5821
1.45	3.5667	3.6263	3.6840	3.7396	3.7930
1.50	3.7870	3.8471	3.9051	3.9609	4.0145
1.55	4.0184	4.0789	4.1373	4.1933	4.2471
1.60	4.2616	4.3225	4.3812	4.4375	4.4913
1.65	4.5170	4.5784	4.6374	4.6939	4.7479
1.70	4.7855	4.8472	4.9065	4.9633	5.0174
1.75	5.0675	5.1297	5.1893	5.2462	5.3004
1.80	5.3639	5.4265	5.4864	5.5435	5.5979
1.85	5.6753	5.7383	5.7985	5.8558	5.9102
1.90	6.0027	6.0661	6.1265	6.1841	6.2386
1.95	6.3466	6.4103	6.4711	6.5288	6.5834
2.00	6.7083	6.7723	6.8333	6.8912	6.9460

Table I. (Continued).

t/α	**0.00**	**-0.05**	**-0.10**	**-0.15**	**-0.20**
0.0001	0.11372	0.17619	0.27197	0.41812	0.63999
0.001	0.15297	0.21175	0.29205	0.40123	0.54905
0.005	0.20087	0.25723	0.32827	0.41742	0.52874
0.01	0.23174	0.28709	0.35449	0.43620	0.53474
0.02	0.27325	0.32762	0.39157	0.46646	0.55370
0.03	0.30478	0.35859	0.42062	0.49178	0.57303
0.04	0.33167	0.38509	0.44578	0.51442	0.59167
0.05	0.35581	0.40892	0.46859	0.53534	0.60962
0.06	0.37810	0.43097	0.48983	0.55507	0.62701
0.07	0.39909	0.45175	0.50993	0.57392	0.64394
0.08	0.41909	0.47157	0.52917	0.59210	0.66050
0.09	0.43834	0.49066	0.54775	0.60976	0.67677
0.10	0.45699	0.50918	0.56581	0.62701	0.69279
0.15	0.54460	0.59625	0.65119	0.70935	0.77062
0.20	0.62746	0.67874	0.73250	0.78860	0.84686
0.25	0.70882	0.75981	0.81257	0.86723	0.92326
0.30	0.79033	0.84109	0.89324	0.94656	1.0008
0.35	0.87303	0.92359	0.97513	1.0274	1.0803
0.40	0.95760	1.0080	1.0590	1.1105	1.1621
0.45	1.0446	1.0948	1.1454	1.1961	1.2467
0.50	1.1345	1.1846	1.2348	1.2848	1.3344
0.55	1.2276	1.2776	1.3274	1.3768	1.4256
0.60	1.3243	1.3742	1.4237	1.4726	1.5207
0.65	1.4250	1.4747	1.5240	1.5724	1.6198
0.70	1.5299	1.5796	1.6285	1.6765	1.7233
0.75	1.6394	1.6890	1.7376	1.7852	1.8314
0.80	1.7537	1.8033	1.8517	1.8989	1.9446
0.85	1.8733	1.9228	1.9710	2.0178	2.0630
0.90	1.9984	2.0478	2.0958	2.1423	2.1871
0.95	2.1294	2.1787	2.2265	2.2727	2.3170
1.00	2.2665	2.3158	2.3634	2.4093	2.4533
1.05	2.4103	2.4595	2.5069	2.5525	2.5961
1.10	2.5609	2.6101	2.6574	2.7027	2.7459
1.15	2.7189	2.7680	2.8151	2.8602	2.9031
1.20	2.8846	2.9336	2.9806	3.0255	3.0680
1.25	3.0584	3.1074	3.1543	3.1989	3.2412
1.30	3.2408	3.2898	3.3365	3.3809	3.4229
1.35	3.4323	3.4812	3.5277	3.5719	3.6136
1.40	3.6332	3.6821	3.7285	3.7725	3.8140
1.45	3.8442	3.8930	3.9393	3.9831	4.0243
1.50	4.0657	4.1145	4.1607	4.2043	4.2453
1.55	4.2983	4.3470	4.3931	4.4366	4.4773
1.60	4.5426	4.5913	4.6373	4.6806	4.7211
1.65	4.7992	4.8478	4.8937	4.9368	4.9771
1.70	5.0687	5.1173	5.1631	5.2061	5.2461
1.75	5.3518	5.4004	5.4461	5.4889	5.5287
1.80	5.6493	5.6978	5.7434	5.7861	5.8257
1.85	5.9617	6.0102	6.0557	6.0982	6.1377
1.90	6.2901	6.3386	6.3840	6.4264	6.4656
1.95	6.6350	6.6834	6.7287	6.7709	6.8100
2.00	6.9976	7.0460	7.0912	7.1333	7.1723

Table I. (Continued).

t/α	-0.25	-0.30	-0.35	-0.40	-0.45
0.0001	0.97500	1.4778	2.2272	3.3336	4.9622
0.001	0.74793	1.0140	1.3674	1.8335	2.4428
0.005	0.66696	0.83757	1.0467	1.3012	1.6082
0.01	0.65294	0.79384	0.96073	1.1569	1.3854
0.02	0.65478	0.77119	0.90435	1.0555	1.2257
0.03	0.66528	0.76938	0.88608	1.0159	1.1591
0.04	0.67812	0.77429	0.88056	0.99710	1.1238
0.05	0.69183	0.78227	0.88110	0.98828	1.1035
0.06	0.70591	0.79192	0.88506	0.98515	1.0918
0.07	0.72015	0.80259	0.89117	0.98563	1.0855
0.08	0.73445	0.81391	0.89873	0.98857	1.0829
0.09	0.74878	0.82571	0.90733	0.99329	1.0831
0.10	0.76311	0.83784	0.91671	0.99935	1.0852
0.15	0.83480	0.90164	0.97077	1.0417	1.1140
0.20	0.90702	0.96877	1.0317	1.0954	1.1594
0.25	0.98047	1.0385	1.0971	1.1557	1.2138
0.30	1.0557	1.1110	1.1662	1.2209	1.2747
0.35	1.1333	1.1862	1.2387	1.2904	1.3408
0.40	1.2135	1.2646	1.3148	1.3640	1.4116
0.45	1.2968	1.3462	1.3946	1.4416	1.4869
0.50	1.3834	1.4314	1.4781	1.5233	1.5666
0.55	1.4736	1.5203	1.5656	1.6092	1.6508
0.60	1.5676	1.6133	1.6573	1.6996	1.7396
0.65	1.6659	1.7106	1.7535	1.7945	1.8332
0.70	1.7686	1.8124	1.8543	1.8941	1.9317
0.75	1.8761	1.9191	1.9600	1.9988	2.0353
0.80	1.9886	2.0308	2.0709	2.1088	2.1443
0.85	2.1065	2.1480	2.1873	2.2243	2.2589
0.90	2.2299	2.2708	2.3094	2.3457	2.3794
0.95	2.3594	2.3996	2.4376	2.4731	2.5061
1.00	2.4952	2.5348	2.5722	2.6070	2.6393
1.05	2.6375	2.6767	2.7134	2.7476	2.7792
1.10	2.7869	2.8256	2.8618	2.8954	2.9264
1.15	2.9437	2.9819	3.0175	3.0506	3.0810
1.20	3.1083	3.1460	3.1812	3.2137	3.2436
1.25	3.2810	3.3183	3.3530	3.3851	3.4144
1.30	3.4624	3.4993	3.5335	3.5651	3.5940
1.35	3.6528	3.6893	3.7232	3.7543	3.7827
1.40	3.8528	3.8890	3.9224	3.9532	3.9811
1.45	4.0628	4.0987	4.1317	4.1621	4.1896
1.50	4.2835	4.3190	4.3517	4.3816	4.4088
1.55	4.5152	4.5504	4.5828	4.6124	4.6391
1.60	4.7587	4.7936	4.8257	4.8549	4.8813
1.65	5.0145	5.0491	5.0808	5.1097	5.1358
1.70	5.2833	5.3176	5.3490	5.3776	5.4033
1.75	5.5656	5.5997	5.6308	5.6590	5.6845
1.80	5.8624	5.8962	5.9270	5.9550	5.9801
1.85	6.1741	6.2076	6.2382	6.2658	6.2907
1.90	6.5019	6.5352	6.5655	6.5929	6.6174
1.95	6.8460	6.8790	6.9091	6.9362	6.9605
2.00	7.2081	7.2409	7.2707	7.2976	7.3216

Table I. (Continued).

t/α	**-0.50**	**-0.55**	**-0.60**	**-0.65**	**-0.70**
0.0001	7.3244	10.716	15.521	22.214	31.348
0.001	3.2317	4.2414	5.5163	7.0995	9.0255
0.005	1.9750	2.4081	2.9128	3.4912	4.1405
0.01	1.6492	1.9501	2.2888	2.6637	3.0702
0.02	1.4154	1.6245	1.8517	2.0948	2.3494
0.03	1.3156	1.4845	1.6645	1.8531	2.0468
0.04	1.2603	1.4055	1.5581	1.7156	1.8751
0.05	1.2262	1.3554	1.4896	1.6267	1.7638
0.06	1.2024	1.3217	1.4425	1.5648	1.6861
0.07	1.1901	1.2984	1.4089	1.5200	1.6293
0.08	1.1811	1.2820	1.3844	1.4866	1.5865
0.09	1.1759	1.2708	1.3664	1.4614	1.5536
0.10	1.1735	1.2632	1.3533	1.4422	1.5281
0.15	1.1868	1.2594	1.3307	1.3997	1.4650
0.20	1.2229	1.2853	1.3457	1.4034	1.4572
0.25	1.2710	1.3265	1.3796	1.4298	1.4761
0.30	1.3271	1.3775	1.4254	1.4701	1.5111
0.35	1.3895	1.4360	1.4798	1.5204	1.5573
0.40	1.4573	1.5006	1.5412	1.5786	1.6123
0.45	1.5301	1.5708	1.6088	1.6435	1.6747
0.50	1.6077	1.6463	1.6820	1.7145	1.7435
0.55	1.6901	1.7268	1.7606	1.7912	1.8184
0.60	1.7773	1.8124	1.8445	1.8736	1.8992
0.65	1.8695	1.9031	1.9338	1.9614	1.9858
0.70	1.9667	1.9990	2.0285	2.0548	2.0780
0.75	2.0692	2.1004	2.1287	2.1539	2.1760
0.80	2.1772	2.2073	2.2346	2.2588	2.2800
0.85	2.2909	2.3201	2.3464	2.3697	2.3900
0.90	2.4105	2.4388	2.4643	2.4868	2.5064
0.95	2.5364	2.5640	2.5886	2.6104	2.6292
1.00	2.6688	2.6957	2.7196	2.7407	2.7589
1.05	2.8081	2.8343	2.8576	2.8781	2.8956
1.10	2.9547	2.9802	3.0029	3.0227	3.0398
1.15	3.1087	3.1336	3.1558	3.1751	3.1916
1.20	3.2707	3.2951	3.3167	3.3355	3.3516
1.25	3.4410	3.4649	3.4860	3.5043	3.5200
1.30	3.6201	3.6435	3.6641	3.6820	3.6972
1.35	3.8084	3.8313	3.8515	3.8690	3.8838
1.40	4.0063	4.0288	4.0486	4.0657	4.0802
1.45	4.2144	4.2365	4.2559	4.2726	4.2867
1.50	4.4332	4.4549	4.4739	4.4903	4.5041
1.55	4.6632	4.6845	4.7031	4.7191	4.7327
1.60	4.9050	4.9259	4.9442	4.9600	4.9732
1.65	5.1591	5.1797	5.1977	5.2131	5.2261
1.70	5.4263	5.4466	5.4643	5.4795	5.4922
1.75	5.7071	5.7271	5.7445	5.7594	5.7718
1.80	6.0025	6.0222	6.0393	6.0539	6.0662
1.85	6.3128	6.3322	6.3490	6.3633	6.3753
1.90	6.6392	6.6584	6.6750	6.6891	6.7009
1.95	6.9820	7.0009	7.0172	7.0310	7.0426
2.00	7.3429	7.3615	7.3776	7.3913	7.4027

Table I. (Continued).

t/α	-0.75	-0.80	-0.85	-0.90	-0.95
0.0001	43.480	59.010	77.832	98.677	117.89
0.001	11.306	13.909	16.723	19.503	21.793
0.005	4.8500	5.5974	6.3432	7.0237	7.5418
0.01	3.4994	3.9362	4.3572	4.7285	5.0017
0.02	2.6090	2.8643	3.1020	3.3046	3.4487
0.03	2.2403	2.4267	2.5969	2.7390	2.8381
0.04	2.0322	2.1814	2.3156	2.4261	2.5022
0.05	1.8975	2.0230	2.1347	2.2256	2.2875
0.06	1.8032	1.9122	2.0083	2.0858	2.1381
0.07	1.7340	1.8307	1.9154	1.9831	2.0285
0.08	1.6815	1.7687	1.8446	1.9049	1.9450
0.09	1.6419	1.7205	1.7893	1.8437	1.8797
0.10	1.6089	1.6823	1.7454	1.7950	1.8276
0.15	1.5253	1.5789	1.6241	1.6589	1.6813
0.20	1.5061	1.5490	1.5846	1.6117	1.6289
0.25	1.5178	1.5538	1.5834	1.6057	1.6197
0.30	1.5476	1.5789	1.6044	1.6234	1.6353
0.35	1.5900	1.6178	1.6402	1.6569	1.6671
0.40	1.6420	1.6670	1.6872	1.7019	1.7110
0.45	1.7019	1.7247	1.7430	1.7563	1.7645
0.50	1.7687	1.7898	1.8066	1.8187	1.8261
0.55	1.8420	1.8616	1.8770	1.8882	1.8950
0.60	1.9213	1.9396	1.9540	1.9644	1.9706
0.65	2.0065	2.0238	2.0373	2.0469	2.0527
0.70	2.0977	2.1140	2.1266	2.1357	2.1411
0.75	2.1948	2.2102	2.2222	2.2307	2.2352
0.80	2.2979	2.3125	2.3239	2.3320	2.3368
0.85	2.4072	2.4212	2.4320	2.4396	2.4442
0.90	2.5228	2.5362	2.5465	2.5538	2.5581
0.95	2.6451	2.6579	2.6678	2.6747	2.6788
1.00	2.7742	2.7865	2.7960	2.8026	2.8065
1.05	2.9104	2.9222	2.9313	2.9377	2.9414
1.10	3.0540	3.0655	3.0742	3.0803	3.0839
1.15	3.2054	3.2165	3.2249	3.2308	3.2342
1.20	3.3649	3.3756	3.3838	3.3894	3.3927
1.25	3.5329	3.5433	3.5512	3.5566	3.5598
1.30	3.7098	3.7199	3.7275	3.7328	3.7359
1.35	3.8961	3.9059	3.9132	3.9183	3.9213
1.40	4.0921	4.1016	4.1088	4.1137	4.1166
1.45	4.2984	4.3076	4.3146	4.3194	4.3221
1.50	4.5155	4.5245	4.5312	4.5359	4.5386
1.55	4.7438	4.7526	4.7591	4.7637	4.7663
1.60	4.9841	4.9926	4.9991	5.0035	5.0060
1.65	5.2367	5.2450	5.2513	5.2556	5.2580
1.70	5.5026	5.5107	5.5169	5.5210	5.5234
1.75	5.7820	5.7899	5.7959	5.8000	5.8022
1.80	6.0761	6.0840	6.0898	6.0938	6.0960
1.85	6.3851	6.3927	6.3984	6.4022	6.4044
1.90	6.7105	6.7180	6.7236	6.7274	6.7295
1.95	7.0520	7.0593	7.0647	7.0683	7.0704
2.00	7.4120	7.4192	7.4245	7.4281	7.4302

Table I. (Continued).

t/α	-1.00	-1.05	-1.10	-1.15	-1.20
0.0001					
0.001				1.3443	
0.005	7.7556	7.4628	6.3836	4.1381	0.22173
0.01	5.1110	4.9681	4.4571	3.4293	1.6972
0.02	3.5046	3.4348	3.1928	2.7223	1.9562
0.03	2.8759	2.8300	2.6737	2.3759	1.9005
0.04	2.5331	2.4967	2.3821	2.1667	1.8278
0.05	2.3105	2.2834	2.1933	2.0259	1.7652
0.06	2.1574	2.1350	2.0611	1.9246	1.7143
0.07	2.0451	2.0261	1.9633	1.8487	1.6733
0.08	1.9596	1.9430	1.8887	1.7901	1.6401
0.09	1.8927	1.8780	1.8302	1.7438	1.6133
0.10	1.8394	1.8262	1.7835	1.7069	1.5916
0.15	1.6893	1.6806	1.6530	1.6045	1.5329
0.20	1.6350	1.6285	1.6083	1.5732	1.5222
0.25	1.6246	1.6195	1.6036	1.5763	1.5370
0.30	1.6393	1.6351	1.6220	1.5998	1.5681
0.35	1.6706	1.6670	1.6560	1.6373	1.6109
0.40	1.7141	1.7110	1.7014	1.6853	1.6627
0.45	1.7673	1.7645	1.7560	1.7419	1.7223
0.50	1.8286	1.8261	1.8186	1.8061	1.7887
0.55	1.8973	1.8950	1.8882	1.8770	1.8615
0.60	1.9727	1.9707	1.9645	1.9543	1.9403
0.65	2.0547	2.0528	2.0471	2.0378	2.0250
0.70	2.1429	2.1412	2.1359	2.1274	2.1157
0.75	2.2375	2.2358	2.2310	2.2231	2.2123
0.80	2.3384	2.3368	2.3323	2.3249	2.3149
0.85	2.4457	2.4442	2.4400	2.4331	2.4238
0.90	2.5595	2.5582	2.5542	2.5477	2.5390
0.95	2.6801	2.6789	2.6751	2.6690	2.6609
1.00	2.8078	2.8065	2.8030	2.7973	2.7896
1.05	2.9426	2.9415	2.9381	2.9327	2.9254
1.10	3.0851	3.0840	3.0807	3.0756	3.0687
1.15	3.2353	3.2343	3.2312	3.2263	3.2198
1.20	3.3938	3.3928	3.3899	3.3852	3.3790
1.25	3.5608	3.5599	3.5571	3.5526	3.5467
1.30	3.7369	3.7359	3.7333	3.7290	3.7233
1.35	3.9223	3.9214	3.9188	3.9147	3.9093
1.40	4.1175	4.1167	4.1142	4.1103	4.1051
1.45	4.3230	4.3222	4.3198	4.3160	4.3111
1.50	4.5394	4.5386	4.5363	4.5327	4.5279
1.55	4.7671	4.7663	4.7641	0.7606	4.7560
1.60	5.0068	5.0060	5.0039	5.0005	4.9961
1.65	5.2588	5.2580	5.2559	5.2527	5.2484
1.70	5.5242	5.5235	5.5215	5.5183	5.5142
1.75	5.8030	5.8023	5.8003	5.7972	5.7932
1.80	6.0968	6.0961	6.0942	6.0912	6.0874
1.85	6.4051	6.4044	6.4025	6.3996	6.3958
1.90	6.7302	6.7296	6.7278	6.7250	6.7214
1.95	7.0710	7.0703	7.0685	7.0658	7.0622
2.00	7.4308	7.4302	7.4286	7.4259	7.4225

Table I. (Continued).

t/α	-1.00	-1.05	-1.10	-1.15	-1.20
0.0001	127.31	110.57	37.047	-148.18	-540.30
0.001	22.807	21.261	15.135	1.3443	-24.702
0.005	7.7556	7.4628	6.3836	4.1381	0.22173
0.01	5.1110	4.9681	4.4571	3.4293	1.6972
0.02	3.5046	3.4348	3.1928	2.7223	1.9562
0.03	2.8759	2.8300	2.6737	2.3759	1.9005
0.04	2.5331	2.4967	2.3821	2.1667	1.8278
0.05	2.3105	2.2834	2.1933	2.0259	1.7652
0.06	2.1574	2.1350	2.0611	1.9246	1.7143
0.07	2.0451	2.0261	1.9633	1.8487	1.6733
0.08	1.9596	1.9430	1.8887	1.7901	1.6401
0.09	1.8927	1.8780	1.8302	1.7438	1.6133
0.10	1.8394	1.8262	1.7835	1.7069	1.5916
0.15	1.6893	1.6806	1.6530	1.6045	1.5329
0.20	1.6350	1.6285	1.6083	1.5732	1.5222
0.25	1.6246	1.6195	1.6036	1.5763	1.5370
0.30	1.6393	1.6351	1.6220	1.5998	1.5681
0.35	1.6706	1.6670	1.6560	1.6373	1.6109
0.40	1.7141	1.7110	1.7014	1.6853	1.6627
0.45	1.7673	1.7645	1.7560	1.7419	1.7223
0.50	1.8286	1.8261	1.8186	1.8061	1.7887
0.55	1.8973	1.8950	1.8882	1.8770	1.8615
0.60	1.9727	1.9707	1.9645	1.9543	1.9403
0.65	2.0547	2.0528	2.0471	2.0378	2.0250
0.70	2.1429	2.1412	2.1359	2.1274	2.1157
0.75	2.2375	2.2358	2.2310	2.2231	2.2123
0.80	2.3384	2.3368	2.3323	2.3249	2.3149
0.85	2.4457	2.4442	2.4400	2.4331	2.4238
0.90	2.5595	2.5582	2.5542	2.5477	2.5390
0.95	2.6801	2.6789	2.6751	2.6690	2.6609
1.00	2.8078	2.8065	2.8030	2.7973	2.7896
1.05	2.9426	2.9415	2.9381	2.9327	2.9254
1.10	3.0851	3.0840	3.0807	3.0756	3.0687
1.15	3.2353	3.2343	3.2312	3.2263	3.2198
1.20	3.3938	3.3928	3.3899	3.3852	3.3790
1.25	3.5608	3.5599	3.5571	3.5526	3.5467
1.30	3.7369	3.7359	3.7333	3.7290	3.7233
1.35	3.9223	3.9214	3.9188	3.9147	3.9093
1.40	4.1175	4.1167	4.1142	4.1103	4.1051
1.45	4.3230	4.3222	4.3198	4.3160	4.3111
1.50	4.5394	4.5386	4.5363	4.5327	4.5279
1.55	4.7671	4.7663	4.7641	0.7606	4.7560
1.60	5.0068	5.0060	5.0039	5.0005	4.9961
1.65	5.2588	5.2580	5.2559	5.2527	5.2484
1.70	5.5242	5.5235	5.5215	5.5183	5.5142
1.75	5.8030	5.8023	5.8003	5.7972	5.7932
1.80	6.0968	6.0961	6.0942	6.0912	6.0874
1.85	6.4051	6.4044	6.4025	6.3996	6.3958
1.90	6.7302	6.7296	6.7278	6.7250	6.7214
1.95	7.0710	7.0703	7.0685	7.0658	7.0622
2.00	7.4308	7.4302	7.4286	7.4259	7.4225

Table I. (Continued).

t/α	**-1.25**	**-1.30**	**-1.35**	**-1.40**	**-1.45**
0.0001	-1299.8	-2690.8	-5138.7	-9314.4	-16252
0.001	-69.694	-143.17	-258.45	-433.76	-693.40
0.005	-6.0230	-15.436	-29.065	-48.193	-74.335
0.01	-0.97102	-4.8564	-10.291	-17.659	-27.388
0.02	0.81602	-0.78794	-2.9554	-5.7940	-9.4143
0.03	1.2070	0.25101	-1.0150	-2.6397	-4.6705
0.04	1.3405	0.67830	-0.18620	-1.2798	-2.6273
0.05	1.3946	0.89649	0.25342	-0.55110	-1.5314
0.06	1.4180	1.0232	0.51827	-0.10776	-0.86363
0.07	1.4280	1.1037	0.69210	0.18571	-0.42103
0.08	1.4319	1.1584	0.81360	0.39219	-0.10937
0.09	1.4330	1.1978	0.90279	0.54442	0.12039
0.10	1.4331	1.2275	0.97095	0.66093	0.29605
0.15	1.4366	1.3140	1.1641	0.98664	0.78193
0.20	1.4545	1.3695	1.2672	1.1477	1.0119
0.25	1.4855	1.4216	1.3455	1.2576	1.1588
0.30	1.5269	1.4763	1.4165	1.3481	1.2719
0.35	1.5768	1.5352	1.4864	1.4311	1.3699
0.40	1.6338	1.5987	1.5579	1.5118	1.4612
0.45	1.6972	1.6670	1.6321	1.5930	1.5502
0.50	1.7667	1.7403	1.7099	1.6760	1.6392
0.55	1.8419	1.8185	1.7917	1.7620	1.7299
0.60	1.9227	1.9018	1.8779	1.8515	1.8232
0.65	2.0091	1.9902	1.9687	1.9451	1.9198
0.70	2.1011	2.0839	2.0645	2.0431	2.0204
0.75	2.1989	2.1831	2.1654	2.1460	2.1254
0.80	2.3025	2.2881	2.2718	2.2540	2.2352
0.85	2.4123	2.3989	2.3838	2.3675	2.3502
0.90	2.5283	2.5158	2.5019	2.4868	2.4709
0.95	2.6508	2.6392	2.6262	2.6122	2.5975
1.00	2.7802	2.7693	2.7571	2.7744	2.7304
1.05	2.9165	2.9063	2.8949	2.8827	2.8700
1.10	3.0603	3.0507	3.0400	3.0286	3.0166
1.15	3.2119	3.2027	3.1928	3.1819	3.1707
1.20	3.3715	3.3628	3.3533	3.3432	3.3326
1.25	3.5395	3.5313	3.5223	3.5127	3.5027
1.30	3.7165	3.7087	3.7002	3.6911	3.6816
1.35	3.9028	3.8953	3.8872	3.8785	3.8696
1.40	4.0988	4.0917	4.0840	4.0758	4.0673
1.45	4.3051	4.2983	4.2909	4.2830	4.2749
1.50	4.5222	4.5157	4.5086	4.5012	4.4935
1.55	4.7504	4.7442	4.7374	4.7302	4.7229
1.60	4.9908	4.9848	4.9783	4.9715	4.9645
1.65	5.2433	5.2375	5.2312	5.2246	5.2179
1.70	5.5093	5.5037	5.4977	5.4914	5.4850
1.75	5.7884	5.7831	5.7773	5.7712	5.7649
1.80	6.0828	6.0777	6.0721	6.0663	6.0603
1.85	6.3913	6.3863	6.3809	6.3752	6.3694
1.90	6.7171	6.7123	6.7071	6.7017	6.6962
1.95	7.0580	7.0532	7.0481	7.0428	7.0374
2.00	7.4185	7.4140	7.4092	7.4041	7.3990

Table I. (Continued).

t/α	**-1.50**	**-1.55**	**-1.60**	**-1.65**	**-1.70**
0.0001	-27508	-45365	-73059		
0.001	-1069.0	-1600.1	-2334.3	-3325.2	-4626.7
0.005	-109.23	-154.77	-212.87	-285.23	-372.93
0.01	-39.931	-55.744	-75.233	-98.679	-126.13
0.02	-13.924	-19.415	-25.953	-33.551	-42.146
0.03	-7.1493	-10.107	-13.559	-17.489	-21.847
0.04	-4.2486	-6.1558	-8.3493	-10.812	-13.503
0.05	-2.6978	-4.0547	-5.5980	-7.3114	-9.1630
0.06	-1.7549	-2.7824	-3.9403	-5.2143	-6.5786
0.07	-1.1310	-1.9431	-2.8514	-3.8430	-4.8968
0.08	-0.69233	-1.3548	-2.0908	-2.8890	-3.7315
0.09	-0.36957	-0.92310	-1.5344	2.1936	-2.8852
0.10	-0.12337	-0.59472	-1.1126	-1.6679	-2.2477
0.15	0.55135	0.29741	0.0024	-0.26334	-0.55730
0.20	0.86105	0.69731	0.5235	0.34349	0.16193
0.25	1.0503	0.93376	0.8115	0.68620	0.56126
0.30	1.1890	1.1008	1.0090	0.91586	0.82380
0.35	1.3039	1.2341	1.1622	1.0896	1.0185
0.40	1.4070	1.3501	1.2918	1.2334	1.1766
0.45	1.5046	1.4571	1.4086	1.3604	1.3137
0.50	1.6002	1.5597	1.5187	1.4781	1.4389
0.55	1.6960	1.6610	1.6257	1.5909	1.5575
0.60	1.7933	1.7627	1.7319	1.7017	1.6728
0.65	1.8933	1.8662	1.8390	1.8125	1.7873
0.70	1.9966	1.9724	1.9482	1.9247	1.9024
0.75	2.1039	2.0821	2.0604	2.0394	2.0196
0.80	2.2157	2.1959	2.1763	2.1574	2.1396
0.85	2.3324	2.3144	2.2966	2.2794	2.2633
0.90	2.4545	2.4380	2.4217	2.4061	2.3914
0.95	2.5823	2.5671	2.5522	2.5379	2.5245
1.00	2.7163	2.7023	2.6885	2.6753	2.6631
1.05	2.8569	2.8438	2.8341	2.8189	2.8076
1.10	3.0044	2.9922	2.9804	2.9691	2.9586
1.15	3.1592	3.1479	3.1368	3.1263	3.1166
1.20	3.3219	3.3112	3.3009	3.2911	3.2820
1.25	3.4926	3.4826	3.4729	3.4637	3.4553
1.30	3.6721	3.6627	3.6536	3.6450	3.6371
1.35	3.8606	3.8517	3.8431	3.8350	3.8275
1.40	4.0587	4.0503	4.0422	4.0346	4.0276
1.45	4.2668	4.2588	4.2511	4.2439	4.2373
1.50	4.4857	4.4782	4.4709	4.4641	4.4579
1.55	4.7155	4.7083	4.7014	4.6949	4.6890
1.60	4.9574	4.9506	4.9440	4.9379	4.9323
1.65	5.2112	5.2046	5.1983	5.1924	5.1871
1.70	5.4786	5.4723	5.4664	5.4608	5.4558
1.75	5.7588	5.7527	5.7469	5.7416	5.7367
1.80	6.0544	6.0487	6.0433	6.0382	6.0336
1.85	6.3637	6.3581	6.3528	6.3479	6.3434
1.90	6.6908	6.6855	6.6805	6.6758	6.6717
1.95	7.0321	7.0268	7.0219	7.0173	7.0131
2.00	7.3940	7.3891	7.3845	7.3802	7.3764

Table I. (Continued).

t/α	-1.75	-1.80	-1.85	-1.90	-1.95
0.0001					
0.001	-6281.7	-8300.1	-10622	-13054	-15169
0.005	-475.86	-591.73	-714.76	-833.75	-929.43
0.01	-157.26	-191.11	-225.83	-258.28	-283.50
0.02	-51.558	-61.447	-71.249	-80.100	-86.750
0.03	-26.523	-31.338	-36.015	-40.154	-43.204
0.04	-16.350	-19.239	-22.006	-24.420	-26.174
0.05	-11.100	-13.045	-14.886	-16.475	-17.617
0.06	-7.9930	-9.3999	-10.720	-11.849	-12.654
0.07	-5.9810	-7.0511	-8.0478	-8.8937	-9.4916
0.08	-4.5926	-5.4370	-6.2181	-6.8768	-7.3392
0.09	-3.5880	-4.2731	-4.9032	-5.4344	-5.8001
0.10	-2.8337	-3.4019	-3.9218	-4.3554	-4.6565
0.15	-0.84847	-1.1252	-1.3733	-1.5761	-1.7142
0.20	-0.01535	-0.18142	-0.32821	-0.44650	-0.52592
0.25	0.44062	0.32885	0.23115	0.15328	0.10156
0.30	0.77571	0.65483	0.58477	0.52944	0.49301
0.35	0.95100	0.88949	0.83660	0.79515	0.76807
0.40	1.1229	1.0744	1.0329	1.0007	0.97970
0.45	1.2700	1.2306	1.1971	1.1713	1.1546
0.50	1.4024	1.3698	1.3422	1.3209	1.3073
0.55	1.5265	1.4990	1.4758	1.4580	1.4466
0.60	1.6462	1.6226	1.6028	1.5877	1.5781
0.65	1.7641	1.7435	1.7265	1.7135	1.7052
0.70	1.8820	1.8640	1.8491	1.8378	1.8306
0.75	2.0014	1.9855	1.9723	1.9624	1.9561
0.80	2.1234	2.1092	2.0975	2.0887	2.0831
0.85	2.2487	2.2360	2.2255	2.2176	2.2127
0.90	2.3782	2.3667	2.3572	2.3502	2.3457
0.95	2.5124	2.5020	2.4934	2.4870	2.4831
1.00	2.6520	2.6425	2.6347	2.6289	2.6253
1.05	2.7975	2.7887	2.7816	2.7763	2.7730
1.10	2.9493	2.9412	2.9347	2.9298	2.9268
1.15	3.1079	3.1005	3.0944	3.0900	3.0872
1.20	3.2740	3.2671	3.2615	3.2574	3.2548
1.25	3.4478	3.4414	3.4362	3.4323	3.4300
1.30	3.6300	3.6240	3.6192	3.6157	3.6135
1.35	3.8210	3.8153	3.8108	3.8075	3.8055
1.40	4.0214	4.0162	4.0120	4.0089	4.0070
1.45	4.2315	4.2265	4.2226	4.2197	4.2179
1.50	4.4524	4.4477	4.4440	4.4413	4.4396
1.55	4.6838	4.6793	4.6758	4.6732	4.6716
1.60	4.9274	4.9232	4.9199	4.9175	4.9161
1.65	5.1824	5.1784	5.1752	5.1729	5.1715
1.70	5.4513	5.4476	5.4447	5.4425	5.4412
1.75	5.7324	5.7288	5.7259	5.7238	5.7225
1.80	6.0296	6.0262	6.0236	6.0216	6.0205
1.85	6.3395	6.3361	6.3335	6.3315	6.3302
1.90	6.6680	6.6649	6.6625	6.6608	6.6597
1.95	7.0095	7.0064	7.0039	7.0020	7.0008
2.00	7.3730	7.3702	7.3680	7.3664	7.3655

Table I. (Continued).

t/α	-2.00	-2.05	-2.10	-2.15	-2.20
0.0001					
0.001	-16150	-14551	-7940.4	7613.8	38282
0.005	-970.77	-910.24	-677.41	-171.01	750.98
0.01	-294.08	-279.30	-224.18	-108.31	95.548
0.02	-89.454	-85.845	-72.798	-46.280	-1.2023
0.03	-44.421	-42.839	-37.222	-26.030	-7.3825
0.04	-26.865	-25.984	-22.895	-16.826	-6.8570
0.05	-18.063	-17.5031	-15.560	-11.785	-5.6518
0.06	-12.965	-12.5787	-11.249	-8.6871	-4.5630
0.07	-9.7215	-9.4391	-8.4738	-6.6283	-3.6798
0.08	-7.5161	-7.3008	-6.5694	-5.1803	-2.9755
0.09	-5.9404	-5.7710	-5.1984	-4.1171	-2.4109
0.10	-4.7705	-4.6338	-4.1738	-3.3096	-1.9531
0.15	-1.7655	-1.7056	-1.5075	-1.1427	-0.58151
0.20	-0.55509	-0.52169	-0.41276	-0.21494	0.08513
0.25	0.08276	0.10397	0.17248	0.29555	0.48018
0.30	0.47988	0.49452	0.54142	0.62493	0.74909
0.35	0.75836	0.76907	0.80311	0.86328	0.95205
0.40	0.97224	0.98040	1.0062	1.0515	1.1179
0.45	1.1486	1.1551	1.1753	1.2105	1.2619
0.50	1.3025	1.3077	1.3239	1.3521	1.3929
0.55	1.4426	1.4469	1.4602	1.4832	1.5164
0.60	1.5747	1.5783	1.5894	1.6085	1.6360
0.65	1.7023	1.7054	1.7148	1.7309	1.7360
0.70	1.8282	1.8308	1.8388	1.8526	1.8722
0.75	1.9540	1.9562	1.9632	1.9751	1.9920
0.80	2.0812	2.0832	2.0893	2.0997	2.1144
0.85	2.2110	2.2128	2.2182	2.2273	2.2402
0.90	2.3443	2.3458	2.3506	2.3687	2.3700
0.95	2.4817	2.4831	2.4874	2.4946	2.5047
1.00	2.6240	2.6253	2.6292	2.6356	2.6446
1.05	2.7719	2.7730	2.7765	2.7823	2.7904
1.10	2.9258	2.9268	2.9300	2.9353	2.9426
1.15	3.0863	3.0872	3.0901	3.0949	3.1015
1.20	3.2540	3.2548	3.2575	3.2618	3.2679
1.25	3.4292	3.4300	3.4324	3.4364	3.4420
1.30	3.6128	3.6135	3.6157	3.6194	3.6245
1.35	3.8048	3.8055	3.8075	3.8109	3.8156
1.40	4.0063	4.0070	4.0089	4.0120	4.0164
1.45	4.2173	4.2178	4.2196	4.2225	4.2265
1.50	4.4391	4.4396	4.4413	4.4440	4.4478
1.55	4.6711	4.6716	4.6731	4.6755	4.6790
1.60	4.9156	4.9161	4.9175	4.9199	4.9231
1.65	5.1709	5.1713	5.1726	5.1748	5.1777
1.70	5.4408	5.4412	5.4425	5.4446	5.4474
1.75	5.7220	5.7223	5.7233	5.7252	5.7277
1.80	6.0201	6.0205	6.0216	6.0234	6.0260
1.85	6.3297	6.3299	6.3308	6.3323	6.3345
1.90	6.6594	6.6597	6.6607	6.6624	6.6646
1.95	7.0002	7.0002	7.0009	7.0022	7.0039
2.00	7.3652	7.3655	7.3664	7.3679	7.3699

Table I. (Continued).

t/α	-2.25	-2.30	-2.35	-2.40	-2.45
0.0001					
0.001	93493				
0.005	2283.2	4686.4	8302.3	13567	21025
0.01	422.85	918.81	1639.7	2653.7	4041.0
0.02	68.722	171.08	314.81	510.11	768.23
0.03	20.966	61.636	117.60	192.12	288.64
0.04	8.0830	29.211	57.870	95.489	143.52
0.05	3.4385	16.152	33.206	55.344	83.297
0.06	1.4943	9.8893	21.049	35.404	53.365
0.07	0.61786	6.5285	14.325	24.278	36.636
0.08	0.21688	4.5783	10.293	17.540	26.479
0.09	0.04506	3.3808	7.7264	13.204	19.921
0.10	-0.01063	2.6139	6.0149	10.280	15.482
0.15	0.20612	1.2490	2.5734	4.2010	6.1465
0.20	0.50025	1.0421	1.7204	2.5421	3.0000
0.25	0.73278	1.0588	1.4625	1.9461	2.5096
0.30	0.91742	1.1327	1.3969	1.7105	2.0726
0.35	1.0715	1.2231	1.4077	1.6251	1.8743
0.40	1.2066	1.3185	1.4538	1.6121	1.7923
0.45	1.3301	1.4158	1.5186	1.6383	1.7737
0.50	1.4469	1.5142	1.5948	1.6879	1.7928
0.55	1.5601	1.6143	1.6788	1.7531	1.8364
0.60	1.6720	1.7164	1.7691	1.8295	1.8969
0.65	1.7841	1.8212	1.8649	1.9149	1.9704
0.70	1.8978	1.9291	1.9659	2.0078	2.0574
0.75	2.0139	2.0407	2.0720	2.1076	2.1468
0.80	2.1334	2.1565	2.1834	2.2140	2.2475
0.85	2.2567	2.2769	2.3003	2.3267	2.3557
0.90	2.3846	2.4023	2.4228	2.4459	2.4711
0.95	2.5176	2.5333	2.5514	2.5716	2.5937
1.00	2.6562	2.6701	2.6862	2.7041	2.7236
1.05	2.8008	2.8132	2.8276	2.8436	2.8609
1.10	2.9519	2.9760	2.9760	2.9903	3.0058
1.15	3.1100	3.1201	3.1317	3.1446	3.1585
1.20	3.2756	3.2848	3.2952	3.3070	3.3195
1.25	3.4490	3.4573	3.4669	3.4774	3.4888
1.30	3.6310	3.6386	3.6472	3.6570	3.6673
1.35	3.8215	3.8285	3.8364	3.8452	3.8546
1.40	4.0219	4.0283	4.0357	4.0437	4.0524
1.45	4.2315	4.2374	4.2441	4.2515	4.2594
1.50	4.4524	4.4580	4.4642	4.4711	4.4784
1.55	4.6833	4.6883	4.6940	4.7003	4.7070
1.60	4.9272	4.9320	4.9373	4.9432	4.9495
1.65	5.1814	5.1857	5.1906	5.1960	5.2016
1.70	5.4510	5.4551	5.4598	5.4649	5.4703
1.75	5.7309	5.7346	5.7388	5.7434	5.7482
1.80	6.0291	6.0327	6.0368	6.0413	6.0460
1.85	6.3372	6.3403	6.3439	6.3478	6.3518
1.90	6.6674	6.6706	6.6742	6.6781	6.6823
1.95	7.0062	7.0088	7.0118	7.0151	7.0185
2.00	7.3723	7.3752	7.3784	7.3819	7.3855

Table I. (Continued).

t/α	**-2.50**	**-2.55**	**-2.60**	**-2.65**	**-2.70**
0.0001					
0.001					
0.005	31327	45226	63538	87067	
0.01	5892.4	8305.4	11376	15188	19788
0.02	1101.0	1519.9	2035.0	2652.6	3372.6
0.03	410.58	561.02	742.28	955.25	1198.6
0.04	203.33	276.08	362.47	462.54	575.24
0.05	117.72	159.12	207.75	263.45	325.49
0.06	75.285	101.41	131.81	166.33	204.42
0.07	51.601	69.300	89.742	112.77	137.98
0.08	37.231	49.864	64.357	80.573	98.215
0.09	27.954	37.336	48.037	59.940	72.814
0.10	21.670	28.861	37.018	46.045	55.756
0.15	8.4146	10.997	13.868	16.981	20.264
0.20	4.6230	5.8717	7.2404	8.7033	10.224
0.25	3.1500	3.8608	4.6313	5.4456	6.2827
0.30	2.4804	2.9289	3.4107	3.9153	4.4293
0.35	2.1527	2.4566	2.7805	3.1172	3.5475
0.40	1.9924	2.2093	2.4389	2.6761	2.9142
0.45	1.9232	2.0843	2.2538	2.4279	2.6016
0.50	1.9080	2.0315	2.1607	2.2927	2.4238
0.55	1.9273	2.0244	2.1255	2.2283	2.3299
0.60	1.9703	2.0482	2.1290	2.2108	2.2913
0.65	2.0305	2.0942	2.1600	2.2263	2.2912
0.70	2.1042	2.1570	2.2114	2.2660	2.3193
0.75	2.1891	2.2334	2.2789	2.3244	2.3687
0.80	2.2835	2.3212	2.3597	2.3981	2.4354
0.85	2.3866	2.4190	2.4519	2.4847	2.5164
0.90	2.4980	2.5260	2.5544	2.5826	2.6099
0.95	2.6173	2.6417	2.6664	2.6909	2.7145
1.00	2.7443	2.7658	2.7875	2.8089	2.8294
1.05	2.8793	2.8982	2.9174	2.9362	2.9542
1.10	3.0222	3.0390	3.0560	3.0727	3.0886
1.15	3.1731	3.1882	3.2033	3.2181	3.2322
1.20	3.3327	3.3462	3.3598	3.3731	3.3857
1.25	3.5007	3.5129	3.5251	3.5369	3.5482
1.30	3.6781	3.6892	3.7002	3.7110	3.7211
1.35	3.8644	3.8744	3.8844	3.8941	3.9032
1.40	4.0614	4.0705	4.0797	4.0885	4.0969
1.45	4.2676	4.2759	4.2842	4.2922	4.2997
1.50	4.4860	4.4937	4.5013	4.5087	4.5156
1.55	4.7139	4.7209	4.7278	4.7345	4.7407
1.60	4.9559	4.9625	4.9690	4.9752	4.9812
1.65	5.2075	5.2134	5.2192	5.2248	5.2299
1.70	5.4758	5.4815	5.4870	5.4923	5.4973
1.75	5.7531	5.7581	5.7630	5.7676	5.7719
1.80	6.0508	6.0557	6.0604	6.0650	6.0693
1.85	6.3560	6.3602	6.3642	6.3680	6.3715
1.90	6.6865	6.6907	6.6949	6.6989	6.7025
1.95	7.0219	7.0253	7.0286	7.0317	7.0344
2.00	7.3892	7.3930	7.3966	7.4001	7.4033

Table I. (Continued).

t/α	-2.75	-2.80	-2.85	-2.90	-2.95
0.0001					
0.001					
0.005	45226	63538	87067		
0.01	8305.4	11376	15188	19788	25159
0.02	1519.9	2035.0	2652.6	3372.6	4184.5
0.03	561.02	742.28	955.25	1198.6	1467.5
0.04	276.08	362.47	462.54	575.24	698.02
0.05	159.12	207.75	263.45	325.49	392.33
0.06	101.41	131.81	166.33	204.42	245.08
0.07	69.300	89.742	112.77	137.98	164.70
0.08	49.864	64.357	80.573	98.215	116.78
0.09	37.336	48.037	59.940	72.814	86.284
0.10	28.861	37.018	46.045	55.756	65.865
0.15	10.997	13.868	16.981	20.264	23.612
0.20	5.8717	7.2404	8.7033	10.224	11.753
0.25	3.8608	4.6313	5.4456	6.2827	7.1150
0.30	2.9289	3.4107	3.9153	4.4293	4.9357
0.35	2.4566	2.7805	3.1172	3.5475	3.7902
0.40	2.2093	2.4389	2.6761	2.9142	3.1454
0.45	2.0843	2.2538	2.4279	2.6016	2.7694
0.50	2.0315	2.1607	2.2927	2.4238	2.5497
0.55	2.0244	2.1255	2.2283	2.3299	2.4270
0.60	2.0482	2.1290	2.2108	2.2913	2.3679
0.65	2.0942	2.1600	2.2263	2.2912	2.3528
0.70	2.1570	2.2114	2.2660	2.3193	2.3696
0.75	2.2334	2.2789	2.3244	2.3687	2.4105
0.80	2.3212	2.3597	2.3981	2.4354	2.4704
0.85	2.4190	2.4519	2.4847	2.5164	2.5461
0.90	2.5260	2.5544	2.5826	2.6099	2.6352
0.95	2.6417	2.6664	2.6909	2.7145	2.7364
1.00	2.7658	2.7875	2.8089	2.8294	2.8485
1.05	2.8982	2.9174	2.9362	2.9542	2.9709
1.10	3.0390	3.0560	3.0727	3.0886	3.1033
1.15	3.1882	3.2033	3.2181	3.2322	3.2452
1.20	3.3462	3.3598	3.3731	3.3857	3.3972
1.25	3.5129	3.5251	3.5369	3.5482	3.5585
1.30	3.6892	3.7002	3.7110	3.7211	3.7305
1.35	3.8744	3.8844	3.8941	3.9032	3.9116
1.40	4.0705	4.0797	4.0885	4.0969	4.1045
1.45	4.2759	4.2842	4.2922	4.2997	4.3066
1.50	4.4937	4.5013	4.5087	4.5156	4.5219
1.55	4.7209	4.7278	4.7345	4.7407	4.7463
1.60	4.9625	4.9690	4.9752	4.9812	4.9863
1.65	5.2134	5.2192	5.2248	5.2299	5.2346
1.70	5.4815	5.4870	5.4923	5.4973	5.5018
1.75	5.7581	5.7630	5.7676	5.7719	5.7757
1.80	6.0557	6.0604	6.0650	6.0693	6.0731
1.85	6.3602	6.3642	6.3680	6.3715	6.3746
1.90	6.6907	6.6949	6.6989	6.7025	6.7059
1.95	7.0253	7.0286	7.0317	7.0344	7.0368
2.00	7.3930	7.3966	7.4001	7.4033	7.4062

Table I. (Continued).

t/α	-3.00	-3.05	-3.10	-3.15	-3.20
0.0001					
0.001					
0.005				-1210.0	
0.01	50519	47512	36055	11402	-32980
0.02	7691.5	7324.4	5968.4	3147.6	-1759.1
0.03	2570.8	2463.5	2074.4	1280.7	-72.436
0.04	1183.9	1139.1	978.57	655.79	113.28
0.05	649.57	626.80	548.05	385.42	118.40
0.06	398.11	385.01	338.94	248.12	98.499
0.07	263.35	255.14	226.48	170.40	78.710
0.08	184.22	178.75	159.75	122.81	62.823
0.09	134.51	130.68	117.46	91.902	50.637
0.10	101.59	98.812	89.252	70.870	41.342
0.15	34.881	34.068	31.324	26.151	18.007
0.20	16.724	16.384	15.252	13.149	9.8831
0.25	9.7506	9.5781	9.0086	7.9615	6.3542
0.30	6.5053	6.4060	6.0811	5.4890	4.5883
0.35	4.8029	4.7408	4.5386	4.1730	3.6210
0.40	3.8384	3.7969	3.6629	3.4220	3.0609
0.45	3.2653	3.2363	3.1430	2.9764	2.7279
0.50	2.9173	2.8962	2.8288	2.7089	2.5312
0.55	2.7074	2.6916	2.6413	2.5523	2.4210
0.60	2.5869	2.5747	2.5363	2.4686	2.3689
0.65	2.5273	2.5178	2.4877	2.4350	2.3577
0.70	2.5110	2.5034	2.4794	2.4376	2.3765
0.75	2.5267	2.5205	2.5012	2.4674	2.4184
0.80	2.5672	2.5621	2.5462	2.5187	2.4787
0.85	2.6275	2.6233	2.6102	2.5874	2.5544
0.90	2.7045	2.7010	2.6899	2.6708	2.6434
0.95	2.7958	2.7928	2.7834	2.7673	2.7442
1.00	2.8998	2.8972	2.8893	2.8755	2.8559
1.05	3.0156	3.0134	3.0065	2.9947	2.9779
1.10	3.1425	3.1405	3.1346	3.1244	3.1099
1.15	3.2799	3.2783	3.2731	3.2642	3.2516
1.20	3.4279	3.4264	3.4218	3.4146	3.4030
1.25	3.5861	3.5848	3.5808	3.5740	3.5643
1.30	3.7549	3.7537	3.7501	3.7441	3.7355
1.35	3.9341	3.9331	3.9299	3.9246	3.9170
1.40	4.1242	4.1233	4.1205	4.1157	4.1089
1.45	4.3255	4.3246	4.3221	4.3178	4.3118
1.50	4.5382	4.5374	4.5352	4.5313	4.5259
1.55	4.7628	4.7622	4.7601	4.7566	4.7517
1.60	4.9999	4.9992	4.9974	4.9992	4.9898
1.65	5.2510	5.2492	5.2475	5.2446	5.2406
1.70	5.5132	5.5126	5.5111	5.5085	5.5048
1.75	5.7906	5.7901	5.7887	5.7863	5.7830
1.80	6.0828	6.0824	6.0811	6.0789	6.0758
1.85	6.3904	6.3900	6.3888	6.3868	6.3840
1.90	6.7142	6.7138	6.7138	6.7109	6.7083
1.95	7.0549	7.0545	7.0535	7.0518	7.0495
2.00	7.4134	7.4130	7.4121	7.4105	7.4084

Table I. (Continued).

t/α	-3.25	-3.30	-3.35	-3.40	-3.45
0.0001					
0.001					
0.005					
0.01					
0.02	-9544.1	-21195	-37913	-61116	-92428
0.03	-2176.5	-5262.5	-9601.9	-15504	-23310
0.04	-718.34	-1920.7	-3587.3	-5822.0	-8735.0
0.05	-286.39	-865.18	-1658.6	-2710.6	-4066.8
0.06	-126.28	-444.76	-877.40	-1445.9	-2172.1
0.07	-57.981	-250.18	-509.27	-847.09	-1275.3
0.08	-26.021	-150.11	-316.29	-531.52	-802.55
0.09	-10.118	-94.483	-206.79	-351.41	-532.45
0.10	-1.9040	-61.642	-140.75	-242.08	-368.27
0.15	6.3174	-9.5073	-30.044	-55.823	-87.283
0.20	5.2624	-0.90387	-8.7922	-18.553	-30.295
0.25	4.1048	1.1364	-2.6192	-7.2148	-12.682
0.30	3.3392	1.7057	-0.34226	-2.8257	-5.7533
0.35	2.8614	1.8755	0.64900	-0.8269	-2.5535
0.40	2.5671	1.9306	1.1439	0.20354	-0.8892
0.45	2.3903	1.9575	1.4259	0.79394	0.06397
0.50	2.2908	1.9843	1.6098	1.1671	0.6582
0.55	2.2443	2.0200	1.7472	1.4263	1.0592
0.60	2.2354	2.0668	1.8626	1.6233	1.3508
0.65	2.2546	2.1249	1.9683	1.7857	1.5786
0.70	2.2953	2.1936	2.0712	1.9290	1.7683
0.75	2.3534	2.2722	2.1750	2.0623	1.9354
0.80	2.4259	2.3602	2.2818	2.1912	2.0895
0.85	2.5110	2.4571	2.3930	2.3192	2.2365
0.90	2.6073	2.5626	2.5096	2.4487	2.3807
0.95	2.7139	2.6764	2.6321	2.5814	2.5249
1.00	2.8302	2.7985	2.7612	2.7185	2.6712
1.05	2.9559	2.9290	2.8972	2.8610	2.8210
1.10	3.0910	3.0678	3.0406	3.0097	2.9755
1.15	3.2352	3.2152	3.1917	3.1651	3.1358
1.20	3.3888	3.3714	3.3510	3.3279	3.3026
1.25	3.5518	3.5366	3.5188	3.4987	3.4766
1.30	3.7245	3.7111	3.6955	3.6779	3.6586
1.35	3.9072	3.8954	3.8816	3.8661	3.8492
1.40	4.1003	4.0898	4.0776	4.0638	4.0489
1.45	4.3040	4.2948	4.2838	4.2716	4.2584
1.50	4.5189	4.5105	4.5008	4.4900	4.4782
1.55	4.7455	4.7379	4.7292	4.7195	4.7090
1.60	4.9841	4.9774	4.9695	4.9608	4.9513
1.65	5.2355	5.2294	5.2223	5.2144	5.2059
1.70	5.5002	5.4946	5.4882	5.4811	5.4734
1.75	5.7788	5.7737	5.7679	5.7615	5.7545
1.80	6.0720	6.0674	6.0621	6.0562	6.0499
1.85	6.3805	6.3763	6.3714	6.3661	6.3603
1.90	6.7051	6.7012	6.6968	6.6919	6.6866
1.95	7.0465	7.0429	7.0389	7.0344	7.0296
2.00	7.4056	7.4023	7.3986	7.3945	7.3901

Table I. (Continued).

t/α	**-3.50**	**-3.55**	**-3.60**	**-3.65**	**-3.70**
0.0001					
0.001					
0.005					
0.01					
0.02					
0.03	-33373	-46040	-61602	-80242	
0.04	-12437	-17031	-22594	-29162	-36700
0.05	-5771.4	-7862.8	-10368	-13294	-16612
0.06	-3076.5	-4176.2	-5481.4	-6991.5	-8689.9
0.07	-1804.6	-2443.2	-3195.4	-4058.9	-5022.7
0.08	-1135.3	-1534.1	-2000.7	-2532.9	-3122.9
0.09	-753.42	-1016.7	-1322.9	-1670.1	-2052.8
0.10	-521.48	-703.07	-913.17	-1150.1	-1410.0
0.15	-124.72	-168.19	-217.49	-271.98	-330.52
0.20	-44.068	-59.836	-77.460	-96.662	-117.00
0.25	-19.023	-26.204	-34.140	-42.693	-51.648
0.30	-9.1183	-12.894	-17.029	-21.445	-26.028
0.35	-4.5228	-6.7154	-9.0986	-11.624	-14.225
0.40	-2.1274	-3.4967	-4.9752	-6.5314	-8.1235
0.45	-0.75824	-1.6623	-2.6326	-3.6480	-4.6808
0.50	0.0881	-0.5354	-1.2012	-1.8942	-2.5954
0.55	0.6498	0.2042	-0.2693	-0.7599	-1.2539
0.60	1.0483	0.72039	0.37346	0.01561	-0.69110
0.65	1.3495	1.1022	0.84165	0.57393	0.04835
0.70	1.5913	1.4009	1.2009	0.99626	0.79264
0.75	1.7961	1.6468	1.4906	1.3312	1.1732
0.80	1.9781	1.8592	1.7352	1.6090	1.4844
0.85	2.1464	2.0503	1.9504	1.8492	1.7494
0.90	2.3068	2.2283	2.1469	2.0646	1.9837
0.95	2.4637	2.3988	2.3317	2.2640	2.1977
1.00	2.6199	2.5658	2.5099	2.4537	2.3988
1.05	2.7777	2.7321	2.6852	2.6381	2.5922
1.10	2.9387	2.9000	2.8603	2.8205	2.7818
1.15	3.1042	3.0711	3.0372	3.0034	2.9706
1.20	3.2754	3.2469	3.2178	3.1888	3.1607
1.25	3.4530	3.4284	3.4032	3.3782	3.3540
1.30	3.6380	3.6166	3.5947	3.5730	3.5521
1.35	3.8311	3.8123	3.7932	3.7743	3.7561
1.40	4.0330	4.0164	3.9997	3.9831	3.9671
1.45	4.2443	4.2296	4.2148	4.2002	4.1862
1.50	4.4656	4.4527	4.4396	4.4267	4.4143
1.55	4.6978	4.6862	4.6746	4.6631	4.6521
1.60	4.9413	4.9310	4.9206	4.9104	4.9006
1.65	5.1969	5.1877	5.1783	5.1692	5.1605
1.70	5.4653	5.4570	5.4486	5.4404	5.4326
1.75	5.7472	5.7396	5.7321	5.7247	5.7177
1.80	6.0432	6.0364	6.0296	6.0229	6.0166
1.85	6.3543	6.3481	6.3419	6.3359	6.3301
1.90	6.6812	6.6755	6.6698	6.6644	6.6592
1.95	7.0246	7.0194	7.0143	7.0093	7.0046
2.00	7.3846	7.3808	7.3761	7.3715	7.3672

Table I. (Continued).

t/α	**-3.75**	**-3.80**	**-3.85**	**-3.90**	**-3.95**
0.0001					
0.001					
0.005					
0.01					
0.02					
0.03					
0.04	-45066	-53961	-62874	-71002	-77164
0.05	-20255	-24086	-27882	-31307	-33874
0.06	-10537	-12462	-14352	-16042	-17287
0.07	-6062.9	-7138.5	-8186.7	-9116.6	-9802.4
0.08	-3755.4	-4405.1	-5034.1	-5588.4	-5994.6
0.09	-2460.7	-2877.2	-3278.1	-3629.3	-3885.2
0.10	-1685.4	-1965.3	-2233.2	-2466.7	-2635.9
0.15	-391.35	-451.91	-508.73	-557.27	-591.76
0.20	-137.83	-158.27	-177.18	-193.11	-204.27
0.25	-60.720	-69.525	-77.580	-84.289	-88.939
0.30	-30.628	-35.052	-39.063	-42.374	-44.648
0.35	-16.815	-19.288	-21.512	-23.334	-24.576
0.40	-9.6989	-11.193	-12.527	-13.614	-14.349
0.45	-5.6966	-6.6542	-7.5049	-8.1932	-8.6567
0.50	-3.2815	-3.9249	-4.4935	-4.9511	-5.2577
0.55	-1.7350	-2.1839	-2.5788	-2.8952	-3.1061
0.60	-0.69110	-1.0144	-1.2975	-1.5233	-1.6733
0.65	0.04835	-0.19063	-0.39910	-0.56472	-0.67427
0.70	0.59671	0.41605	0.25903	0.13474	0.05282
0.75	1.0217	0.88243	0.76182	0.66668	0.60419
0.80	1.3652	1.2561	1.1619	1.0878	1.0393
0.85	1.6544	1.5675	1.4928	1.4342	1.3960
0.90	2.5490	1.8369	1.7769	1.7299	1.6993
0.95	2.1349	2.0778	2.0290	1.9910	1.9662
1.00	2.3469	2.2999	2.2598	2.2286	2.2083
1.05	2.5490	2.5099	2.4765	2.4507	2.4340
1.10	2.7454	2.7126	2.6848	2.6632	2.6493
1.15	2.9397	2.9120	2.8885	2.8703	2.8586
1.20	3.1344	3.1108	3.0908	3.0754	3.0655
1.25	3.3314	3.3112	3.2941	3.2809	3.2725
1.30	3.5325	3.5151	3.5004	3.4891	3.4818
1.35	3.7391	3.7239	3.7112	3.7014	3.6952
1.40	3.9523	3.9391	3.9280	3.9195	3.9195
1.45	4.1732	4.1617	4.1520	4.1446	4.1399
1.50	4.4028	4.3926	4.3841	4.3776	4.3735
1.55	4.6420	4.6330	4.6255	4.6198	4.6161
1.60	4.8916	4.8836	4.8770	4.8719	4.8687
1.65	5.1524	5.1453	5.1394	5.1349	5.1321
1.70	5.4254	5.4190	5.4138	5.4098	5.4072
1.75	5.7112	5.7055	5.7008	5.6972	5.6950
1.80	6.0108	6.0057	6.0014	5.9982	5.9962
1.85	6.3249	6.3203	6.3165	6.3136	6.3118
1.90	6.6544	6.6503	6.6468	6.6442	6.6426
1.95	7.0003	6.9965	6.9934	6.9910	6.9895
2.00	7.3633	7.3599	7.3570	7.3549	7.3536

Table I. (Continued).

t/α	-4.00	-4.05	-4.10	-4.15	-4.20
0.0001					
0.001					
0.005					
0.01					
0.02					
0.03				-97710	45586
0.04	-79686	-76287	-63934	-38698	4391.1
0.05	-34915	-33533	-28561	-18514	-1548.4
0.06	-17802	-17140	-14776	-10042	-2120.4
0.07	-10076	-9720.7	-8460.2	-5954.7	-1793.7
0.08	-6155.9	-5948.3	-5217.2	-3773.3	-1391.0
0.09	-3986.2	-3857.2	-3404.9	-2516.9	-1060.4
0.10	-2702.5	-2618.1	-2323.8	-1749.0	-810.97
0.15	-605.08	-588.65	-532.33	-424.49	-252.03
0.20	-208.52	-203.37	-185.95	-153.06	-101.19
0.25	-90.694	-88.600	-81.589	-68.492	-48.069
0.30	-45.500	-44.496	-41.162	-34.991	-25.454
0.35	-25.038	-24.500	-22.722	-19.455	-14.445
0.40	-14.621	-14.307	-13.276	-11.393	-8.5249
0.45	-8.8272	-8.6317	-7.994	-6.8359	-5.0822
0.50	-5.3700	-5.2421	-4.8270	-4.0775	-2.9480
0.55	-3.1831	-3.0960	-2.8146	-2.3088	-1.5503
0.60	-1.7278	-1.6665	-1.4691	-1.1159	-0.58849
0.65	-0.71392	-0.66555	-0.52715	-0.27331	0.10424
0.70	0.02327	0.05617	0.16143	0.34839	0.62544
0.75	0.58171	0.60666	0.68607	0.82670	1.0344
0.80	1.0219	1.0411	1.1021	1.2099	1.3685
0.85	1.3823	1.3973	1.4450	1.5289	1.6521
0.90	1.6884	1.7004	1.7382	1.8044	1.9014
0.95	1.9574	1.9670	1.9973	2.0503	2.1277
1.00	2.2012	2.2090	2.2336	2.2765	2.3389
1.05	2.4281	2.4345	2.4547	2.4897	2.5407
1.10	2.6443	2.6497	2.6663	2.6953	2.7372
1.15	2.8545	2.8589	2.8728	2.8969	2.9318
1.20	3.0620	3.0657	3.0774	3.0976	3.1268
1.25	3.2695	3.2727	3.2826	3.2997	3.3243
1.30	3.4793	3.4820	3.4904	3.5050	3.5258
1.35	3.6930	3.6954	3.7026	3.7150	3.7329
1.40	3.9122	3.9143	3.9205	3.9312	3.9465
1.45	4.1382	4.1400	4.1454	4.1546	4.1679
1.50	4.3721	4.3736	4.3783	4.3864	4.3978
1.55	4.6149	4.6162	4.6203	4.6274	4.6374
1.60	4.8676	4.8688	4.8724	4.8785	4.8873
1.65	5.1311	5.1322	5.1354	5.1408	5.1485
1.70	5.4064	5.4073	5.4101	5.4149	5.4217
1.75	5.6942	5.6950	5.6975	5.7018	5.7078
1.80	5.9955	5.9962	5.9985	6.0023	6.0076
1.85	6.3111	6.3118	6.3138	6.3172	6.3220
1.90	6.6420	6.6426	6.6444	6.6474	6.6517
1.95	6.9890	6.9896	6.9912	6.9939	6.9977
2.00	7.3531	7.3536	7.3551	7.3575	7.3610

Table II. Values of the Volterra functions $\mu(t,1,\alpha)$ for $t \leq 2.0$ and $-1.0 \leq \alpha \leq 1.0$

t/α	-1.00	-0.75	-0.50	-0.25	0.00
0.001	6.6573	2.1980	0.54690	0.11729	0.02281
0.005	2.9017	1.2582	0.44134	0.13663	0.03878
0.010	2.1702	1.05199	0.42528	0.15352	0.05111
0.015	1.8770	0.96926	0.42486	0.16752	0.06114
0.02	1.7155	0.92561	0.42923	0.18002	0.07009
0.03	1.5433	0.88453	0.44331	0.20253	0.08628
0.04	1.4561	0.87016	0.46019	0.22315	0.10123
0.05	1.4074	0.86808	0.47823	0.24269	0.11553
0.06	1.3795	0.87283	0.49687	0.26156	0.12945
0.07	1.3646	0.88183	0.51588	0.27999	0.14316
0.08	1.3583	0.89374	0.53516	0.29816	0.15677
0.09	1.3581	0.90774	0.55487	0.31617	0.17034
0.10	1.3624	0.92334	0.57439	0.33409	0.18394
0.11	1.3702	0.94020	0.59431	0.35199	0.19760
0.12	1.3807	0.95811	0.61444	0.36992	0.21136
0.13	1.3935	0.97690	0.63477	0.38791	0.22523
0.14	1.4082	0.99646	0.65533	0.40599	0.23923
0.15	1.4245	1.0167	0.67610	0.42419	0.25340
0.20	1.5240	1.1265	0.78365	0.51765	0.32700
0.25	1.6453	1.2484	0.89805	0.61650	0.40615
0.30	1.7833	1.3814	1.0202	0.72220	0.49180
0.35	1.9361	1.5253	1.1511	0.83500	0.58473
0.40	2.1030	1.6806	1.2914	0.95689	0.68564
0.45	2.2841	1.8479	1.4421	1.0880	0.79526
0.50	2.4798	2.0278	1.6040	1.2292	0.91430
0.60	2.9176	2.4287	1.9647	1.5455	1.1836
0.70	3.4226	2.8907	2.3810	1.9129	1.5001
0.80	4.0033	3.4221	2.8611	2.3392	1.8707
0.90	4.6694	4.0325	3.4143	2.8330	2.3036
1.00	5.4318	4.7326	4.0508	3.4039	2.8078
1.10	6.3034	5.5347	4.7822	4.0631	3.3936
1.20	7.2986	6.4525	5.6216	4.8228	4.0726
1.30	8.4335	7.5014	6.5838	5.6971	4.8579
1.40	9.7266	8.6991	7.6854	6.7016	5.7645
1.50	11.198	10.065	8.9452	7.8542	6.8092
1.60	12.872	11.622	10.384	9.1750	8.0109
1.70	14.774	13.394	12.027	10.687	9.3911
1.80	16.933	15.410	13.899	12.415	10.974
1.90	19.383	17.702	16.032	14.388	12.787
2.00	22.161	20.305	18.459	16.639	14.862

Table II. (Continued).

t/α	**0.25**	**0.50**	**0.75**	**1.00**
0.001	0.00413	0.00071	0.00012	0.00002
0.005	0.01030	0.00260	0.00063	0.00015
0.010	0.01598	0.00475	0.00135	0.00037
0.015	0.02101	0.00687	0.00216	0.00065
0.02	0.02574	0.00901	0.00303	0.00098
0.03	0.03476	0.01337	0.00494	0.00176
0.04	0.04352	0.01788	0.00707	0.00270
0.05	0.05220	0.02257	0.00940	0.00379
0.06	0.06090	0.02745	0.01192	0.00501
0.07	0.06967	0.03251	0.01463	0.00638
0.08	0.07855	0.03777	0.01752	0.00787
0.09	0.08755	0.04322	0.02059	0.00951
0.10	0.09671	0.04886	0.02384	0.01128
0.11	0.10602	0.05470	0.02727	0.01319
0.12	0.11551	0.06075	0.03088	0.01523
0.13	0.12519	0.06699	0.03467	0.01742
0.14	0.13505	0.07344	0.03467	0.01974
0.15	0.14512	0.08010	0.04279	0.02220
0.20	0.19864	0.11657	0.06632	0.03669
0.25	0.25797	0.15858	0.09465	0.05499
0.30	0.32367	0.20650	0.12808	0.07741
0.35	0.39629	0.26075	0.16698	0.10430
0.40	0.47639	0.32177	0.21174	0.13602
0.45	0.56456	0.39007	0.26282	0.17301
0.50	0.66139	0.46617	0.32071	0.21570
0.60	0.88374	0.64417	0.45905	0.32024
0.70	1.1492	0.86096	0.63151	0.45400
0.80	1.4642	1.1225	0.84359	0.62205
0.90	1.8362	1.4356	1.1016	0.83021
1.00	2.2737	1.8081	1.4128	1.0851
1.10	2.7861	2.2489	1.7853	1.3945
1.20	3.3845	2.7682	2.2287	1.7670
1.30	4.0810	3.3773	2.7537	2.2125
1.40	4.8897	4.0896	3.3725	2.7426
1.50	5.8264	4.9197	4.0990	3.3701
1.60	6.9091	5.8846	4.9489	4.1097
1.70	8.1580	7.0033	5.9403	4.9782
1.80	9.5961	8.2977	7.0935	5.9946
1.90	11.249	9.7919	8.4314	7.1807
2.00	13.147	11.514	9.9803	8.5608

Table III. Values of the Volterra functions $\mu(t,1/2,\alpha)$ for $t \leq 2.0$ and $-1.0 \leq \alpha \leq 1.0$

t/α	-1.00	-0.75	-0.50	-0.25	0.00
0.001	13.102	5.0587	1.3387	0.29743	0.05924
0.005	5.0494	2.5188	0.94292	0.30375	0.08864
0.010	3.5475	1.9618	0.84744	0.31901	0.10942
0.015	2.9478	1.7275	0.80953	0.33327	0.12548
0.02	2.6141	1.5944	0.79067	0.34654	0.13932
0.03	2.2476	1.4477	0.77607	0.37100	0.16341
0.04	2.0491	1.3702	0.77506	0.39361	0.18481
0.05	1.9258	1.3245	0.78037	0.41502	0.20464
0.06	1.8432	1.2965	0.78920	0.43560	0.22346
0.07	1.7856	1.2795	0.80022	0.45559	0.24159
0.08	1.7444	1.2698	0.81274	0.47515	0.25923
0.09	1.7149	1.2655	0.82634	0.49439	0.27651
0.10	1.6938	1.2651	0.84075	0.51341	0.29355
0.11	1.6791	1.2676	0.85581	0.53226	0.31041
0.12	1.6695	1.2726	0.87141	0.55100	0.32715
0.13	1.6640	1.2794	0.88747	0.57773	0.34382
0.14	1.6617	1.2879	0.90393	0.58830	0.36044
0.15	1.6621	1.2977	0.92074	0.60693	0.37706
0.20	1.6934	1.3618	1.0094	0.70072	0.46078
0.25	1.7558	1.4430	1.1046	0.79696	0.54690
0.30	1.8380	1.5364	1.2060	0.89694	0.63668
0.35	1.9350	1.6400	1.3138	1.0015	0.73095
0.40	2.0445	1.7528	1.4281	1.1115	0.83039
0.45	2.1652	1.8745	1.5494	1.2274	0.93559
0.50	2.2967	2.0051	1.6780	1.3499	1.0471
0.60	2.5910	2.2939	1.9593	1.6167	1.2911
0.70	2.9287	2.6219	2.2760	1.9163	1.5667
0.80	3.3127	2.9929	2.6325	2.2535	1.8784
0.90	3.7472	3.4115	3.0339	2.6333	2.2310
1.00	4.2376	3.8833	3.4856	3.0612	2.6297
1.10	4.7898	4.4143	3.9940	3.5433	3.0805
1.20	5.4110	5.0115	4.5658	4.0863	3.5900
1.30	6.1089	5.6827	5.2088	4.6978	4.1653
1.40	6.8924	6.4367	5.9316	5.3861	4.8146
1.50	7.7716	7.2831	6.7436	6.1605	5.5470
1.60	8.7575	8.2329	7.6556	7.0315	6.3725
1.70	9.8624	9.2983	8.6795	8.0107	7.3024
1.80	11.100	10.493	9.8285	9.1108	8.3494
1.90	12.487	11.831	11.117	10.347	9.5275
2.00	14.039	13.331	12.563	11.734	10.852

Table III (continued)

t/α	**0.25**	**0.50**	**0.75**	**1.00**
0.001	0.01091	0.00189	0.00031	0.00005
0.005	0.02405	0.00616	0.00151	0.00035
0.010	0.03500	0.01060	0.00306	0.00085
0.015	0.04416	0.01473	0.00470	0.00144
0.02	0.05244	0.01872	0.00640	0.00210
0.03	0.06757	0.02654	0.00998	0.00362
0.04	0.08162	0.03429	0.01381	0.00536
0.05	0.09507	0.04206	0.01785	0.00731
0.06	0.10817	0.04991	0.02211	0.00945
0.07	0.12104	0.05785	0.02656	0.01178
0.08	0.13378	0.06592	0.03122	0.01428
0.09	0.14645	0.07411	0.03606	0.01696
0.10	0.15910	0.08245	0.04110	0.01981
0.11	0.17176	0.09093	0.04633	0.02283
0.12	0.18446	0.09956	0.05175	0.02602
0.13	0.19722	0.10836	0.05735	0.02937
0.14	0.21006	0.11731	0.06314	0.03290
0.15	0.22298	0.12644	0.06912	0.03658
0.20	0.28938	0.17466	0.10180	0.05752
0.25	0.35944	0.22748	0.13923	0.08270
0.30	0.43388	0.28156	0.18156	0.11227
0.35	0.51325	0.34819	0.22900	0.14644
0.40	0.59804	0.41671	0.28180	0.18546
0.45	0.68868	0.49111	0.34025	0.22958
0.50	0.78563	0.57176	0.40465	0.27912
0.60	1.0003	0.75335	0.55272	0.39578
0.70	1.2457	0.96480	0.72908	0.53840
0.80	1.5261	1.2099	0.93724	0.71033
0.90	1.8459	1.4928	1.1812	0.91544
1.00	2.2101	1.8183	1.4655	1.1581
1.10	2.6245	2.1918	1.7953	1.4431
1.20	3.0952	2.6192	2.1762	1.7761
1.30	3.6293	3.1074	2.6148	2.1633
1.40	4.2345	3.6637	3.1183	2.6117
1.50	4.9196	4.2966	3.6949	3.1290
1.60	5.6946	5.0157	4.3536	3.7242
1.70	6.5701	5.8315	5.1048	4.4070
1.80	7.5585	6.7558	5.9598	5.1885
1.90	8.6735	7.8018	6.9315	6.0812
2.00	9.9302	8.9845	8.0342	7.0989

Table IV. Values of the Volterra functions $\mu(t,-1/4,\alpha)$ for $t \leq 2.0$ and $-1.0 \leq \alpha \leq 1.0$

t/α	-1.00	-0.75	-0.50	-0.25	0.00
0.001	27.587	16.664	4.9908	1.18190	0.24524
0.005	8.8034	6.5997	2.8328	0.98304	0.30128
0.010	5.6156	4.5682	2.2758	0.92801	0.33574
0.015	4.3858	3.7283	2.0217	0.90477	0.36039
0.02	3.7114	3.2489	1.8686	0.89265	0.38049
0.03	2.9737	2.7051	1.6868	0.88242	0.41348
0.04	2.5696	2.3968	1.5801	0.88063	0.44103
0.05	2.3113	2.1950	1.5093	0.88290	0.46535
0.06	2.1308	2.0518	1.4589	0.88744	0.48750
0.07	1.9974	1.9446	1.4216	0.89339	0.50811
0.08	1.8947	1.8613	1.3931	0.90029	0.52755
0.09	1.8082	1.7949	1.3710	0.90786	0.54607
0.10	1.7477	1.7409	1.3537	0.91593	0.56387
0.11	1.6936	1.6963	1.3401	0.92193	0.58106
0.12	1.6484	1.6591	1.3295	0.93314	0.59777
0.13	1.6104	1.6276	1.3211	0.94215	0.61406
0.14	1.5792	1.6010	1.3148	0.95136	0.63000
0.15	1.5506	1.5782	1.3101	0.96074	0.64564
0.20	1.4611	1.5052	1.3039	1.0097	0.72066
0.25	1.4199	1.4736	1.3163	1.0612	0.79254
0.30	1.4054	1.4657	1.3399	1.1149	0.86308
0.35	1.4079	1.4731	1.3715	1.1707	0.93336
0.40	1.4223	1.4914	1.4091	1.2285	1.0041
0.45	1.4457	1.5180	1.4518	1.2886	1.0757
0.50	1.4764	1.5515	1.4989	1.3509	1.1488
0.60	1.5554	1.6354	1.6051	1.4830	1.3002
0.70	1.6542	1.7387	1.7264	1.6259	1.4604
0.80	1.7708	1.8596	1.8625	1.7810	1.6316
0.90	1.9045	1.9977	2.0139	1.9495	1.8152
1.00	2.0555	2.1531	2.1814	2.1327	2.0130
1.10	2.2245	2.3267	2.3663	2.3324	2.2269
1.20	2.4125	2.5196	2.5699	2.5501	2.4586
1.30	2.6209	2.7330	2.7938	2.7877	2.7101
1.40	2.8514	2.9689	3.0399	3.0472	2.9035
1.50	3.1058	3.2291	3.3103	3.3308	3.2811
1.60	3.3863	3.5158	3.6072	3.6409	3.6055
1.70	3.6953	3.8313	3.9331	3.9802	3.9592
1.80	4.0354	4.1785	4.2910	4.3516	4.3455
1.90	4.4098	4.5604	4.6836	4.758	4.7674
2.00	4.8216	4.9804	5.1151	5.2039	5.2289

Table IV. (continued).

t/α	**0.25**	**0.50**	**0.75**	**1.00**
0.001	0.04652	0.00825	0.00139	0.00022
0.005	0.08473	0.02232	0.00558	0.00134
0.010	0.11171	0.03487	0.01034	0.00294
0.015	0.13222	0.04555	0.01491	0.00468
0.02	0.14957	0.05525	0.01940	0.00653
0.03	0.17904	0.07293	0.02827	0.01051
0.04	0.20443	0.08922	0.03708	0.01478
0.05	0.22733	0.10465	0.04589	0.01932
0.06	0.24852	0.11948	0.05474	0.02408
0.07	0.26848	0.13387	0.06365	0.02906
0.08	0.28749	0.14794	0.07255	0.03424
0.09	0.30576	0.16175	0.08166	0.03961
0.10	0.32343	0.17537	0.09077	0.04518
0.11	0.34061	0.18884	0.09997	0.05089
0.12	0.35738	0.20218	0.10926	0.05678
0.13	0.37381	0.21543	0.11864	0.06284
0.14	0.38994	0.22861	0.12810	0.06906
0.15	0.40584	0.24174	0.13766	0.07544
0.20	0.48269	0.30698	0.18691	0.10961
0.25	0.55704	0.37243	0.23868	0.14745
0.30	0.63045	0.43880	0.29307	0.18885
0.35	0.70386	0.50655	0.35020	0.23376
0.40	0.77794	0.57605	0.41017	0.28219
0.45	0.85314	0.64755	0.47309	0.33418
0.50	0.92986	0.72131	0.53907	0.38980
0.60	1.0890	0.87638	0.68068	0.51216
0.70	1.2576	1.0428	0.83603	0.65012
0.80	1.4373	1.2222	1.0063	0.80462
0.90	1.6301	1.4159	1.1927	0.97685
1.00	1.8375	1.6254	1.3967	1.1681
1.10	2.0613	1.8527	1.6197	1.3800
1.20	2.3034	2.0993	1.8637	1.6141
1.30	2.5659	2.3673	2.1304	1.8723
1.40	2.8508	2.6588	2.4220	2.1568
1.50	3.1605	2.9761	2.7407	2.4698
1.60	3.4975	3.3218	3.0890	2.8139
1.70	3.8647	3.6986	3.4698	3.1919
1.80	4.2649	4.1095	3.8862	3.6069
1.90	4.7015	4.5579	4.3414	4.0622
2.00	5.1784	5.0476	4.8393	4.5618

LIST OF SYMBOLS AND DEFINITIONS OF SPECIAL FUNCTIONS.

Mathematical constants

$\pi = 3.14159265....$

$\gamma = 0.577215664...$ Euler's constant

$G = 0.915965594...$ Catalan's constant

Exponential integral and related functions

$$Ei(x) = -\int_{-x}^{\infty} \frac{e^{-t}}{t} dt = li(e^{x}) \quad , \quad x > 0$$

$$Ei(-x) = -\int_{x}^{\infty} \frac{e^{-t}}{t} dt = \gamma + \ln x + \sum_{n=1}^{\infty} (-1)^{n} \frac{x^{n}}{n!n} \quad , \quad x > 0$$

$$E_{n}(x) = \int_{1}^{\infty} \frac{e^{-xt}}{t^{n}} dt \quad , \quad x > 0 \quad , \quad n = 0,1,2,...$$

$$li(x) = \int_{0}^{x} \frac{1}{\ln t} dt \quad , \quad x > 0$$

$$li(x) = \gamma + \ln(\ln x) + \sum_{n=1}^{\infty} \frac{(\ln x)^{n}}{n!n} = Ei(\ln x) \quad , \quad x > 1$$

$$li(x) = \gamma + \ln(-\ln x) + \sum_{n=1}^{\infty} \frac{(\ln x)^{n}}{n!n} \quad , \quad 0 < x < 1$$

Sine and cosine integrals.

$$si(x) = -\int_{x}^{\infty} \frac{\sin t}{t} dt = -\frac{\pi}{2} + \sum_{n=1}^{\infty} \frac{(-1)^{n+1} x^{2n-1}}{(2n-1)!(2n-1)}$$

$$Si(x) = \int_{0}^{x} \frac{\sin t}{t} dt = si(x) + \frac{\pi}{2}$$

$$ci(x) = -\int_x^{\infty} \frac{\cos t}{t} dt = \gamma - \ln x + \sum_{n=1}^{\infty} \frac{(-1)^n x^{2n}}{(2n)!2n}$$

$$Ci(x) = \gamma + \ln x + \int_0^x \frac{\cos t - 1}{t} dt = ci(x)$$

Gamma function and related functions.

$$\Gamma(z) = \int_0^{\infty} t^{z-1} e^{-t} dt \quad , \quad \operatorname{Re} z > 0$$

$$\Gamma(n+1) = n!$$

$$\Gamma(\alpha, x) = \int_x^{\infty} t^{\alpha-1} e^{-t} dt \quad , \quad \operatorname{Re} \alpha > 0$$

$$\gamma(\alpha, x) = \int_0^x t^{\alpha-1} e^{-t} dt$$

$$B(x, y) = \int_0^1 t^{x-1} (1-t)^{y-1} dt = \frac{\Gamma(x)\Gamma(y)}{\Gamma(x+y)} = B(y, x)$$

$$\psi(z) = \frac{d \ln \Gamma(z)}{dz}$$

Error functions and Fresnel integrals.

$$erf(z) = \frac{2}{\sqrt{\pi}} \int_0^z e^{-t^2} dt$$

$$erfc(z) = 1 - erf(z)$$

$$i^n erfc(z) = \int_z^\infty i^{n-1} erfc(t)dt \quad , \quad n = 0,1,2...$$

$$i^{-1} erfc(z) = \frac{2}{\sqrt{\pi}} e^{-z^2}$$

$$i^0 erfc(z) = erfc(z)$$

$$C(z) = \int_0^z \cos(\frac{\pi}{2}t^2)dt$$

$$S(z) = \int_0^z \sin(\frac{\pi}{2}t^2)dt$$

Legendre functions.

$$P_\nu^\mu(x) = \frac{1}{\Gamma(1-\mu)}\left(\frac{x+1}{x-1}\right)^{\mu/2} {}_2F_1(-\nu,\nu+1;1-\mu;\frac{1-x}{2}) \quad , \quad |x| < 1$$

$$Q_\nu^\mu(x) = \frac{\pi}{2\sin(\pi\mu)}[P_\nu^\mu(x)\cos(\pi\mu) - \frac{\Gamma(\nu+\mu+1)}{\Gamma(\nu-\mu+1)} P_\nu^{-\mu}(x)]$$

Bessel, Airy, Kelvin and Lommel functions.

$$J_\nu(z) = \left(\frac{z}{2}\right)^\nu \sum_{n=0}^\infty \frac{(-\frac{z^2}{4})^n}{n!\Gamma(\nu+n+1)} \quad , \quad |\arg z| < \pi$$

$$Y_\nu(z) = \frac{J_\nu(z)\cos(\pi\nu) - J_{-\nu}(z)}{\sin(\pi\nu)}$$

$$I_\nu(z) = \left(\frac{z}{2}\right)^\nu \sum_{n=0}^\infty \frac{(\frac{z^2}{4})^n}{n!\Gamma(\nu+n+1)}$$

$$K_\nu(z) = \frac{\pi}{2}[\frac{I_{-\nu}(z) - J_\nu(z)}{\sin(\pi\nu)}]$$

$$Ai(z) = \frac{\sqrt{z}}{3}[I_{-1/3}(\xi) - I_{1/3}(\xi)] = \sqrt{\frac{z}{3}}K_{1/3}(\xi) \quad , \quad \xi = \frac{2}{3}z^{3/2}$$

$$Bi(z) = \sqrt{\frac{z}{3}}[I_{-1/3}(\xi) + I_{1/3}(\xi)]$$

$$ber_\nu(z) = \frac{1}{2}[J_\nu(ze^{3\pi i/4}) + J_\nu(ze^{-3\pi i/4})]$$

$$= \left(\frac{z}{2}\right)^\nu \sum_{n=0}^{\infty} \frac{\cos[(\frac{3\nu+2n}{4})\pi]}{n!\Gamma(\nu+n+1)} \left(\frac{z^2}{4}\right)^n$$

$$bei_\nu(z) = \frac{1}{2i}[J_\nu(ze^{3\pi i/4}) - J_\nu(ze^{-3\pi i/4})]$$

$$= \left(\frac{z}{2}\right)^\nu \sum_{n=0}^{\infty} \frac{\sin[(\frac{3\nu+2n}{4})\pi]}{n!\Gamma(\nu+n+1)} \left(\frac{z^2}{4}\right)^n$$

$$ker_\nu(z) = \frac{\pi}{2\sin(\pi\nu)}[ber_{-\nu}(z) - \cos(\pi\nu)ber_\nu(z) - \sin(\pi\nu)bei_\nu(z)]$$

$$kei_\nu(z) = \frac{\pi}{2\sin(\pi\nu)}[bei_{-\nu}(z) + \sin(\pi\nu)ber_\nu(z) - \cos(\pi\nu)bei_\nu(z)]$$

$$ber(z) = ber_0(z) \quad , \quad bei(z) = bei_0(z)$$
$$ker(z) = ker_0(z) \quad , \quad kei(z) = kei_0(z)$$

$$s_{\mu,\nu}(z) = \frac{\pi}{2}[Y_\nu(z)\int_0^z t^\mu J_\nu(t)dt - J_\nu(z)\int_0^z t^\mu Y_\nu(t)dt]$$

$$= \frac{z^{\mu+1}}{(\mu-\nu+1)(\mu+\nu+1)}\,{}_1F_2(1;\frac{\mu-\nu+3}{2},\frac{\mu+\nu+3}{2};-\frac{z^2}{4})$$

$$S_{\mu,\nu}(z) = s_{\mu,\nu}(z) +$$

$$\frac{2^{\mu-1}\Gamma(\frac{\mu-\nu+1}{2})\Gamma(\frac{\mu+\nu+1}{2})}{\sin(\pi\nu)}\{\cos[(\frac{\mu-\nu}{2})\pi]J_{-\nu}(z) - \cos[(\frac{\mu+\nu}{2})\pi]J_\nu(z)\}$$

Struve, Anger and Weber functions.

$$H_\nu(z) = \left(\frac{z}{2}\right)^{\nu+1} \sum_{n=0}^{\infty} \frac{(-1)^n (\frac{z^2}{4})^n}{\Gamma(n+\frac{3}{2})\Gamma(\nu+n+\frac{3}{2})}$$

$$L_\nu(z) = \left(\frac{z}{2}\right)^{\nu+1} \sum_{n=0}^{\infty} \frac{(\frac{z^2}{4})^n}{\Gamma(n+\frac{3}{2})\Gamma(\nu+n+\frac{3}{2})}$$

$$\boldsymbol{J}_\nu(z) = \frac{1}{\pi} \int_0^{\pi} \cos(\nu\theta - z\sin\theta) d\theta$$

$$\boldsymbol{E}_\nu(z) = \frac{1}{\pi} \int_0^{\pi} \sin(\nu\theta - z\sin\theta) d\theta$$

Hypergeometric, confluent and generalized hypergeometric functions.

$${}_2F_1(a,b;c;z) = \frac{\Gamma(c)}{\Gamma(a)\Gamma(b)} \sum_{n=0}^{\infty} \frac{\Gamma(a+n)\Gamma(b+n)}{\Gamma(c+n)} \frac{z^n}{n!}$$

$${}_1F_1(a;b;z) = \frac{\Gamma(b)}{\Gamma(a)} \sum_{n=0}^{\infty} \frac{\Gamma(a+n)}{\Gamma(b+n)} \frac{z^n}{n!}$$

$${}_mF_n(a_1, a_2, ...a_m; b_1, b_2, ...b_n; z) = \sum_{k=0}^{\infty} \frac{(a_1)_k (a_2)_k ... (a_m)_k}{(b_1)_k (b_2)_k ... (b_n)_k} \frac{z^k}{k!}$$

$$(a)_k = \frac{\Gamma(a+k)}{\Gamma(a)} = a(a+1)(a+2)....(a+k-1)$$

Parabolic cylinder and Whittaker functions.

$$D_\mu(z) = 2^{\mu/2+1/4} W_{\mu/2+1/4,-1/4}(\frac{z^2}{2})$$

$$M_{\nu,\mu}(z) = z^{\mu+1/2} e^{-z/2} {}_1F_1(\mu - \nu + \frac{1}{2}; 2\mu + 1; z)$$

$$W_{\nu,\mu}(z) = \frac{\Gamma(-2\mu)}{\Gamma(\frac{1}{2} - \mu - \nu)} M_{\nu,\mu}(z) + \frac{\Gamma(2\mu)}{\Gamma(\frac{1}{2} + \mu - \nu)} M_{\nu,-\mu}(z)$$

Jacobi, Gugenbauer, Chebyshev, Legendre, Laguerre and Hermite orthogonal polynomials

$$P_n^{\alpha,\beta}(x) = \frac{(-1)^n}{2^n n!(1-x)^\alpha (1+x)^\beta} \frac{d^n}{dx^n} \left\{ (1-x)^\alpha (1+x)^\beta (1-x^2)^n \right\}$$

$$C_n^\alpha(x) = \frac{(-1)^n \Gamma(\alpha + \frac{1}{2}) \Gamma(2\alpha + n)}{2^n n! \Gamma(2\alpha) \Gamma(\alpha + n + \frac{1}{2})(1-x^2)^{\alpha-1/2}} \frac{d^n}{dx^n} \left\{ (1-x^2)^{\alpha-1/2} (1-x^2)^n \right\}$$

$$T_n(x) = \frac{(-1)^n \sqrt{\pi} (1-x^2)^{1/2}}{2^n \Gamma(n + \frac{1}{2})} \frac{d^n}{dx^n} \left\{ (1-x^2)^{n-1/2} \right\}$$

$$U_n(x) = \frac{(-1)^n \sqrt{\pi} (n+1)}{2^{n+1} \Gamma(n + \frac{3}{2})(1-x^2)^{1/2}} \frac{d^n}{dx^n} \left\{ (1-x^2)^{n+1/2} \right\}$$

$$P_n(x) = \frac{(-1)^n}{2^n n!} \frac{d^n}{dx^n} \left\{ (1-x^2)^n \right\}$$

$$L_n^\alpha(x) = \frac{x^{-\alpha} e^x}{n!} \frac{d^n}{dx^n} \left\{ x^{n+\alpha} e^{-x} \right\}$$

$$H_n(x) = (-1)^n e^{x^2} \frac{d^n}{dx^n} \left\{ e^{-x^2} \right\}$$

$$He_n(x) = (-1)^n e^{x^2/2} \frac{d^n}{dx^n} \left\{ e^{-x^2/2} \right\} \quad He_n(x) = (-1)^n e^{x^2/2} \frac{d^n}{dx^n} \left\{ e^{-x^2/2} \right\}$$

Riemann zeta, Lerch, Jonquière, Liouville, Mittag-Leffler and Möbius functions.

$$\zeta(z)=\sum_{n=1}^{\infty}\frac{1}{n^{z}} \quad , \quad \operatorname{Re} z>1$$

$$\zeta(z,\alpha)=\sum_{n=1}^{\infty}\frac{1}{(\alpha+n)^{z}} \quad , \quad \operatorname{Re} z>1$$

$$\Phi(z,s,\alpha)=\sum_{n=0}^{\infty}\frac{z^{n}}{(\alpha+n)^{s}} \quad , \quad |z|<1 \quad , \quad \alpha\neq 0,-1$$

$$F(z,s)=\sum_{n=1}^{\infty}\frac{z^{n}}{n^{s}}=z\Phi(z,s,1)\ F(z,s)=\sum_{n=1}^{\infty}\frac{z^{n}}{n^{s}}=z\Phi(z,s,1)$$

$$\lambda(n)=\sum_{m=0}^{\infty}\frac{1}{(2m+1)^{n}}=(1-\frac{1}{2^{n}})\zeta(n) \quad , \quad n=2,3,...$$

$$E_{\alpha}(z)=\sum_{n=0}^{\infty}\frac{z^{n}}{\Gamma(\alpha n+1)}$$

$$E_{\alpha,\beta}(z)=\sum_{n=0}^{\infty}\frac{z^{n}}{n!\Gamma(\alpha n+\beta)}$$

$${}_{m}E_{\alpha,\beta}(z)=\sum_{n=0}^{\infty}\frac{z^{n}}{(n!)^{m}\Gamma(\alpha n+\beta)} \quad , \quad m=1,2,3...$$

the **Möbius function** $\mu(n)$ is defined by:

$\mu(n) = 1$ if $n = 1$.
$\mu(n) = (-1)^{k}$ if n is product of k distinct primes.
$\mu(n) = 0$ if n is divisible by a square bigger than unity.

Bernoulli and Euler polynomials and numbers

$$\frac{te^{zt}}{e^{t}-1}=\sum_{n=0}^{\infty}B_{n}(x)\frac{t^{n}}{n!} \quad , \quad |t|<2\pi$$

$$B_{n}=B_{n}(0)$$

$$\frac{2e^{zt}}{e^t+1} = \sum_{n=0}^{\infty} E_n(x)\frac{t^n}{n!}$$

$$E_n = 2^n E_n(\frac{1}{2})$$

REFERENCES

[1] ABRAMOWITZ, A. and STEGUN, I. E. (Eds.) *Handbook of Mathematical Functions*. U.S. National Bureau of Standards, Washington DC, 1965.

[2] AGARWAL, R. P. A propos d'une note de Pierre Humbert. *C. R. Acad. Sciences,* 236 (1953) 2031-2032.

[3] APELBLAT, A. Some integrals of gamma, polygamma and Volterra functions. *IMA J. Appl. Math.,* 34 (1985) 173-186.

[4] APELBLAT, A. and KRAVITSKY, N. Integral representations of derivatives and integrals with respect to the order of the Bessel functions $J_\nu(t)$, $I_\nu(t)$, the Anger function $\mathbf{J}_\nu(t)$ and the integral Bessel function $Ji_\nu(t)$. *IMA J. Appl. Math.,* 34 (1985) 187-210.

[5] APELBLAT, A. and KRAVITSKY, N. Improper integrals associated with the inverse Laplace transforms of $[\ln(s^2 + a^2)]^{-n}$, n = 1,2. ZAMM *Z. Angew. Math. Mech.*, 78 (1998) 565-571

[6] APELBLAT, A. Derivatives and integrals with respect to the order of the Struve functions $H_\nu(x)$ and $L_\nu(x)$. *J. Math. Anal. Appl.,* 137 (1989) 17-36.

[7] APELBLAT, A. *Tables of Definite and Infinite Integrals.* Elsevier Sci. Publ. Co., Amsterdam, 1983.

[8] APELBLAT, A. *Tables of Integrals and Series*. Verlag Harri Deutsch, Frankfurt am Main, 1996.

[9] APELBLAT, A. The integral representation of functions as the asymptotic limit of infinite integral of the Bessel function $J_\nu(\nu x)$. *Intern. J. Appl. Math.*, 1 (1999) 19-27.

[10] APELBLAT, A. The asymptotic limit of infinite integral of the Bessel function $J_\nu(\nu x)$ as integral representation of elementary and special functions. *Intern. J. Appl. Math.*, 2 (2000) 743-762.

[11] APELBLAT, A. The application of the Dirac delta function $\delta(x\text{-}1)$ to the evaluation of limits and integrals of elementary and special functions.. *Intern. J. Appl. Math.*, 16 (2004) 323-339.

[12] BARRUCAND, R. Sur les suites réciproques. *C. R. Acad. Sciences,* 230 (1950) 1727-1728.

[13] BARRUCAND, P. Sur les fonctions de S. Colombo. *C. R. Acad. Sciences,* 232 (1951) 1058-1060.

[14] BARRUCAND, P. A. and COLOMBO, S. Sur la fonction $\nu(t,n)$. *C. R. Acad. Sciences,* 230 (1950) 1335-1337.

[15] BERG, L. *Introduction to the Operational Calculus*. North-Holland Publ. Co., Amsterdam, 1967.

[16] BOUWKAMP, C. J. Note on an asymptotic expansion. *Indiana Univ. Math. J.*, 21 (1971) 547-549.

[17] BUTZER, P. L. Singular integral equations of Volterra type and the finite part of divergent integrals. *Arch. Rational Mech. Anal.*, 3 (1959) 194-205.

[18] CHAE, S.B., *Lebesgue Integration*. 2nd ed., Springer Verlag, 1995.

[19] CARSLAW, H. S. and JAEGER, J. C. *Operational Methods in Applied Mathematics*. Dover Publ. Inc., New York, 1963.

[20] CARSON, J. R. Electric Circuit Theory and the Operational Calculus. McGraw-Hill Book Co., New York, 1926.

[21] CARSTOIU, I. Une formule générale opératoire dans le calcul symbolique. *C. R. Acad. Sciences*, 232 (1951) 721-723.

[22] CHURCHILL, R. V. *Operational Mathematics*. 2nd ed. McGraw-Hill Book Co., New York, 1958.

[23] COLOMBO, S. Sur quelques correspondances symboliques. *C. R. Acad. Sciences*, 216 (1943) 368-369.

[24] COLOMBO, S. Sur quelques nouvelles correspondances symboliques. *Bull. Sci. Math.*, 67 (1943) 104-108.

[25] COLOMBO, S. Sur la fonction $\nu(t,n)$. *C. R. Acad. Sciences*, 226 (1948) 1235-1236.

[26] COLOMBO, S. Sur les équations intégrales de Volterra à noyaux logarithmiques. *C. R. Acad. Sciences*, 235 (1952) 928-929.

[27] COLOMBO, S. Sur quelques transcendantes introduites par la resolution des équations intégrales de Volterra a noyaux logarithmiques. *Bull. Sci. Math.*, 77 (1953) 89-104.

[28] COLOMBO, S. Sur la fonction $\nu(x,n)$ et $\mu(x,m,n)$. *Bull. Sci. Math.*, 79 (1955) 72-78.

[29] COLOMBO, S. *Les Transformations de Mellin et de Hankel*. CNRS, Paris, 1959.

[30] COLOMBO, S. *Étude des Transcendantes Intervenant Dans la Résolution des Équations Intégrales de Volterra a Noyaux Logarithmiques*. Studia Universitatis "Lovanium" No.18, Gauthier-Villars, Paris, 1964.

[31] COLOMBO, S and LAVOINE, J. *Transformations de Laplace et de Mellin. Formulaires. Mode D'Utilisation*. Gauthier-Villars, Paris, 1972.

[32] COLOMBO, S. *Holomorphic Functions of One Variable*. Gordon and Breach Sci. Publ., New York, 1983.

[33] DAVIES, B. *Integral Transforms and Their Applications*. Springer-Verlag, Berlin, 1978.

[34] DELAVAULT, H. *Les Transformations Intégrales a Plusieurs Variables et Leurs Applications*. Gauthier-Villars, Paris, 1961.

[35] DELAVAULT, H. Application de la transformation de Laplace et de la transformation de Hankel à la détermination de solutions de l'équation de la chaleur et des équations de Maxwell en coordonnées cylindriques. Publ. Scientifique et Technique du Ministére de l'Air, S.D.I.T. Note technique No 71, Dec. 1957.

[36] DITKIN, V. A. and PRUDNIKOV, A. P. *Operational Calculus in Two Variables and its Application*. Pergamon Press, Oxford, 1962.

[37] DORNING, J. J., NICOLAENCO, B. and THURBER, J. K. An integral identity due to Ramanujan which occurs in neutron transport theory. *J. Math. Mech.*, 19 (1969) 429-438.

[38] DOETSCH, G. *Theorie und Anwendung der Laplace Transformation.* Springer-Verlag, Berlin, 1937.

[39] DZRBASHYAN, M. M. On the theory of integral transforms with the Volterra kernels. *Dokl. Akad. Nauk SSSR, Math. Ser.*, 124 (1959) 22-25.

[40] DZRBASHYAN, M. M. Integral transforms with the Volterra kernels. *Izv. Akad. Nauk SSSR, Math. Ser.*, 24 (1960) 387-420.

[41] ERDÉLYI, A. *Asymptotic Expansions.* Dover Publ., Inc., New York, 1956.

[42] ERDÉLYI, A. *Operational Calculus and Generalized Functions.* Holt, Rinehart and Winston, New York, 1966.

[43] ERDÉLYI, A., MAGNUS, W., OBERHETTINGER, F. and TRICOMI, F. G. *Higher Transcendental Functions.*Vol. 3, Chapter 18.3, pp. 217-227, McGraw-Hill Book Co., Inc., New York, 1955.

[44] ERDÉLYI, A., MAGNUS, W., OBERHETTINGER, F. and TRICOMI, F. G. *Tables of Integral Transforms.* Vol.1 and 2. McGraw-Hill Book Co., Inc., New York, 1954.

[45] FAN, K. Exposé sur le calcul symbolique de Heaviside. *Rev. Sci.*, 80 (1942) 147-163.

[46] FOCK, V. A. On a certain Volterra integral equation. *Mat. Sbornik,* 31 (1924) 519-528.

[47] FOCK, V. A. Über eine Klasse von Integralgleichungen. *Mat. Z.* 21 (1924) 160-173.

[48] FRASER, W. C. G. The factorial transform. *Duke Math. J.*, 11 (1944) 469-486.

[49] FRASER, W. C. G. Inversion formulae for the factorial transform. *Duke Math. J.*, 13 (1946) 239-246.

[50] GHIZZETTI, A. *Calcolo Simbolico.* N. Zanichelli, Bologna, 1943.

[51] GHIZZETTI, A. and OSSICINI, A. *Transformate di Laplace e Calcolo Simbolico.* Unione Tipografico-Editrice Torinese, Torino, 1971.

[52] GRADSTEIN, I. and RYZHIK, I. *Tables of Series, Products and Integrals.*Verlag Harri Deutsch, Thun-Frankfurt, 1981.

[53] HARDY, G.H. A formula of Ramanujan in the theory of primes. *J. London Math. Soc.,* 12 (1937) 94-98.

[54] HARDY, G.H. Another formula of Ramanujan. *J. London Math. Soc.,* 12 (1937) 314-318.

[55] HARDY, G. H. *Ramanujan, Twelve Lectures on Subjects Suggested by His Life and Work.* Chelsea Publ. Co., New York, 1959.

[56] HLADIK, J. *La Transformation de Laplace a Plusieurs Variables.* Masson et Cie Éditeurs. Paris, 1969.

[57] HILLE, E. *Methods in Classical and Functional Analysis.* Addison-Wesley, Reading, Ma., 1972.

[58] HILLE, E. *Analytical Function Theory.* Vol. 2, Ginn and Company, Boston, 1962.

[59] HILLE, E. *Analysis.* Vol. 2, Blaisdell Publ. Co., Waltham, Masschusetts, 1966.

[60] HOSKINS, R. F. *Generalised Functions.* Ellis Horwood Ltd., Chichester, 1979.

[61] HUMBERT, P. Une nouvelle correspondance symbolique. *C. R. Acad. Sciences,* 218 (1944) 99-100.

[62] HUMBERT, P. Nouvelles correspondance symboliques. *Bull. Sci. Math.,* 69 (1945) 121-129.

[63] HUMBERT, P. Image nouvelle pour la fonction de Bessel. *C. R. Acad. Sciences,* 230 (1950) 504-505.

[64] HUMBERT, P. Quelques resultats rélatif à la fonction de Mittag-Leffler. *C. R. Acad. Sciences,* 236 (1953) 1467-1468.

[65] HUMBERT, P. and AGARWAL, R. P. Sur la fonction de Mittag-Leffler et quelques-unes de ses généralisations. *Bull. Sci. Math.,* 77 (1953) 180-185.

[66] HUMBERT, P. and COLOMBO, S. *Le Calcul Symbolique et Ses Applications a la Physique Mathématique.* 2nd ed., Gauthier-Villars, Paris, 1965.

[67] HUMBERT, P. and POLI, L. Sur certaines transcendantes liées au calcul symbolique. *Bull. Sci. Math.,* 68 (1944) 204-214.

[68] INCE, E.L. *Ordinary Differential Equations.* Dover Publ., Inc., New York, 1959.

[69] KÖLBIG, K. S. On the integral $\int_0^\infty e^{-\mu t} t^{\nu-1} \log^m t dt$. *Math. Comp.*, 40 (1983) 171-182.

[70] KOLMOGOROV, A. and FOMINE, S. *Éléments de la Théorie des Fonctions et de l'Analyse Fonctionnelle.* Editions Mir, Moscou, 1977.

[71] KRASNOV, M., KISSÉLEV, A. and MAKARENKO, G. *Équations Integrales.* Éditions Mir, Moscou, 1977.

[72] LAMBORN, B. N. A. An expression of the Dirac delta function. *SIAM Review,* 11 (1969) 603.

[73] LANDAU, E. Über einige ältere Vermutungen und Behauptungen in der Primzahlentheorie. *Mat. Z.* 1 (1918) 213-219.

[74] LePAGE, W. R. *Complex Variables and the Laplace Transform for Engineers.* McGraw-Hill, New York, 1961.

[75] LINDELÖF E. *Calcul des Résidus et Ses Applications a la Théorie des Fonctions.* pp. 52-69, Chelsea Publ. Co., New York, 1947.

[76] McLACHLAN, N. W. *Laplace Transforms and Their Applications to Differential Equations.* Dover Publ., Inc., New York, 1962.

[77] McLACHLAN, N. W. *Complex Variable and Transform Calculus.* 2nd. ed., Cambridge University Press, Cambridge, 1963.

[78] LINZ, P. *Analytical and Numerical Methods for Volterra Equations.* SIAM Studies in Applied Mathematics, Philadelphia, 1985.

[79] LIOUVILLE, J. *Journ. Ecole Polytechnique,* Paris, 21 (1832).

[80] MAGNUS, W., OBERHETTINGER, F. and SONI, R. P. *Formulas and Theorems for the Special Functions of Mathematical Physics.* 3rd ed. Springer-Verlag, Berlin, 1966.

[81] MIKUSINSKI, J. G. Sur les fondements du calcul opèratoire. *Studia Math.*, 11 (1949) 41-70.

[82] MIKUSINSKI, J. G. Sur les équations différentialles du calcul opèratoire et leurs applications. *Studia Math.*, 12 (1951) 227-270.

[83] MILES, J. W. *Integral Transforms in Applied Mathematics.* Cambridge University Press, Cambridge, 1971.

[84] MURNAGHAM, F. D. *The Laplace Transformation.* Spartan Book, Washington, D.C. 1962.

[85] OBERHETTINGER, F. *Tables of Bessel Transforms.* Springer-Verlag, Berlin, 1972.

[86] OBERHETTINGER, F. *Fourier Expansions,* Academic Press, New York 1973.

[87] OBERHETTINGER, F. *Tables of Mellin Transforms.* Springer-Verlag, Berlin, 1974.

[88] OBERHETTINGER, F. and BADII, L. *Table of Laplace Transforms*. Springer-Verlag, Berlin, 1973.

[89] PARODI, M. Relations intégrales entre les fonctions $\nu(t)$ et $\mu(t,x)$ de Serge Colombo. *Rev. Sci.*, 85 (1947) 360.

[90] PARODI, M. *Applications Physique de la Transformation de Laplace*. CNRS, Paris, 1948.

[91] PARODI, M. Sur un type d'équations intégrales de seconde espèce résolubles par calcul symbolique. *J. Math. Pures et Appl.*, 28 (1949) 35-62.

[92] PARODI, M. *Équations Intégrales et Transformation de Laplace*. Publ. Scientifique et Technique du Ministére de l'Air No. 242, pp. 54-55, 83-89, 112-114, 1950.

[93] PARODI, M. *Introduction à l'Etude de l'Analyse Symbolique*. Gauthier-Villars, Paris, 1957.

[94] PARODI, M. Équations integrals et calcul symbolique a deux variables. *Rev. Sci.*, 85 (1947) 550-551.

[95] PÉRES, J. Sur les Fonctions Permutable de Premiére Espéce de M. Vito Volterra. Thése-Faculté des Sciences de l'Universite de Paris No.1367. Gauthier-Villars, Paris, 1915.

[96] van der POL, B. and BREMMER, H. *Operational Calculus Based on the Two-Sided Laplace Integral*. Cambridge University Press, Cambridge, 1964.

[97] POLI, L. Équations integrales et calcul symbolique. *Ann. Soc. Sci. Bruxelles,* Ser. I, 55 (1935) 111-119.

[98] POLI, L. and DELERNE, P. *Le Calcul Symbolique à Deux Variables et Ses Applications*. Gauthier-Villars, Paris, 1954.

[99] POLYANIN, A. D. and MANZHIROV, A. V. *Handbook of Integral Equations*. CRC Press, Boca Raton 1998.

[100] REED, I. S. The Mellin type of double integral. *Duke Math. J.*, 2 (1944) 565-572.

[101] RIESZ, M. L'intégrale de Riemann-Liouville. *Acta Math.*, 81 (1949) 1-222.

[102] ROBERTS, G. E. and KAUFMAN, H. *Table of Laplace Transforms*.W.B. Saunders Co., Philadelphia, 1966.

[103] ROONEY, P. G. On an integral equation of Sub-Sizenenko. *Glasgow Math. J.*, 24 (1983) 207-210.

[104] RÜHS, F. Über die Anwendung der Operatorenrechnung auf lineare Integralgleichungen vom Faltungstypus. *Z. Angew. Math. Mech.*, 42 (1962) 203-209.

[105] SANSONE, G. and GERRETSEN, J. *Lectures on the Theory of Functions of Complex Variable*. Springer-Verlag, Berlin, 1968.

[106] SCHWARTZ, L. *Théorie des Distributions*. Hermann, Paris, Vol. 1, 1950, Vol. 2, 1951.

[107] SNEDDON, I. N. *The Use of Integral Transforms*. McGraw-Hill Book Co., New York, 1971.

[108] SRIVASTAVA, H. M. and BUSCHMAN, R. G. *Theory and Applications of Convolution Integral Equations*. Kluwer Acad. Publ., Dordrecht, 1992.

[109] TITCHMARSH, E. C. *The Theory of Functions*. Oxford University Press, Oxford, 1964.

[110] TOUCHARD, J. Sur la fonction gamma. *Bull. Soc. Math. France,* 41 (1913) 234-242.

[111] TRICOMI, F. G. *Integral Equations*. Interscience, New York, 1957.

[112] VALLÉE, O. and SOARES, M. *Airy Functions and Applications to Physics*. Word Scientific, New Jersey, 2004.

[113] VOELKER, D. and DOETSCH, G. *Die Zweidimensionale Laplace-Transformation.* Birkhaüser Verlag, Basel, 1950.

[114] VOLTERRA, V. Theoria delle potenze, dei logaritmi e delle funzioni di composizione. *R. Acc. dei Lincei,* Mem., ser. 5a, vol.11, 1916.

[115] VOLTERRA, V. and PÉRES, J. *Leçons sur la Composition et les Fonctions Permutables.* Gauthier-Villars, Paris, 1924.

[116] VOLTERRA, V. *Theory of Functionals and of Integral and Integro-differential Equations.* Dover Publ., Inc., New York, 1959.

[117] WATSON, E. J. *Laplace Transforms and Applications.* Van Nostrand Reinhold Co., New York, 1981.

[118] WATSON, G. A. *A Treatise on the Theory of Bessel Functions.* Cambridge University Press, Cambridge, 1958.

[119] WHEELON, A. D. *Tables of Summable Series and Integrals of Bessel Functions,* Holden-Day, San Francisco, 1968.

[120] WHITTAKER, E. and WATSON, G. *A Course of Modern Analysis.* Cambridge University Press, Cambridge, 1963.

[121] WIDDER, D. V. *The Laplace Transform.* Princeton Univ. Press, Princeton, 1946.

[122] YOSIDA, K. *Differential and Integral Equations.* Interscience, New York, 1960.

INDEX

W